DIGITAL IMAGE
PROCESSING METHODS

OPTICAL ENGINEERING

Series Editor

Brian J. Thompson
Provost
University of Rochester
Rochester, New York

1. Electron and Ion Microscopy and Microanalysis: Principles and Applications, *Lawrence E. Murr*
2. Acousto-Optic Signal Processing: Theory and Implementation, *edited by Norman J. Berg and John N. Lee*
3. Electro-Optic and Acousto-Optic Scanning and Deflection, *Milton Gottlieb, Clive L. M. Ireland, and John Martin Ley*
4. Single-Mode Fiber Optics: Principles and Applications, *Luc B. Jeunhomme*
5. Pulse Code Formats for Fiber Optical Data Communication: Basic Principles and Applications, *David J. Morris*
6. Optical Materials: An Introduction to Selection and Application, *Solomon Musikant*
7. Infrared Methods for Gaseous Measurements: Theory and Practice, *edited by Joda Wormhoudt*
8. Laser Beam Scanning: Opto-Mechanical Devices, Systems, and Data Storage Optics, *edited by Gerald F. Marshall*
9. Opto-Mechanical Systems Design, *Paul R. Yoder, Jr.*
10. Optical Fiber Splices and Connectors: Theory and Methods, *Calvin M. Miller with Stephen C. Mettler and Ian A. White*
11. Laser Spectroscopy and Its Applications, *edited by Leon J. Radziemski, Richard W. Solarz and Jeffrey A. Paisner*
12. Infrared Optoelectronics: Devices and Applications, *William Nunley and J. Scott Bechtel*
13. Integrated Optical Circuits and Components: Design and Applications, *edited by Lynn D. Hutcheson*
14. Handbook of Molecular Lasers, *edited by Peter K. Cheo*
15. Handbook of Optical Fibers and Cables, *Hiroshi Murata*
16. Acousto-Optics, *Adrian Korpel*
17. Procedures in Applied Optics, *John Strong*
18. Handbook of Solid-State Lasers, *edited by Peter K. Cheo*
19. Optical Computing: Digital and Symbolic, *edited by Raymond Arrathoon*
20. Laser Applications in Physical Chemistry, *edited by D. K. Evans*

21. Laser-Induced Plasmas and Applications, *edited by Leon J. Radziemski and David A. Cremers*

22. Infrared Technology Fundamentals, *Irving J. Spiro and Monroe Schlessinger*

23. Single-Mode Fiber Optics: Principles and Applications, Second Edition, Revised and Expanded, *Luc B. Jeunhomme*

24. Image Analysis Applications, *edited by Rangachar Kasturi and Mohan M. Trivedi*

25. Photoconductivity: Art, Science, and Technology, *N. V. Joshi*

26. Principles of Optical Circuit Engineering, *Mark A. Mentzer*

27. Lens Design, *Milton Laikin*

28. Optical Components, Systems, and Measurement Techniques, *Rajpal S. Sirohi and M. P. Kothiyal*

29. Electron and Ion Microscopy and Microanalysis: Principles and Applications, Second Edition, Revised and Expanded, *Lawrence E. Murr*

30. Handbook of Infrared Optical Materials, *edited by Paul Klocek*

31. Optical Scanning, *edited by Gerald F. Marshall*

32. Polymers for Lightwave and Integrated Optics: Technology and Applications, *edited by Lawrence A. Hornak*

33. Electro-Optical Displays, *edited by Mohammad A. Karim*

34. Mathematical Morphology in Image Processing, *edited by Edward R. Dougherty*

35. Opto-Mechanical Systems Design: Second Edition, Revised and Expanded, *Paul R. Yoder, Jr.*

36. Polarized Light: Fundamentals and Applications, *Edward Collett*

37. Rare Earth Doped Fiber Lasers and Amplifiers, *edited by Michel J. F. Digonnet*

38. Speckle Metrology, *edited by Rajpal S. Sirohi*

39. Organic Photoreceptors for Imaging Systems, *Paul M. Borsenberger and David S. Weiss*

40. Photonic Switching and Interconnects, *edited by Abdellatif Marrakchi*

41. Design and Fabrication of Acousto-Optic Devices, *edited by Akis P. Goutzoulis and Dennis R. Pape*

42. Digital Image Processing Methods, *edited by Edward R. Dougherty*

Additional Volumes in Preparation

Visual Science and Engineering: Models and Applications, *edited by Donald Kelly*

Spatial Light Modulator Technology: Materials, Devices, and Applications, *edited by Uzi Efron*

DIGITAL IMAGE PROCESSING METHODS

EDITED BY
EDWARD R. DOUGHERTY

Rochester Institute of Technology
Rochester, New York

MARCEL DEKKER, INC.

NEW YORK · BASEL

Library of Congress Cataloging-in-Publication Data

Digital image processing methods / edited by Edward R. Dougherty.
     p.   cm. -- (Optical engineering; 42)
    Includes bibliographical references and index.
    ISBN 0-8247-8927-X (acid-free paper)
    1. Image processing--Digital techniques.   I. Dougherty, Edward R.   II. Series:
Optical engineering (Marcel Dekker, Inc.); v. 42.
   TA1637.D52   1994
   621.36'7'0285--dc20   93-43308
                                              CIP

The publisher offers discounts on this book when ordered in bulk
quantities. For more information, write to Special Sales/Professional
Marketing at the address below.

This book is printed on acid-free paper.

Marcel Dekker, Inc.
270 Madison Avenue, New York, New York 10016

Current printing (last digit):
10 9 8 7 6 5 4 3

PRINTED IN THE UNITED STATES OF AMERICA

Series Introduction

The philosophy of the Optical Engineering series is to discuss topics in optical engineering at a level useful to those working in the field or attempting to design subsystems that are based on optical techniques or that have significant optical subsystems. The concept is not to provide detailed monographs on narrow subject areas but to deal with the material at a level that makes it immediately useful to the practicing scientist and engineer. We expect that workers in optical research will also find them extremely valuable. In this volume, Edward Dougherty brings together a very important set of materials that relate to digital image processing methods.

Image processing and pattern recognition have been subjects of considerable interest for a long time. In the early days, of course, the methods were purely optical, stemming from Abbe's theory of vision in a microscope and the experimental illustrations of H. E. Fripp (published in the *Proceedings of the Bristol Naturalists Society* from 1875 to 1877) and the milestone work of A. B. Porter published in the *Philosophical Magazine* in 1906. This work led to the extensive study of coherent methods of optical image processing which gained considerable momentum in the 1960s even though the phase contrast microscope was developed by Zernike in 1935. Incoherent optical processing had its significant development starting in the 1970s.

Coherent optical processing used the fact that a Fourier transform occurs naturally from the diffraction process. With the advent of digital computers it became obvious that processing that uses a Fourier transform can equally well be carried out digitally if the image to be processed is first scanned. The digital method offers the possibility of more flexibility than the essentially analogous methods of optics. It is important to realize that while the manipulation may be the same from a mathematical point of view, there are significant differences. Optical methods are real time and parallel; digital methods have the ability to carry out multiple processes on the image data, with a variety of transformations possible.

Two other aspects of this field should not be ignored. First is the field of hybrid systems, which uses optical methods combined with electronic (digital) methods, and second is the field of so-called optical computing, which uses optical methods to do digital computing.

Edward Dougherty has brought together a leading set of experts to contribute to this integrated volume. Image data flow and processing steps are well captured in these pages.

Brian J. Thompson
University of Rochester
Rochester, New York

Preface

The literature on image processing continues to grow at an ever-increasing rate. This is reflected by the number of both books and new journals concentrating on the digital processing of images. For the most part, depth of coverage for particular methods remains polarized between the glancing coverage of general texts, designed for students, and the tightly focused coverage of academic journals, designed for specialists. On the other hand, many practitioners require a degree of depth that is deeper than that of a typical textbook but wider and more fully explained than that normally given in a journal article. Our goal in producing this book has been to reach members of the latter audience, practicing scientists and engineers, and to do this with expository chapters written by researchers with expertise in the particular subject covered.

Since each chapter is self-contained and sufficiently developed that one can go to the literature after reading it, the number of subjects treated must be small, as compared to a standard textbook on image processing. Thus, the scope of the book reflects the interest of the editor; nevertheless, we believe the chosen methods are both topical and important. Moreover, many of them are nowhere else so fully explained at such an accessible mathematical level. Except for the second chapter's dependence on a few basic definitions from the first chapter, they are logically independent.

Overall, the book treats several basic image processing topics: filtering, segmentation, thinning, pattern recognition, compression, image processing architecture, and digital halftoning. Not only are these treated in greater depth than is typically done in a standard text, but the subjects are also considered from the perspectives of persons intimately involved in their development. Consequently, the book provides both historical understanding and insight relevant to current trends.

In any survey book, it is important to see the manner in which a specific methodology fits into the overall program. Thus, an introduction has been included to provide a general framework for the material covered in the book, especially concerning the manner in which methods for filtering, segmentation, thinning, compression, and so on, fit into a coherent imaging paradigm. I encourage you to read the introduction before going to an individual chapter so that its place in the universal scheme is appreciated.

Edward R. Dougherty

Contents

Series Introduction   iii
Preface   v
Contributors   ix
Introduction   xi

1   Nonlinear Filters   **1**
        Jaakko Astola and Edward R. Dougherty

2   Morphological Segmentation for Textures and Particles   **43**
        Luc Vincent and Edward R. Dougherty

3   Multispectral Image Segmentation in Magnetic Resonance Imaging   **103**
        Joseph P. Hornak and Lynn M. Fletcher

4   Thinning and Skeletonizing   **143**
        Jennifer L. Davidson

5   Syntactic Image Pattern Recognition   **167**
        Edward K. Wong

6   Heuristic Parallel Approach for 3D Articulated Line-Drawing   **197**
        Object Pattern Representation and Recognition
        P. S. P. Wang

vii

7 Handwritten Character Recognition 223
Paul Gader, Andrew Gillies, and Daniel Hepp

8 Digital Image Compression 261
Paul W. Jones and Majid Rabbani

9 Image-Processing Architectures 327
Stephen S. Wilson

10 Digital Halftoning 363
Paul G. Roetling and Robert P. Loce

11 Glossary of Computer Vision Terms 415
Robert M. Haralick and Linda G. Shapiro

Index 469

Contributors

Jaakko Astola *Tampere University of Technology, Tampere, Finland*

Jennifer L. Davidson *Iowa State University, Ames, Iowa*

Edward R. Dougherty *Rochester Institute of Technology, Rochester, New York*

Lynn M. Fletcher *Rochester Institute of Technology, Rochester, New York*

Paul Gader *University of Missouri at Columbia, Columbia, Missouri*

Andrew Gillies *Environmental Research Institute of Michigan, Ann Arbor, Michigan*

Robert M. Haralick *University of Washington, Seattle, Washington*

Daniel Hepp *Environmental Research Institute of Michigan, Ann Arbor, Michigan*

Joseph P. Hornak *Rochester Institute of Technology, Rochester, New York*

Paul W. Jones *Eastman Kodak Company, Rochester, New York*

Robert P. Loce   *Xerox Corporation, Webster, New York*

Majid Rabbani   *Eastman Kodak Company, Rochester, New York*

Paul G. Roetling   *Xerox Corporation, Webster, New York*

Linda G. Shapiro   *University of Washington, Seattle, Washington*

Luc Vincent   *Xerox Imaging Systems, Peabody, Massachusetts*

P. S. P. Wang   *Northeastern University, Boston, Massachusetts*

Stephen S. Wilson   *Applied Intelligent Systems, Inc., Ann Arbor, Michigan*

Edward K. Wong   *Polytechnic University, Brooklyn, New York*

Introduction

During the last two decades, digital image processing has grown into a subject in its own right with applications spanning all areas of human endeavor, from documents to medicine to astronomy, and theory utilizing all branches of mathematics, from differential geometry to probability to lattice theory. Methods applying to any given application area are apt to apply to numerous areas and any given algorithm might use numerous branches of mathematics. With this scope and interpenetration, one might wonder whether there is any central core to digital image processing. One potential answer to this question comes from looking at the practice of digital image processing, that is, perhaps the nature of the subject lies in the kinds of methods employed in processing images.

At the outset, prior to digitization, an image must be captured by a sensor. The sensor (or sensors) may be one of any number of modalities. Typically, one thinks of optical devices, but images can be generated in many other ways, such as nuclear magnetic resonance (NMR), infrared (IR), sonar, radar, ultrasound, synthetic aperture radar (SAR), scanning tunneling microscopy (STM), or atomic force microscopy (AFM). Figure 1 shows an NMR image of a wrist bone affected by osteoporosis and a noncontact AFM image of blood cells. Within a particular modality there might be numerous methods employed to enable image formation. For instance, in biological or astronomical imaging one may be dealing with very low light levels and these require their own special hardware. The kind of information contained in an image is dependent upon the physics involved in its formation, so that the specifics of processing are often related to the

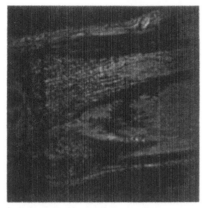

(a)

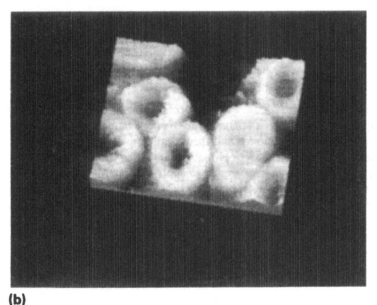

(b)

Figure 1   (a) NMR image of a wrist bone; (b) AFM image of blood cells.

formation modality. Yet the image processing stages from filtering to classification/decision tend to be rather universal, as do the mathematical principles governing the main algorithmic approaches.

Once captured, an image needs to be sampled; that is, it must be put into numerical form for digital computation. Assuming a continuous tone capture,

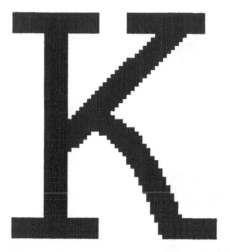

Figure 2 Binary bit map of the character K.

the image must go through a spatial digitization and a gray-tone quantization. It must be placed into computation in a functional form $f(x,y)$, where (x,y) is a discrete pixel (picture element) in the grid and $f(x,y)$ is a discrete gray value. A digital image may be binary (1 bit), have multiple levels of gray (for instance, 8 bits), or be color. Figure 2 shows a binary bit map of the character K. The amount of information from the physical analog image that is available for digital processing depends on the sampling rates (both spatial digitization and quantization). We shall assume that capture and sampling form a block of activity prior to digital processing and that digital processing commences with a sampled (digitized and quantized) image stored in a computer.

Typically, we view a specific captured and stored digital image as representative of some underlying ideal image. In the vernacular, often the image obtained is degraded by some form of noise. The noise could have occurred in capture or transmission, or perhaps the physical image itself was degraded prior to capture. A first stage of processing (often called preprocessing) involves filtering the image in an attempt to restore it back to a closer version of the underlying ideal image. Various filtering procedures might be employed and these typically depend upon the image type (print, biological), imaging modality (NMR, STM, low-light detector), and imaging purpose (visual quality, automatic character recognition). Restoration often depends upon some statistical model of the ideal and observed (degraded) image, and it might depend upon some known deterministic degradation in the imaging device (blurring, gray-level illumination gradient). Filters are of many and varied kinds, and the best filter to employ depends upon all of the above-mentioned considerations. Linear

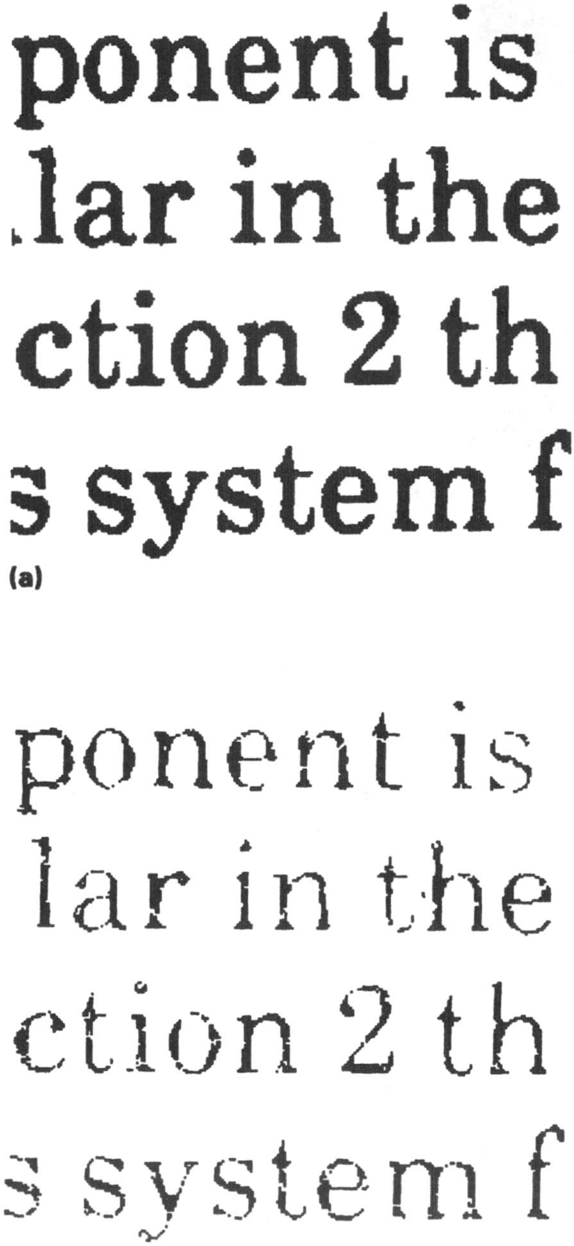

ponent is
lar in the
ction 2 th
s system f

(a)

ponent is
lar in the
ction 2 th
s system f

(b)

Figure 3   (a) Text fragment; (b) thinned and broken text fragment; (c) morphologically restored text fragment.

ponent is
lar in the
ction 2 th
s system f

(c)

filters have historically been most commonly employed and studied because of their long use in signal processing and the simplicity of their mathematical characterization. Nonlinear filters are becoming more popular because they represent less of a constraint in the design of optimal filters and because they often have properties conducive to image understanding, such as the preservation of edges by medians and the incorporation of pattern information by morphological filters. Figure 3(a) shows a page fragment, 3(b) a thinned and broken version of the page fragment, and 3(c) the degraded page fragment restored to an estimate of the original by a morphological filter. Figure 4(a) shows an original image, 4(b) an impulse-noise (5%) degraded version, and 4(c) the degraded version filtered by a 5-by-5 median.

Figure 5 shows images corresponding to those of Figure 4. Notice that the effect of the median (good or bad) depends upon the underlying image: the highly textured animal is severely degraded by the median, whereas the less-textured road scene suffers much less from application of the median. The goodness of median filtering cannot be measured simply by its effect on impulse noise; statistical analysis is required.

A second stage of preprocessing involves enhancement. Whether we wish to employ a processed image for visualization, measurement, or automatic decision, certain information in the image is more salient than other information. Typically, edges of objects within a gray-level image delineate object boundaries and shapes, whereas slowly varying gray areas tend to indicate regions of ho-

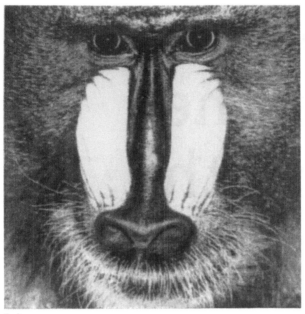

(a)

Figure 4 (a) Original animal image; (b) impulse-noise degraded animal image; (c) median-filtered degraded animal image.

mogeneity. One might wish to filter the image in a manner that accentuates the edges (edge enhancement) and flattens both intraobject and background regions (smoothing). The kinds of filters employed for enhancement are the same as those employed for restoration; here, however, the purpose is not to bring an image back to a form we believe is more faithful to the underlying real-world image of interest, but rather it is to alter the image in such a way as to make it more visually pleasing, more amenable for measurement, or more useful for decision. Figure 6 shows a low-contrast chest X-ray and a processed version of the X-ray that has been contrast-enhanced by employing the full gray range.

Our understanding of an image is not pixel-based; that is, image understanding relates to objects or regions of interest, not a particular value at this or that pixel. For instance, in an image of electrophoresis gels, both measurement and visual appreciation depend on the global perception of the gels as entities. A similar comment applies to character, target, or biological information. More globally, various regions of interest may often need to be segmented from the full image, if only for the purpose of reducing the amount of further processing. Also, various portions of a large image are subject to different processing so that an image may need to be segmented into smaller images, each of which belongs

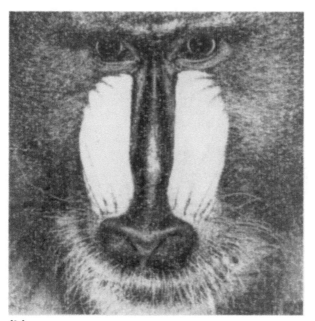

(b)

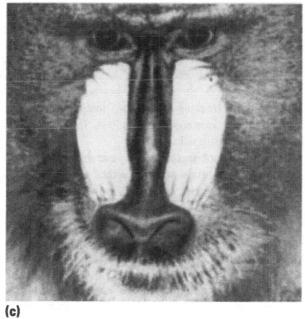

(c)

(a)

Figure 5   (a) Original road scene; (b) impulse-noise degraded road scene; (c) median-filtered degraded road scene.

to a certain class of images. For instance, page segmentation can involve segmentation into text, figures, and photographs, or more finely into small, medium, and large text; a medical image may require segmentation of various kinds of tissue in the image. Figure 7 shows an unsegmented binarized coffee-bean image and a processed version segmented into individual beans. Figure 8 shows segmentation of the NMR wristbone (rectangular region) image from Figure 1(a) into osteoporotic (outlined) and nonosteoporotic regions. In both images, segmentation has been accomplished automatically.

Prior to, following, or in the midst of segmentation, one might choose to drastically reduce the complexity of an image so that further processing takes place on a much simpler image. We distinguish reduction from enhancement because here we are considering a drastic change of image content, such as thresholding to form a binary image from a gray-scale image. We also choose to distinguish it from compression (although there exists no clear demarcation) because here the goal is not the suppression of redundant information while maintaining image integrity; rather, it is the elimination of information not useful to subsequent processing. In addition to simple thresholded images, examples of reduced images are edge images and skeletons. One might also include various

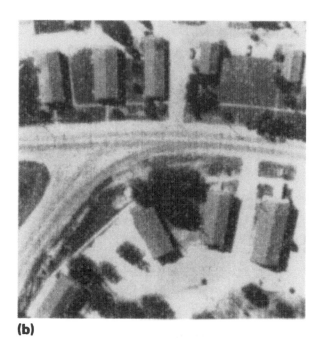

(b)

(c)

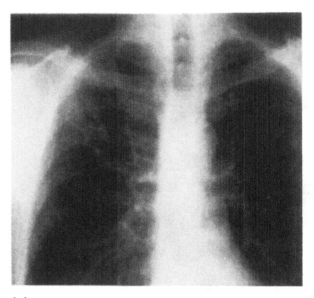

(a)

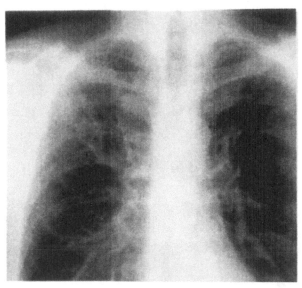

(b)

Figure 6   (a) Low-contrast chest X-ray; (b) contrast-enhanced chest X-ray.

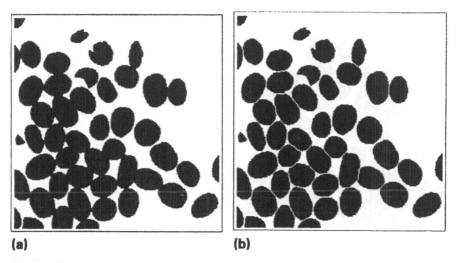

(a)   **(b)**

Figure 7   (a) Unsegmented binary coffee beans; (b) segmented coffee beans.

simplified representations such as mosaics. Figure 9(a) shows the skeleton by maximal balls for the coffee-bean image of Figure 7(a); Fig. 9(b) shows a connected version of the skeleton, and Figure 9(c) a thinned, connected skeleton.

If an image is to be used for an automatic decision, then relevant information in the image must be organized. Such information takes many forms. It might be the area and geometry of particles within the image, it might be some algebraic representation of shapes conducive to object recognition, or it might be a col-

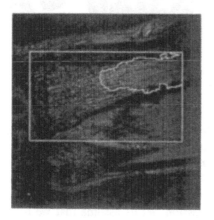

Figure 8   Segmentation of NMR wristbone image into osteoporotic and nonosteoporotic regions.

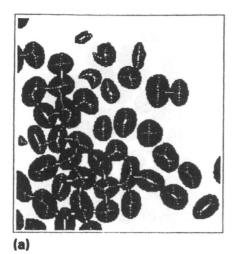

(a)

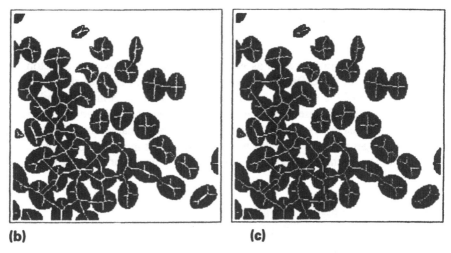

(b)   **(c)**

Figure 9   (a) Skeleton by maximal balls of coffee-bean image; (b) connected skeleton; (c) thinned, connected skeleton.

lection of transformation-like numerical features representative of the image, such as Fourier descriptors or texture features. Actually, if one takes a sufficiently abstract view of features, one might claim that all such schemes involve features, whether or not they are numeric. For instance, a text character may be represented in various ways (syntactically, numerically, graphically) to facilitate automatic recognition. For the sake of discussion, let us assume that we have associated a vector of numerical features with an image. Then these features can be used to classify the image or objects within an image. In the latter case, they may be part of a segmentation procedure; in either case, they may form part of a classification procedure leading to detection and/or decision. Since images, restored or otherwise, are random processes (shapes and textures possessing variability) feature vectors are random in nature and therefore classification by comparing numerical features of an image (or region of interest) with some archetypal stored feature vector typically involves statistical analysis. Decision routines reflect the resulting uncertainty.

Digital images contain large amounts of information, and a standard functional form, or bit map, may not be the most efficient way of representing an image. From the standpoints of data storage and transmission, one would like to have an image stored in a way that requires the fewest bits. If one is processing large numbers of images, then efficient representation is necessary in order not to overwhelm memory; if images are to be transmitted, there are bandwidth limitations so that timely transmission requires efficient representation. Thus, data need to be compressed, or coded, in such a way as to facilitate storage and transmission. Furthermore, various image representations enhance the speed of various algorithms. A key here is elimination of redundancy, and various transformations serve to reduce various types of redundancy. More than simply finding an efficient image representation, one often alters the image in a noninvertible manner. Since this involves the loss of information, such compression is termed "lossy," as opposed to invertible encodings, which are lossless. Should certain image data be highly correlated to other data, this correlation can be used to estimate the discarded data in some optimal statistical manner, thereby reconstructing the discarded data. When performing a noninvertible compression, one must keep in mind the end purpose of the processing. If certain information is not useful to a decision procedure or measurement process, it need not be kept. If certain information does not affect the human visual system and the purpose of processing and transmission is to supply images for human visual consumption, then it can be suppressed without harm. In sum, given transmission limitations, we would like to transmit as many images in a given amount of time as possible while at the same time keeping necessary information. Also taken into account must be the complexity of encoding the data at the source and decoding at the destination. Figure 10 shows an original 8-bit/pixel monochrome image and a 0.5-bit/pixel compressed version. To the eye, the effect of compression is not overly visible; however, loss of information is visible when comparing the enlarged sections of the images shown in Figure 11.

All of the above processing requires a computer (or computers). Certain architectures are suitable for certain kinds of image processing. Whether or not one can use a conventional single-instruction-single-data (SISD) machine depends on the processing demands. Is processing for off-line analysis or on-line decisions? Is real-time processing necessary? When there are many instructions to be executed on each datum, then multiple-instruction-single-data (MISD)

(a)

(b)

Figure 10 (a) Original 512 × 512, 8-bit/pixel image; (b) 0.5-bit/pixel compressed image.

pipeline architectures may be conducive to concurrent processing; if, on the other hand, a great amount of data needs to be fed to a small number of instructions, a single-instruction–multiple-data (SIMD) vector processor may be called for. For extreme compute-bound types of applications where the number of repeated operations in a computation is large compared to the amount of input–output instructions, a systolic array may be appropriate. One might think of convolutions where there are repeated multiplications and additions or morphological operations where there are repeated minima and maxima. Since these kinds of operations are commonly employed in image processing, the utility of vector or array processors is evident. Since image processing algorithms tend to

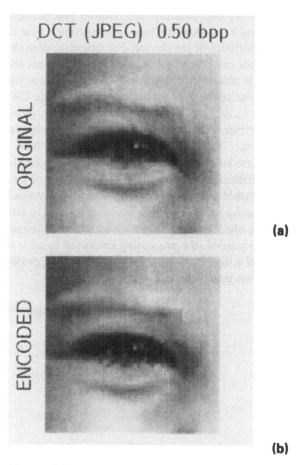

(a)

(b)

Figure 11 (a) Enlarged section of 8-bit/pixel image; (b) enlarged section of 0.5-bit/pixel compressed image.

be rather complex, the role of special-purpose hardware needs to examined in the context of the overall computer system. For instance, much processing now takes place with reduced instruction sets simply to gain better overall CPU performance. As a general rule of thumb, one needs to know his or her application and particular algorithms of interest before opting for very narrowly usable hardware. This choice must be weighed against the possibility of finding some very efficient algorithmic implementation conducive to more general-purpose hardware.

If a digital image, processed or otherwise, is to be rendered visible, then it is the goal of the rendering device to accurately represent the digital image in spatial patterns of varying light intensity and possibly hue. Digital processing is typically required prior to rendering to account for tone reproduction, color fidelity, and spatial filtering characteristics of the device. For example, consider rendering a color image on a cathode ray tube (CRT) possessing red, green, and blue phosphors. The digital image is converted to temporal analog electrical signals that drive electron guns that excite the appropriate phosphors on the CRT screen. The phosphors collectively radiate a color image. Now consider printing this same image on a digital film printer, where the film employs cyan, magenta, and yellow dyes to achieve the color. For the CRT and film images to have a similar appearance, processing must be performed, prior to rendering, on the digital image to account for the difference in color media. On the issue of tone reproduction, consider that xerographic printers typically produce binary black dots on a white page. Image processing halftone techniques must be used to render a print that is made up of these black dots but is perceived to be continuous gray scale by a human observer. In general, the processing employed to achieve an accurate visual representation of a digital image must account for the device physics, as well as human eye–brain psychophysics.

Edward R. Dougherty

1

Nonlinear Filters

Jaakko Astola

Tampere University of Technology
Tampere, Finland

Edward R. Dougherty

Rochester Institute of Technology
Rochester, New York

I. MEDIAN FILTERING

Semantically, nonlinear filtering concerns all image-to-image operators that are nonlinear, and since digital images do not form a vector space, all image filtering. Nonetheless, insofar as classical linear techniques are adapted to image filtering, linear methods do compose a large segment of image filtering. Moreover, certain types of inherently nonlinear filters have been studied extensively and these concern us in the present chapter.

Linear filters are attractive for several reasons: they possess useful algebraic properties; their operation is easy to understand; via Fourier transform they have a direct relation to frequency representation; their statistical properties are well understood; and there exist elegant, closed-form solutions for finding statistically optimal linear filters. Yet requiring linearity imposes a strong constraint on filter design. Although the linear constraint might be appropriate for some image models, for many it is certainly disadvantageous. The example cited most often is the manner in which linear filters blur edges, which in images often contain key information; on the other hand, median filters, which are nonlinear, leave edges invariant. There any many other instances where linearity is a poor filter requirement, albeit one that is mathematically attractive. As a result, more recently much attention has been focused on the analysis and design of nonlinear filters for accomplishing various image-processing tasks.

1

The present chapter is broken into two parts. The first considers median filters, and the second, morphological filters. As intuitively conceived, median filters are numerically based, and morphological filters are shape based. Median filters arise from classical maximum-likelihood estimation and from certain operations on logical variables; morphological filters arise from fitting shape probes within larger shapes. Nevertheless, there is a close relation between the types of filtering and they form a unified, coherent theory. We begin with median filters, proceed to stack filters, then to shape-based morphological filters, and then to more general morphological filters. In the end we complete the circle and see the unity by showing the manner in which all of the filters discussed possess morphological representations.

Median filtering in signal processing was introduced by Tukey [40] to smooth economic time series. The one-dimensional median filter is implemented by sliding a window of odd length over the input signal one sample at a time. At each window position the samples within the window are sorted by magnitude, and the centermost value, the median of the samples within the window, is the filter output. We denote window size by N and, as it is required to be an odd integer, write $N = 2k + 1$. Thus the median filter of length N can be expressed as

$$y(n) = \text{MED}[x(n - k), \ldots, x(n), \ldots, x(n + k)] \tag{1}$$

where $x(n)$ and $y(n)$ are the input and output sequences. The operation defined by Eq. (1) is often called the running median. In image processing, median filtering is usually applied by moving a square or cross-shaped window and choosing the median of the pixel values within the current window as output.

Example.   Consider an $M \times M$ image $x(m, n)$, $1 \leq m, n \leq M$, and its filtering with a median filter having a cross-shaped window of size 5. The filtered image is then

$$y(m, n) = \text{MED}\{x(m - 1, n), x(m, n), x(m + 1, n), x(m, n - 1),$$
$$x(m, n + 1)\}, \quad 2 \leq m, n \leq M$$

Depending on the type of image-processing application, there are very many different median image-processing algorithms. The methods depend heavily on the goal or the constraints imposed by the application. If speed is essential, only simple and fast algorithms can be applied, which often leads to inevitable loss of information. In off-line processing one can do very extensive computations, perhaps first segmenting the image and then applying different methods on segments. Also, noise can be analyzed carefully and methods tailored using this information. A comprehensive exposition of median image-processing methods can be found, for example, in [33]. Our discussion is not aimed at providing methods for designing optimal median filtering algorithms but to reveal broad underlying principles of median filters and to develop general methods of analyzing them.

We look at the median filter defined by Eq. (1) from two different points of view: statistical and algebraic. We will see that both viewpoints give new insight into the median filtering operation and also lead to different useful generalizations.

A. Statistical Approach

Many signal-processing tasks can be formulated in the following way. Suppose that we have the observations

$$x_i = \theta s_i + n_i, \qquad i = 1, \ldots, N \tag{2}$$

where $(s_1, \ldots, s_N)$ is a known signal waveform, θ is an unknown "amplitude" parameter to be estimated, and $(n_1, \ldots, n_N)$ is a sequence of independent, identically distributed (i.i.d.) random variables with a common distribution function $F(t)$.

Assume first that $N = 2k + 1$, $s_1 = \cdots = s_N = 1$, and that n_i, $i = 1, \ldots, N$, has a Laplace (or biexponential) distribution with probability density function

$$f(t) = \frac{\alpha}{2} e^{-\alpha |t|}, \qquad \alpha > 0$$

This means that the observations form a simple random sample from a population having density

$$f(t) = \frac{\alpha}{2} e^{-\alpha |t - \theta|}, \qquad \alpha > 0 \tag{3}$$

The maximum-likelihood (ML) estimation principle says that given a simple random sample $x_1, \ldots, x_N$ we should choose as the estimate for θ the value $\hat{\theta}$ for which the joint density is maximized for this particular sample. The joint density is, by independence,

$$f(t_1, \ldots, t_N) = \prod_{i=1}^{N} f(t_i) = \left(\frac{\alpha}{2}\right)^N \exp\left(-\alpha \sum_{i=1}^{N} |t_i - \theta|\right) \tag{4}$$

Substituting the values x_i for t_i we see that maximizing (4) at $t_i = x_i$, $i = 1, \ldots, N$, is equivalent to minimizing

$$L(\theta) = \sum_{i=1}^{N} |x_i - \theta| \tag{5}$$

It is easy to see that the value of θ minimizing (5), for which we use the short-hand notation

$$\arg\left\{\min_{\theta} \sum_{i=1}^{N} |x_i - \theta|\right\} \tag{6}$$

is exactly the median of $x_1, \ldots, x_N$, that is, $\hat{\theta} = \text{MED}\,[x_1, \ldots, x_N]$. This means that using the median filter is equivalent to finding the ML estimate of the amplitude of a constant signal under the assumption that noise is i.i.d. Laplace distributed. It is interesting that only changing the noise to i.i.d. Gaussian changes (5) to

$$L(\theta) = \sum_{i=1}^{N} (x_i - \theta)^2 \tag{7}$$

which leads to the simplest linear smoother: namely, the moving average of length N,

$$\hat{\theta} = \frac{1}{N} \sum_{i=1}^{N} x_i$$

This estimator, written as a digital filter, is

$$y(n) = \frac{1}{2k + 1} \sum_{i=-k}^{k} x(n + i)$$

The interpretation of the median filter as a device producing the ML estimate for location parameter under Laplace noise partly explains the good behavior of the median filter when impulsive noise is present. To wit, the Laplace distribution is often used to model impulsive or heavy-tailed noise. As an estimator the median belongs to the class of scaled robust estimators [22,23] which have the property of not being sensitive to variations in the distributions of the underlying population. A simple example of the different behavior of median and mean of the same window size is the following. If we let one sample value become arbitrarily large, the mean will also become arbitrarily large. The median either remains the same (if the particular sample value is originally at least the median) or just moves to the next larger sample value. Also, when we compare the variances of sample mean and sample median for large sample sizes we find that they depend on the distribution of the underlying population in radically different ways. For instance, changes in the variance of the underlying distribution need not affect the variance of the sample median at all. A good survey of robust methods in linear and nonlinear signal processing is [22].

The preceding ML approach leads to useful generalizations of the median filter in three ways. First, we can assume that the corrupting noise has density of the form

$$f(t) = \alpha e^{-\beta |t|^{\gamma}} \tag{8}$$

where β and γ are positive constants and α is the necessary normalizing factor. This assumption on the noise distribution leads to the filter operation [1]

$$y(n) = \arg\left\{\min_x \sum_{i=-k}^{k} |x(n+i) - x|^{\gamma}\right\} \tag{9}$$

which has several interesting properties. If $\gamma = 1$, it results in the median filter. If $\gamma = 2$, it results in the linear simple moving average. If $\gamma \to \infty$, it will approach a scaled midrange detector, that is,

$$y(n) = \tfrac{1}{2}(\min\{x(n+i) : -k \le i \le k\} + \max\{x(n+i) : -k \le i \le k\})$$

If $\gamma \le 1$, the filter will behave similarly to the median filter in the sense that its impulse response is zero and its step response is ideal. If $\gamma < 1$, the filter has an edge-enhancing property, which means that if used on a gray-scale image, it will increase its contrast. The general appearance of the objective function in Eq. (9) is plotted in Fig. 1 for $\gamma = 0.25, 1, 2,$ and 3.

A second generalization is obtained in the following way. We again assume both that in Eq. (2), $s_1 = \cdots = s_N = 1$, and independence of noise components, but allow them to have centered Laplace densities with different variances. Let the noise component n_i come from a population of density

$$f_i(t) = \frac{\gamma_i}{2}e^{-\gamma_i |t|}, \qquad i = 1, \ldots, N$$

This implies that the ML estimate for θ is

$$\hat{\theta} = \arg\left\{\min_\theta \sum_{i=1}^{N} \gamma_i |x_i - \theta|\right\} \tag{10}$$

To clarify the meaning of Eq. (10), assume that the γ_i are positive integers. Compared to Eq. (6), we see that $\hat{\theta}$ in Eq. (10) is in fact the median of the values

$$x_1, \ldots, x_1, x_2, \ldots, x_2, \ldots, x_N, \ldots, x_N$$

where each sample x_i is repeated γ_i times. The operation defined by Eq. (10) is called the weighted median of $x_1, \ldots, x_N$ with weights $\gamma_1, \ldots, \gamma_N$ and is denoted by

$$\text{WM}[\gamma_1 \Diamond x_1, \ldots, \gamma_N \Diamond x_N] \tag{11}$$

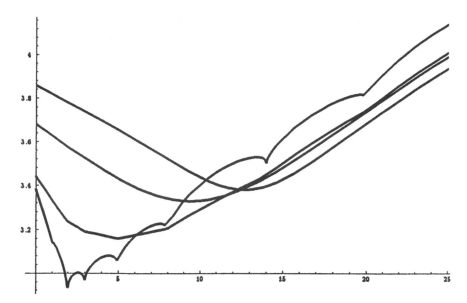

Figure 1 The general appearance of the objective function in Eq. (9) is plotted for $\gamma = 0.25$, 1, 2, and 3. The graphs are logarithmically scaled and vertically adjusted to make comparison easier. It can be seen how the minimum point moves from a cluster of x_i to the midrange as γ increases.

We saw that the median filter is the counterpart (in the median world) of the simple moving average. In the same way, the weighted median filter

$$y(n) = \text{WM}[\gamma_{-k} \diamond x(n - k), \ldots, \gamma_k \diamond x(n + k)] \tag{12}$$

is the counterpart of the linear FIR filter

$$y(n) = \frac{1}{\sum_{i=-k}^{k} \gamma_i} \sum_{i=-k}^{k} \gamma_i x(n + i) \tag{13}$$

It has been shown [42] that weighted median filters can be used successfully in many applications, especially in image processing. Compared to the standard median filter, where the only adjustable parameter is the window size, weighted median filters offer much more design freedom as we can choose any positive weights. There are also effective methods to compute optimal weights adaptively [42]. As we shall see later, even though there are an infinite number of weight combinations, there are only finitely many distinct weighted medians of a fixed window size.

A third generalization can be obtained. We consider again the model (2),

$$x_i = \theta s_i + n_i, \qquad i = 1, \ldots, N \tag{14}$$

At this time we assume that the n_i are independent random variables having the same Laplace distribution,

$$f(t) = \frac{\alpha}{2} e^{-\alpha |t|}, \qquad \alpha > 0 \tag{15}$$

but now we require only that $s_i \neq 0$ for $i = 1, \ldots, N$. Forming the ML estimate $\hat{\theta}$ for θ, we get

$$\hat{\theta} = \arg \left\{ \min_{\theta} \sum_{i=1}^{N} | x_i - \theta s_i | \right\} \tag{16}$$

Writing this in a slightly different way, as

$$\hat{\theta} = \arg \left\{ \min_{\theta} \sum_{i=1}^{N} | s_i | \left| \theta - \frac{x_i}{s_i} \right| \right\} \tag{17}$$

we find that this ML estimate is the weighted median of normalized observed signal values and that the weights come from the underlying signal shape. The filtering operation obtained from Eq. (17) is called *matched median filtering* [3]. It forms the counterpart "in the median world" of the fundamental linear matched filter forming the basis of most communications systems. It is shown in [3] that the matched median filter, when used either for signal detection and/or pulse compression, outperforms the linear matched filter if the noise contains enough impulsive components. In image processing the natural application of matched median filtering is in the search for specific forms buried in impulsive noise.

We can also consider the extension of median operations to vector-valued signals. Depending on the application, vector-valued signals are also called *multispectral, multivariate, multisample,* or *multichannel* signals. Typical examples of vector-valued signals are multispectral satellite images and standard color images in television systems. In these examples, the various components contain information from different parts of the spectrum of the underlying signal. Another example of a vector-valued signal is the signal representing the instantaneous velocity of an object.

Let the vector-valued signal have d components $x_1(n), \ldots, x_d(n)$, $d \geq 1$, forming the signal

$$\mathbf{x}(n) = (x_1(n), \ldots, x_d(n))^T \tag{18}$$

where the superscript T denotes the transpose. A natural approach to processing $\mathbf{x}$ is to design filters T_i, $i = 1, \ldots, d$, for each scalar signal $x_i(n)$, $i = 1, \ldots,$ d, separately, and apply them to components separately. Hence the filter output is

$$\mathbf{y}(n) = (T_1[x_1(n)], \ldots, T_d[x_d(n)])^T$$

This method, however, has some drawbacks. The signal components in real applications are in general dependent, and if each component is processed separately, this dependence is not utilized. The previous approach, where the median was interpreted as the ML estimate of the location parameter, gives a direct way to extend the median filter to vector-valued signals. We shall discuss it briefly for two-dimensional signals. Suppose that we have observations $(x_1, y_1), \ldots,$ (x_N, y_N) from a two-dimensional population which is characterized by the density

$$f(s, t) = \rho e^{-\beta | (s,t) - (u,v) |} \tag{19}$$

where ρ and β are constants, the center (u, v) of the distribution is not known, and $| \cdot |$ a denotes the Euclidean distance. This distribution is the two-dimensional equivalent of the Laplace distribution. As the Laplace distribution is used to model noise having impulsive characteristics, this distribution naturally models two-dimensional impulsive noise with circular symmetry. The ML estimate for (u, v) is the vector $(\hat{u}, \hat{v})$, minimizing

$$\sum_{i=1}^{N} | (x_i, y_i) - (u, v) |$$

that is,

$$(\hat{u}, \hat{v}) = \arg\left\{ \min_{(u,v)} \sum_{i=1}^{N} | (x_i, y_i) - (u, v) | \right\} \tag{20}$$

In the statistical literature $(\hat{u}, \hat{v})$ is sometimes called the *two-dimensional median*. Notice that $(\hat{u}, \hat{v})$ has the following robustness property, which resembles the insensitivity of the one-dimensional median to variations in one (or more) sample value. Let $(x_1, y_1), \ldots, (x_N, y_N)$ be fixed vectors and $(\hat{u}, \hat{v})$ be as given by Eq. (20). Take any of the vectors $(x_1, y_1), \ldots, (x_N, y_N)$, say (x_p, y_p), and consider the ray through (x_p, y_p) from $(\hat{u}, \hat{v})$. The minimizing point $(\hat{u}, \hat{v})$ will not change if the point (x_p, y_p) moves along the ray. The proof of this fact is straightforward. If instead of the ML rule (13) we estimate (u, v) by the observation (x_i, y_i), minimizing Eq. (20), we obtain the vector median of $(x_1, y_1), \ldots, (x_N, y_N)$. That is, the vector median of $(x_1, y_1), \ldots, (x_N, y_N)$ is the vector (x_k, y_k) among $(x_1, y_1), \ldots, (x_N, y_N)$, which minimizes

$$\sum_{i=1}^{N} | (x_i, y_i) - (x_k, y_k) | \tag{21}$$

The fact that the vector median is defined to be one of the samples within the window is important in color signal processing, as it guarantees that filtering will not introduce false colors into noiseless images. Again, changing the density (19) to a Gaussian density will change the ML estimate to a simple average of $(x_1, y_1), \ldots, (x_N, y_N)$. This can be processed componentwise, which is to be expected, as the components of (x_i, y_i) are then independent. It is worth noticing that although the components of the random vector characterized by Eq. (19) are uncorrelated, they are not independent.

B. Algebraic Approach

In this part we look at some deterministic properties of the median filter and show that this leads to another very useful generalization of the median filter, the stack filter. The stack filter expression in turn provides us with a general method to analyze the statistical properties of weighted median filters and even some morphological filters. If we consider the median filter as a function μ: $R^{2k+1} \rightarrow R$, we see that it has the following properties: (a) For any $a \in R$,

$$\mu(x_1 + a, \ldots, x_{2k+1} + a) = \mu(x_1, \ldots, x_{2k+1}) + a \tag{22}$$

(b) For any $a \in R$,

$$\mu(ax_1, \ldots, ax_{2k+1}) = a\mu(x_1, \ldots, x_{2k+1}) \tag{23}$$

On the other hand, in general,

$$\mu(x_1 + y_1, \ldots, x_{2k+1} + y_{2k+1}) \neq \mu(x_1, \ldots, x_{2k+1})$$
$$+ \mu(y_1, \ldots, y_{2k+1}) \tag{24}$$

and thus the mapping μ is, of course, not linear. Property (a) says that the median operation is spatially translation invariant, and property (b) says that it is invariant to scaling by any real number. In fact, a far stronger invariance holds. Let $\xi: R \rightarrow R$ be monotonic. Then

$$\mu(\xi(x_1), \ldots, \xi(x_{2k+1})) = \xi(\mu(x_1, \ldots, x_{2k+1})) \tag{25}$$

Let ξ_a be defined by

$$\xi_a(t) = \begin{cases} 1 & \text{if } t > a \\ 0 & \text{otherwise} \end{cases} \tag{26}$$

Applying Eq. (25) with ξ_a we find that

$$\mu(\xi_a(x_1), \ldots, \xi_a(x_{2k+1})) = \xi_a(\mu(x_1, \ldots, x_{2k+1})) \tag{27}$$

The interpretation of Eq. (27) is obvious; if we want to know whether $\mu(x_1, \ldots, x_{2k+1})$ is greater than a, it is enough to check if at least $k + 1$ of the x_i are greater than a. The fact that Eq. (27) can be understood as a Boolean function leads to an important generalization of the median operation, the stack filter.

It is very difficult to analyze or optimally design nonlinear filters. Almost all the powerful tools and elegant methods of linear digital signal processing are completely useless when we are dealing with nonlinear methods. However, the output of the median filter is decided solely on the basis of the ranks of the samples. This property makes it possible to use a powerful technique, called *threshold decomposition*, to divide the analysis into smaller and simpler parts. Using threshold decomposition we can, by Eq. (25), derive all properties of median filters by just studying their effect on binary signals.

Threshold decomposition of an *M*-valued signal $x(n)$, where the samples are integer valued and $0 \leq x(n) \leq M - 1$, means decomposing it into $M - 1$ binary signals $x^{(1)}(n), \ldots x^{(M-1)}(n)$ according to the thresholding rule,

$$x^{(m)}(n) = \begin{cases} 1 & \text{if } x(n) \geq m \\ 0 & \text{otherwise} \end{cases} \tag{28}$$

Let **u** and **v** be binary signals (sequences) of fixed length. Define

$$\mathbf{u} \leq \mathbf{v} \quad \text{iff } u(n) \leq v(n) \quad \text{for all } n \tag{29}$$

As the relation defined by Eq. (29) is reflexive, antisymmetric, and transitive, it defines a partial order in the set of binary signals of fixed length. Now, consider a signal **x** and its thresholded binary signals $\mathbf{x}^{(1)}, \ldots, \mathbf{x}^{(M-1)}$. It is obvious that

$$\mathbf{x}^{(i)} \leq \mathbf{x}^{(j)} \quad \text{if } i \geq j \tag{30}$$

Thus the binary signals $\mathbf{x}^{(1)}, \ldots, \mathbf{x}^{(M-1)}$ are nonincreasing in the sense of the partial ordering (29).

Example.   Consider a five-level ($M = 5$) integer-valued signal $x(n)$ (i.e., x is a mapping $x: Z \rightarrow \{0, 1, 2, 3, 4\}$ and its section $0 \leq n \leq 10$:

$$x(n) = \ldots 0 \ 0 \ 1 \ 3 \ 2 \ 4 \ 4 \ 0 \ 0 \ 4 \ 0 \ \ldots \tag{31}$$

Its thresholded binary signals are given by

$$\begin{aligned}
x^{(4)}(n) &= \ldots 0 \ 0 \ 0 \ 0 \ 0 \ 1 \ 1 \ 0 \ 0 \ 1 \ 0 \ \ldots \\
x^{(3)}(n) &= \ldots 0 \ 0 \ 0 \ 1 \ 0 \ 1 \ 1 \ 0 \ 0 \ 1 \ 0 \ \ldots \\
x^{(2)}(n) &= \ldots 0 \ 0 \ 0 \ 1 \ 1 \ 1 \ 1 \ 0 \ 0 \ 1 \ 0 \ \ldots \\
x^{(1)}(n) &= \ldots 0 \ 0 \ 1 \ 1 \ 1 \ 1 \ 1 \ 0 \ 0 \ 1 \ 0 \ \ldots
\end{aligned} \tag{32}$$

We see in Eq. (31) that (29) is true and clearly

$$x(n) = \sum_{i=0}^{M-1} x^{(i)}(n) \tag{33}$$

This illustrates the fact which follows from Eq. (25) that median filtering of (31) can be done by separately filtering the threshold signals and then adding them. We can write this as the following three-stage procedure:

1. Decompose the signal into $M - 1$ binary signals according to Eq. (28).
2. Apply the filter to the binary signals separately.
3. Add the filtered signals.

From threshold decomposition we know that the thresholded binary signals satisfy

$$\mathbf{x}^{(i)} \leq \mathbf{x}^{(j)} \qquad \text{if} \quad i \geq j$$

Property (27) implies also that the binary output signals $\mathbf{y}^i$ satisfy

$$\mathbf{y}^{(i)} \leq \mathbf{y}^{(j)} \qquad \text{if} \quad i \geq j \tag{34}$$

Median filtering of a binary signal is essentially a Boolean function of the $2k + 1$ variables inside the filter window. For example, the binary median filtering with window length 3 can be written as

$$y(n) = x(n - 1)x(n) + x(n - 1)x(n + 1) + x(n)x(n + 1)$$

where $+$ and $\cdot$ denote Boolean OR and AND operations.

At this point a natural question arises: Can we use any Boolean function to define a filtering operation via threshold decomposition? It turns out that to obtain useful properties the class of Boolean functions must be restricted in the way that the binary output signals $\mathbf{y}^{(i)}$ satisfy Eq. (34), which means essentially that the "stack" of output binary signals corresponds to the threshold decomposition of some M-valued signal. This suitably restricted class of Boolean functions, that is, positive Boolean functions, form the basis of stack filters first introduced in [41].

Let f be a Boolean function with arguments indexed from $-k$ to k, k a nonnegative integer, and use the notation $\mathbf{x} = (x_{-k}, \ldots, x_k)$ for the elements of $\{0, 1\}^{2k + 1}$. Now f is called a positive Boolean function if it satisfies

$$\mathbf{u} \leq \mathbf{v} \Rightarrow f(\mathbf{u}) \leq f(\mathbf{v})$$

It is well known (cf. [24,32]) that a Boolean function is positive if and only if it contains no complemented variables in its minimum sum-of-products form.

Example. Let $k = 1$. The functions

$$f(\mathbf{x}) = x_{-1}x_0 + x_{-1}x_1 + x_0x_1$$

and

$$f(\mathbf{x}) = x_0 + x_{-1}x_1$$

are positive Boolean functions.

An M-valued stack filter is defined using threshold decomposition and a positive Boolean function $f(\mathbf{x}) = f(x_{-k}, \ldots, x_k)$ as follows. Let $\mathbf{x}$ be an M-valued input signal, that is, $x(n) \in \{0, 1, \ldots, M - 1\}$ for all $n \in Z$. Form the thresholded binary signals $\mathbf{x}^{(1)}, \ldots, \mathbf{x}^{(M-1)}$ by

$$x^{(i)}(n) = \begin{cases} 1 & \text{if } x(n) \geq i \\ 0 & \text{otherwise} \end{cases} \tag{35}$$

The output of filtering the ith thresholded signal $\mathbf{x}^{(i)}$ at point n is defined by

$$y^{(i)}(n) = f(x^{(i)}(n - k), \ldots, x^{(i)}(n + k)) \tag{36}$$

where $x^{(i)}(n + j)$, $-k \leq j \leq k$, are understood as Boolean variables. The output of the stack filter defined by f at point n is now

$$y(n) = \sum_{i=1}^{M-1} y^{(i)}(n) \tag{37}$$

where the values $y^{(i)}(n)$ of Boolean functions are now understood as real 0's and 1's.

Example. Consider the Boolean function $f(\mathbf{x}) = x_{-1}x_1 + x_0$ and a five-valued signal segment $\ldots 00212340100 \ldots$ The stack filtering can now be expressed schematically as in Fig. 2.

Earlier we defined the weighted median of $x_{-k}, \ldots, x_k$ as the value minimizing the expression (10) and noticed that it is also the standard median of a suitably extended set of numbers. In the following we see that the weighted median filter is, in fact, a stack filter where the positive Boolean function is of a special type: self-dual and linearly separable.

Consider the computation of the weighted median of length $2k + 1$ with positive integer weights $w_{-k}, \ldots, w_k$. By Eq. (10)

$$\mathrm{WM}[w_{-k} \lozenge x_{-k}, \ldots, w_k \lozenge x_k] = \arg\left\{\min_t \sum_{i=-k}^{k} w_i \,|\, x_i - t \,|\right\}$$

The objective function

$$s(t) = \sum_{-k}^{k} w_i \,|\, x_i - t \,| \tag{38}$$

is everywhere continuous, and differentiable if $t \not\in \{x_{-k}, \ldots, x_k\}$. Its derivative is

$$s'(t) = - \sum_{(x_i < t)} w_i + \sum_{(x_i > t)} w_i \tag{39}$$

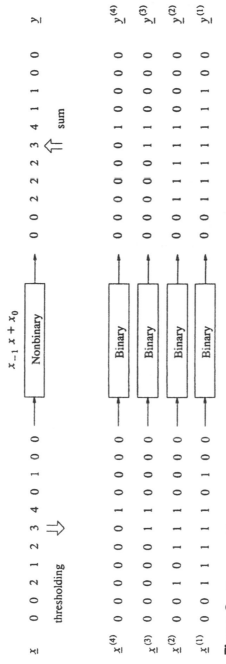

Figure 2   How a nonbinary stack filter is defined via threshold decomposition and binary filtering.

From Eq. (39) we see that $s(t)$ has unique minimum for any choise of $\{x_{-k}, \ldots, x_k\}$ if and only if there is no partition

$$\{x_{-k}, \ldots, x_k\} = A \cup B, \qquad A \cap B = \varnothing \tag{40}$$

such that

$$\sum_{i \in A} w_i = \sum_{i \in B} w_i \tag{41}$$

Writing Eq. (41) as

$$2 \sum_{i \in A} w_i = \sum_{i \in A} w_i + \sum_{i \in B} w_i = \sum_{i=-k}^{k} w_i \tag{42}$$

we see that a sufficient condition for a weighted median with positive integer weights to be unique is that the sum of weights is odd.

Equation (39) gives a straightforward method to compute WM $[w_{-k} \diamond x_{-k}, \ldots, w_k \diamond x_k]$: Denote by $((-k), (-k+1), \ldots, (k))$ a permutation of $(-k, -k+1, \ldots, k)$ that arranges $x_{-k}, \ldots, x_k$ into an increasing order, that is,

$$x_{(-k)} \leq x_{(-k+1)} \leq \ldots \leq x_{(k)}$$

Compute the sums

$$\sum_{i=-k}^{l} w_{(i)}, \qquad l = -k, -k+1, \ldots \tag{43}$$

and let l_0 be the smallest index such that

$$\sum_{i=-k}^{l} w_{(i)} \geq \frac{1}{2} \sum_{i=-k}^{k} w_{(i)} \tag{44}$$

Then $x_{l_0} = \text{WM}[w_{-k} \diamond x_{-k}, \ldots, w_k \diamond x_k]$.

Example. Compute WM $[1 \diamond 3, 2 \diamond 2, 3 \diamond 1, 2 \diamond 5, 1 \diamond 4]$. The permutation that orders the elements is $((-2), (-1), (0), (1), (2)) = (0, -1, -2, 2, 1)$ and

$$\sum_{i=-2}^{-2} w_{(i)} = 3, \qquad \sum_{i=-2}^{-1} w_{(i)} = 5 \geq \frac{9}{2}$$

Thus WM $[1 \diamond 2, 2 \diamond 3, 3 \diamond 1, 2 \diamond 5, 1 \diamond 4] = x_{(-1)} = x_{-2} = 2$.

Consider the Boolean function $f(x)$ that will realize WM $[w_{-k} \diamond x_{-k}, \ldots, w_k \diamond x_k]$ on each level of threshold decomposition. From (43) and (44) we see that it can be written in the form

$$f(x_{-k}, \ldots, x_k) = \begin{cases} 1 & \text{if } \sum_{i=-k}^{k} w_i x_i > \frac{1}{2} \sum_{i=-k}^{k} w_i \\ 0 & \text{otherwise} \end{cases} \tag{45}$$

A Boolean function that can be expressed in the form

$$f(x_{-k}, \ldots, x_k) = \begin{cases} 1 & \text{if } \sum_{i=-k}^{k} w_i x_i > T \\ 0 & \text{otherwise} \end{cases}$$

is called *linearly separable*. A Boolean function $f(\mathbf{x}) = f(x_{-k}, \ldots, x_k)$ satisfying

$$\bar{f}(\bar{\mathbf{x}}) = f(\mathbf{x}) \qquad \text{for all } \mathbf{x} \tag{46}$$

where $\bar{\mathbf{x}}$ means the negation of $\mathbf{x}$, is called *self-dual*. By direct computation we
see that any Boolean function of the form (45), where w_i are positive integers
with Σw_i odd, is self-dual: Let $\mathbf{x} \in \{0,1\}^{2k+1}$. The following chain of equiva-
lences holds:

$$f(\mathbf{x}) = 1 \Leftrightarrow \sum_{i=-k}^{k} w_i x_i > \frac{1}{2} \sum_{i=-k}^{k} w_i$$

$$\Leftrightarrow \sum_{i=-k}^{k} w_i(1 - x_i) < \frac{1}{2} \sum_{i=-k}^{k} w_i$$

$$\Leftrightarrow f(\bar{\mathbf{x}}) = 0 \tag{47}$$

proving the assertion.

There are two very important advantages of the stack filter expression of median filters. First, the analysis of iterated or cascaded operations becomes in a
sense trivial because we only need to compute the Boolean function obtained by
substituting a shifted version of the function into each variable.

Example. Consider the operation that results when a three-point median is ap-
plied twice to a signal $x(n)$. The corresponding Boolean function is $f(x_{-1}, x_0, x_1) = x_{-1}x_0 + x_{-1}x_1 + x_0x_1$. Thus after one pass of the three-point median the
resulting signal can be expressed (with Boolean operations) as

$$y(n) = x(n - 1)x(n) + x(n - 1)x(n + 1) + x(n)x(n + 1)$$

and thus the final result is

$$\begin{aligned} z(n) &= y(n - 1)y(n) + y(n - 1)y(n + 1) + y(n)y(n + 1) \\ &= x(n - 2)x(n - 1)x(n + 1) + x(n - 2)x(n)x(n + 2) \\ &\quad + x(n - 1)x(n) \\ &\quad + x(n - 1)x(n + 1)x(n - 2) + x(n)x(n + 1) \end{aligned}$$

It is interesting that this particular Boolean function is linearly separable and the resulting stack filter is a weighted median

$$z(n) = \text{WM}[1 \diamond x(n - 2), 2 \diamond x(n - 1), 3 \diamond x(n),$$
$$2 \diamond x(n + 1), 1 \diamond x(n + 2)] \tag{48}$$

Example. Consider the weighted median filter

$$y(n) = \text{WM}[1 \diamond x(n - k), \ldots, 1 \diamond x(n - 1),$$
$$(2k - 1) \diamond x(n), 1 \diamond x(n + 1), \ldots, 1 \diamond x(n + k)] \tag{49}$$

The corresponding Boolean function can be written as

$$f(x_{-k}, \ldots, x_k) = x_0 \sum_{i=1}^{k} (x_{-i} + x_i) + \prod_{i=1}^{k} (x_{-i} x_i) \tag{50}$$

Straightforward calculation shows that if appropriately shifted versions of this Boolean function are substituted into the variables, the resulting function collapses back to its original form (50). This shows that the center weighted median (49) is an idempotent operation. The result holds also for two- and higher-dimensional center weighted median filters of the foregoing type [17]. A more detailed exposition of cascading weighted median and stack filters can be found in [43].

There is a simple expression involving the defining positive Boolean function for the output distribution of a stack filter in the case of i.i.d. inputs [43]. Using this formula, certain statistical properties of stack and median filters can be reduced to the properties of Boolean functions. To derive these results in such a form that they also hold for weighted median filters over real-valued signals, we first define the continuous stack filter. Let $f(x_{-k}, \ldots, x_k)$ be a positive Boolean function (not identically $= 1$) and $\mathbf{x} = \ldots, x(n - 1), x(n), x(n + 1), \ldots$ a real-valued signal. Define a threshold function $T(x, \beta)$ by

$$T(x, \beta) = \begin{cases} 1 & \text{if } x \geq \beta \\ 0 & \text{otherwise} \end{cases} \tag{51}$$

Then the output of the continuous stack filter defined by f with input $\mathbf{x}$ at time instant n is

$$y(n) = \max\{\beta : f(T(x(n - k), \beta), \ldots, T(x(n + k), \beta)) = 1\} \tag{52}$$

Example. If $f(x_{-1}, x_0, x_1) = x_{-1}x_0 + x_{-1}x_1 + x_0x_1$, we have just the three-point median for real-valued signals.

We now investigate the random variable defined by the output of a continuous stack filter when the inputs are real-valued random variables. We shall consider the case when the input components (samples) of the input signal are identically distributed independent random variables. As we are interested only in the statistical properties of the output, we shall adopt the more standard indexing of the variables of the positive Boolean function. That is, we consider the positive Boolean function $f(\mathbf{x}) = f(x_1, \ldots, x_n)$. Let the inputs $X_1, \ldots, X_n$ be independent, identically distributed random variables with a common distribution function $\Phi(t)$ and define the random variable Y by

$$Y = \max\{\beta \mid f(T(X_1, \beta), \ldots, T(X_n, \beta)) = 1\} \tag{53}$$

Now, the distribution function $\Psi(t)$ of Y can be written as

$$\Psi(t) = \sum_{i=0}^{n} A_i (1 - \Phi(t))^i \Phi(t)^{n-i} \tag{54}$$

where the numbers A_i are defined by

$$A_i = |\{\mathbf{x} \in \{0,1\}^n : f(\mathbf{x}) = 0, w_H(\mathbf{x}) = i\}|$$

where $|S|$ means the cardinality of the set S and $w_H(\mathbf{x})$ denotes the number of 1's in $\mathbf{x}$, that is, its Hamming weight. Now

$$\Psi(t) = P\{Y \leq t\}$$

and the input space can be divided into 2^n mutually exclusive events

$$(-\infty, t] \times (-\infty, t] \times \cdots \times (-\infty, t]$$
$$(t, \infty) \times (-\infty, t] \times \cdots \times (-\infty, t]$$
$$\cdots$$
$$(t, \infty) \times (t, \infty) \times \cdots \times (t, \infty)) \tag{55}$$

A typical event having i terms of type (t, ∞) and $n - i$ terms of type $(-\infty, t]$ has probability $(1 - \Phi(t))^i \Phi(t)^{n-i}$. It is easy to see that the event $\{Y \leq t\}$ is the union of exactly those events in (55) whose terms of type (t, ∞) match with 1's in some $\mathbf{x} \in \{0,1\}^n$ with $f(\mathbf{x}) = 0$. As the events in (55) are mutually exclusive, we can write the probability of the event $\{Y \leq t\}$ as the sum (54).

Example. Compute the output distribution function of the weighted median WM [1 ◊ x(n − 2), 2 ◊ x(n − 1), 3 ◊ x(n), 2 ◊ x(n + 1), 1 ◊ x(n + 2)] when the input signal values are independent and identically distributed with a common distribution function $\Phi(t)$. Formula (54) shows that we only need to compute the number of distinct Boolean vectors of each Hamming weight $i = 0, 1, 2, 3, 4, 5$ such that $f(\mathbf{x}) = 0$, where f is the linearly separable Boolean function

$$f(x_1, x_2, x_3, x_4, x_5) = \begin{cases} 1 & \text{if } x_1 + 2x_2 + 3x_3 + 2x_4 + x_5 > 3 \\ 0 & \text{otherwise} \end{cases}$$

By listing the possibilities we see that the numbers A_i are $A_0 = 1$, $A_1 = 5$, $A_2 = 6$, $A_3 = 4$, $A_4 = A_5 = 0$. Thus

$$\Psi(t) = \Phi(t)^5 + 5(1 - \Phi(t))\Phi(t)^4 + 6(1 - \Phi(t))^2\Phi(t)^3$$
$$+ 4(1 - \Phi(t))^3\Phi(t)^2$$
$$= 4\Phi(t)^2 - 6\Phi(t)^3 + 5\Phi(t)^4 - 2\Phi(t)^5$$

In the actual computation of the numbers A_i for a weighted median it is not necessary to list the vectors $x \in \{0, 1\}^n$ for which $f(x) = 0$. The computation can be done extremely fast using generating functions. Consider the weighted median WM $[w_1 \diamond x_1, \ldots, w_n \diamond x_n]$, where $\sum_{i=0}^n w_i = 2T + 1$. We form the polynomial

$$P(\xi, \eta) = \prod_{i=1}^n (1 + \xi\eta^{w_i}) \tag{56}$$

which can be written in the form

$$P(\xi, \eta) = \sum_{x \in \{0,1\}^n} \prod_{i=1}^n (\xi\eta^{w_i})^{x_i}$$

$$= \sum_{x \in \{0,1\}^n} \xi^{\sum x_i}\eta^{\sum w_i x_i} \tag{57}$$

From Eq. (57) we see that $P(\xi, \eta)$ has one term for each $x \in \{0, 1\}^n$, the exponent of ξ is the Hamming weight, and the exponent of η is $\sum w_i x_i$. Deleting all terms whose power of η is at least $T + 1$, we obtain exactly the vectors x for which WM $[w_1 \diamond x_1, \ldots, w_n \diamond x_n] = 0$. Thus the procedure for computing A_i can be written as follows:

1. Form $P(\xi, \eta) = \prod_{i=1}^n (1 + \xi\eta^{w_i})$.
2. Expand $P(\xi, \eta) = \sum \xi^{\sum x_i}\eta^{\sum w_i x_i}$.
3. Collect the powers of η: $P(\xi, \eta) = \sum_{k=0}^{2T+1} S_k(\xi)\eta^k$.
4. Truncate with respect to η at T: $Q(\xi, \eta) = \sum_{k=0}^T S_k(\xi)\eta^k$.
5. Now $R(\xi) = Q(\xi, 1) = \sum_{i=0}^n A_i\xi^i$.

Notice that once we have the generating function

$$R(\xi) = \sum_{i=0}^n A_i\xi^i \tag{58}$$

by (54), the output distribution $\Psi(t)$ can simply be written as

$$\Psi(t) = \Phi(t)^n R\left(\frac{1 - \Phi(t)}{\Phi(t)}\right) \tag{59}$$

Example. We compute the output distribution of WM $[1 \diamond x_1, 2 \diamond x_2, 3 \diamond x_3, 2 \diamond x_4, 1 \diamond x_5]$ using generating functions. Now

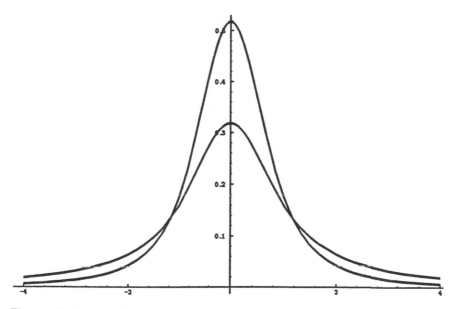

Figure 3 Input and output densities of WM [1 $\lozenge$ x_2, 2 $\lozenge$ x_2, 3 $\lozenge$ x_3, 2 $\lozenge$ x_4, 1 $\lozenge$ x_5] for Cauchy-distributed input.

$$P(\xi,\eta) = (1 + \xi\eta)^2(1 + \xi\eta^2)^2(1 + \xi\eta^3)$$
$$= 1 + 2\xi\eta + (2\xi + \xi^2)\eta^2 + (\xi + 4\xi^2)\eta^3$$
$$+ (3\xi^2 + 2\xi^3)\eta^4 + (2\xi^2 + 3\xi^3)\eta^5 + (4\xi^3 + \xi^4)\eta^6$$
$$+ (\xi^3 + 2\xi^4))\eta^7 + 2\xi^4\eta^8 + \xi^5\eta^9$$

and

$$Q(\xi, \eta) = 1 + 2\xi\eta + (2\xi + \xi^2)\eta^2 + (\xi + 4\xi^2)\eta^3 + (3\xi^2 + 2\xi^3)\eta^4$$

Thus the generating function for A_i is

$$R(\xi) = 1 + 5\xi + 8\xi^2 + 2\xi^3$$

giving

$$\Psi(t) = \Phi(t)^n\, R\!\left(\frac{1 - \Phi(t)}{\Phi(t)}\right)$$
$$= 2\Phi(t)^2 + 2\Phi(t)^3 - 5\Phi(t)^4 + 2\Phi(t)^5 \tag{60}$$

In Fig. 3 the noise attenuation capability of this filter is illustrated by plotting the input and output densities in the case of Cauchy distributed input.

II. MORPHOLOGICAL FILTERS

Conceptually, morphological filters are rooted in planar geometric structure and the desire to quantify or alter that structure by probing with structuring elements. At first sight it might appear that a geometric conception of filtering is quite distinct from numerical-based median filtering; however, the proper mathematical perspective shows that such is not the case. Median filtering is grounded in the ordering of the real numbers (or integers). Stack filters are also based on ordering. The summation expression of Eq. (37) that applies to discrete stack filters is replaced by the maximum expression of Eq. (52) for continuous stack filters, but in fact, the discrete summation is merely historical and is equivalent to the digital counterpart of the latter maximum expression. The key is order. Binary morphological filtering is grounded in the subset relation, and there, too, the key is order. When extended to real-valued functions, the order-induced morphological operations produce a class of filters that contains stack filters, and therefore median filters, as a subclass.

As a discipline within imaging, *mathematical morphology* concerns the application of its basic operators to all aspects of image processing. A number of full-length books are dedicated to mathematical morphology, its theory, and applications: Serra [36,37], Serra and Vincent [39], Dougherty [7,8], and Giardina and Dougherty [16]. Because this chapter concerns nonlinear filtering, we view the fundamental morphological operations as filters and study their properties as filters.

A. Binary Filters

When we say that morphological processing is geometrically based, we mean this in a very specific sense. The basic idea, conceived by Matheron [31] and Serra [36], is to probe an image with a *structuring element* and to quantify the manner in which the structuring element fits (or does not fit) within the image. Figure 4 depicts a binary image and a circular structuring element (probe). The structuring element is placed in two different positions. In one location it fits; in the other it does not fit. By marking the locations at which the structuring element fits within the image, we derive structural information concerning the image. This information depends on both the size and shape of the structuring element. Thus the nature of that information is dependent on the choice of structuring element.

Formal characterization of fitting depends on the subset relation and the *translation*, $A + x$, of a set A by a point x,

$$A + x = \{a + x : a \in A\} \tag{61}$$

Geometrically, $A + x$ is A translated along the vector x. The nature of probing is to mark the positions (translations) of a structuring element where it fits into

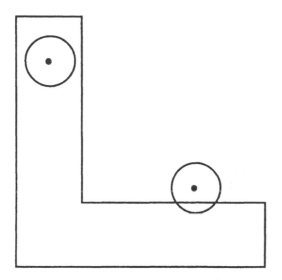

Figure 4 Structuring element fitting and not fitting.

an image. The fundamental operation of mathematical morphology, *erosion*, is defined by

$$A \ominus B = \{x : B + x \subseteq A\} \tag{62}$$

where $\subseteq$ denotes the subset relation. Relative to erosion, A is the input image and B is the structuring element. $A \ominus B$ consists of all points x for which the translation of B by x fits inside A.

If the origin lies inside the structuring element, erosion has the effect of shrinking the input image (see Fig. 5, where the structuring element B is a disk). Formally, if the origin is contained within the structuring element, the eroded image is a subset of the input image. Should the origin not lie within the struc-

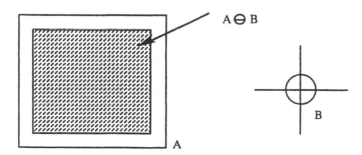

Figure 5 Erosion by a disk.

turing element, it may not be that the eroded image lies within the input image. Many important applications employ structuring elements not containing the origin, in particular, image restoration by filling holes in an image.

Example.   Consider the image

$$S = \begin{pmatrix} 1 & 1 & 1 & 1 & 1 & 1 & 1 \\ 1 & 1 & 0 & 1 & 1 & 1 & 1 \\ 1 & 1 & 1 & 1 & 0 & 1 & 1 \\ 1 & 0 & 1 & 0 & 1 & 1 & 1 \\ \mathbf{1} & 1 & 1 & 1 & 1 & 0 & 1 \end{pmatrix}$$

where 1 and 0 denote activated and nonactivated pixels, respectively, and the bold entry denotes the value at the origin of the Cartesian grid. From the perspective of restoration, S might represent a 5×7 square with some salt holes. Eroding S by $E = (1 \quad \mathbf{0} \quad 1)$ yields

$$S \ominus E = \begin{pmatrix} 0 & 1 & 1 & 1 & 1 & 1 & 0 \\ 0 & 0 & 1 & 0 & 1 & 1 & 0 \\ 0 & 1 & 1 & 0 & 1 & 0 & 0 \\ 0 & 1 & 0 & 1 & 0 & 1 & 0 \\ \mathbf{0} & 1 & 1 & 1 & 0 & 1 & 0 \end{pmatrix}$$

Restoration of the full 5×7 square is given by $S \cup (S \ominus E)$.

Erosion can be formulated in other ways besides the fitting characterization of Eq. (62). Of particular importance is its representation by an intersection of image translates:

$$A \ominus B = \cap \{A - b : b \in B\} \tag{63}$$

The intersection formulation of erosion is related to the classical *Minkowski subtraction* of A by B, which, relative to the intersection of Eq. (63), is defined by $A \ominus (-B)$, where $-B = \{-b : b \in B\}$ is the reflection of B through the origin.

The second basic operation of binary mathematical morphology is *dilation* of A by B. It is *dual* to erosion, in that it is defined via erosion by set complementation. Specifically,

$$A \oplus B = [A^c \ominus (-B)]^c \tag{64}$$

where A^c denotes the complement of A. Dilation is commutative, $A \oplus B = B \oplus A$, and associative, $(A \oplus B) \oplus C = A \oplus (B \oplus C)$. As indicated by Fig. 6, where B is a disk, if B contains the origin, then dilation of A by B results in an expansion of A. Since dilation involves a fitting into the complement of an image, it represents a filtering on the outside, whereas erosion represents a filtering on the inside.

There are equivalent formulations of dilation that deserve mention. The historical *Minkowski addition* formulation is

A ⊕ B

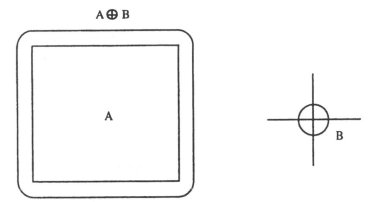

Figure 6 Dilation by a disk.

$$A \oplus B = \cup \{A + b : b \in B\} \tag{65}$$

By commutativity,

$$A \oplus B = \cup \{B + a : a \in A\} \tag{66}$$

The formulation of Eq. (66) is often used for image-processing intuition: dilation by B is formed by translating B to all points of A and then forming the union. A third formulation of dilation involves translates of the reflected structuring element that "hit" (intersect) the input image:

$$A \oplus B = \{x : (-B + x) \cap A \neq \varnothing\} \tag{67}$$

Our central interest is morphological filters. Generically, a binary-image filter is an operator Ψ that takes as input a binary image and outputs a binary image. If A is the input image, $\Psi(A)$ denotes the output. For instance, if Ψ is erosion by B, then $\Psi(A) = A \ominus B$. The study of any class of filters is the study of the properties defining the class. We focus on four properties, two immediately and two subsequently.

A filter Ψ is said to be *translation invariant* if

$$\Psi(A + x) = \Psi(A) + x \tag{68}$$

Translating and then filtering is equivalent to filtering and then translating. Dilation and erosion are translation invariant:

$$(A + x) \oplus B = (A \oplus B) + x \tag{69}$$

$$(A + x) \ominus B = (A \ominus B) + x \tag{70}$$

As a filter property, translation invariance applies to the input image. Nevertheless, because dilation is commutative,

$$A \oplus (B + x) = (A \oplus B) + x \tag{71}$$

so that dilation is also translation invariant relative to the structuring element. Not so for erosion; in fact,

$$A \ominus (B + x) = (A \ominus B) - x \tag{72}$$

so that translating the structuring element by x prior to eroding is equivalent to eroding and then translating by $-x$.

Filter Ψ is said to be *monotonically increasing* if whenever A_1 is a subset of A_2, then $\Psi(A_1)$ is a subset of $\Psi(A_2)$, so that Ψ preserves order. For fixed structuring element, both dilation and erosion are increasing: if $A_1 \subseteq A_2$, then $A_1 \oplus B \subseteq A_2 \oplus B$ and $A_1 \ominus B \subseteq A_2 \ominus B$.

Increasing monotonicity for erosion is relative to a fixed structuring element and input images ordered by set inclusion; a reverse phenomenon occurs if the input image is fixed and two ordered structuring elements are employed. If A is a fixed image and $B_1 \subseteq B_2$, it is easier to fit B_1 inside A than to fit B_2. Hence $A \ominus B_1 \supseteq A \ominus B_2$. This property is central to the design of morphological filters involving parallel erosions. We have mentioned that dilation is dual to erosion. In general, the *dual filter* of a filter Ψ is defined by

$$\Psi^*(A) = \Psi(A^c)^c \tag{73}$$

According to Eq. (64), dilation by B is the dual of erosion by $-B$. If we apply Eq. (64) to the dilation of A^c by $-B$ (instead of A by B) and then take the complement of each side [recognizing that $(A^c)^c = A$ and $-(-B) = B$], we obtain

$$A \ominus B = [A^c \oplus (-B)]^c \tag{74}$$

so that erosion by B is the dual of dilation by $-B$. The equation pair Eqs. (64) and (74) illustrate a general filter property: the dual of the dual is the original filter: $\Psi^{**} = \Psi$.

Two secondary operations play key roles in morphological filtering. The *opening* of image A by structuring element B can be defined as an iteration of erosion and dilation, namely,

$$A \circ B = (A \ominus B) \oplus B \tag{75}$$

However, for filtering applications it is better to view it in terms of an alternative formulation:

$$A \circ B = \cup \{B + x : B + x \subseteq A\} \tag{76}$$

Here the opening results from unioning all translations of the structuring element that fit inside the input image. Each fit is marked and the opening results from taking the union of the structuring-element translations to each marked location. Indeed, this is precisely what is meant by eroding and then dilating. Figure 7 illustrates opening by a disk. In it, one can see how the opening is the

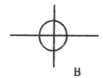

Figure 7 Opening by a disk.

region swept out by "sliding" the disk about the inside of the input image. From the perspective of filtering, opening by a disk yields a filter that smooths from the inside; that is, it rounds corners extending into the background. The roundness of the disk yields a "low-pass" effect; the effect would be quite different with a square structuring element.

The dual operation to opening is *closing*, defined by

$$A \cdot B = [A + (-B)] \ominus (-B) \tag{77}$$

Duality with respect to opening is expressed by

$$A \cdot B = (A^c \circ B)^c \tag{78}$$

Because closing is dual to opening, opening is dual to closing. Replacing A by A^c in Eq. (78) and complementing yields

$$A \circ B = (A^c \cdot B)^c \tag{79}$$

Whereas opening filters from the inside, closing filters from the outside. Moreover, whereas the position of the origin relative to the structuring element plays a role in both erosion and dilation, it plays no role in opening and closing.

Opening and closing are both translation invariant:

$$(A + x) \circ B = (A \circ B) + x \tag{80}$$

$$(A + x) \cdot B = (A \cdot B) + x \tag{81}$$

They are also increasing: if $A_1 \subseteq A_2$, then $A_1 \circ B \subseteq A_2 \circ B$ and $A \cdot B \subseteq A_2 \cdot B$.

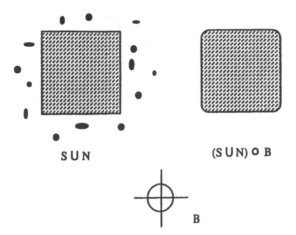

SUN   (SUN) O B

B

Figure 8   Restoration of rectangle degraded by pepper noise.

As noted previously, there are four filter properties that are of key concern in morphological filtering. The two remaining ones are of special import to opening and closing. Filter Ψ is said to be *antiextensive* if $\Psi(A) \subseteq A$ [*extensive* if $\Psi(A) \supseteq A$] for any set A. Opening is antiextensive: $A \circ B \subseteq A$. This follows from the fact that the opening is a union of translates lying within the input image. Closing is extensive: $A \cdot B \supseteq A$.

A filter Ψ is said to be *idempotent* if for any set A, $\Psi(\Psi(A)) = \Psi(A)$. Both opening and closing are idempotent:

$$(A \circ B) \circ B = A \circ B \tag{82}$$

$$(A \cdot B) \cdot B = A \cdot B \tag{83}$$

The import of idempotence is that once an image has been opened (closed), successive openings (closings) produce no further effects. This is quite different from erosion, or if we think of linear processing, moving-average filters.

We examine the kind of restoration that can be accomplished by opening (analogous considerations applying to closing). Consider the rectangle degraded by pepper noise in Fig. 8. Opening by the disk B has a restorative effect because the disk does not fit into the small noise components strewn about the background of the rectangle. Except for the rounded corners, the rectangle has been restored.

Consideration of the fitting formulation of opening [Eq. (76)] shows the manner in which the opening acts as a filter: treating the structuring element as a shape primitive, it passes only those portions of the image that are part of some translate of the shape primitive that fits inside the image. If the image is a union of such translates, it is fully passed by the opening; otherwise, the filter eliminates some of the image.

An image-noise model can help to clarify the filter action. Suppose that there is an underlying uncorrupted image S, a noise image N, and a corrupted image $S \cup N$ formed by the union of S with N. Because of increasing montonicity and antiextensivity,

$$S \circ B \subseteq (S \cup N) \circ B \subseteq S \cup N \tag{84}$$

The filtered image lies between the opened uncorrupted image and the noisy image. Should the filter pass the entire uncorrupted image ($S \circ B = S$), the filtered image lies between the uncorrupted and noisy images. We desire a structuring element B that passes little noise and for which $S \circ B$ is close to S. In the event that $S \circ B = S$, S is invariant under opening by B and we say that S is *B-open*. It can be shown that S is B-open if and only if S is a union of translations of B.

Filtering by openings and closings is more problematic if there is both union noise and subtractive noise (salt-and-pepper noise). One strategy is to open to eliminate the pepper noise in the background and then to follow with a closing to eliminate the salt noise in the foreground, the latter elimination working because the salt in the foreground is part of the complement and the structuring element does not fit into those portions of the complement, as it must if they are to be passed by the closing. Such a filter is known as a *open-close*. The difficulty with a open-close filter is that it may not be possible to choose a suitable size structuring element, one sufficiently large to eliminate background noise components but not too large to fit "between" salt holes in the foreground so as not to destroy the image during opening. A way around the size dilemma is to employ an *alternating sequential filter*. Here a sequence of close-open (or open-close) filters are performed iteratively, beginning with a very small structuring element and then proceeding with ever-increasing structuring elements. The strategy is first to eliminate small salt-and-pepper components, thus causing the larger structuring elements to be more likely to fit when they are eventually applied in the process.

An opening passes only those portions of an image conforming to the shape of the structuring element. Suppose that one wishes to pass portions of an image conforming to any one of a number of primitive shapes, not simply a single primitive. Such an effect can be accomplished by using a filter composed of a number of parallel openings, one for each desired shape primitive. The final filter output is the union of the individual openings. As defined by Matheron [31], a filter Ψ is called a *τ-opening* if there exists some class **B** of structuring elements such that

$$\Psi(A) = \cup \{A \circ B : B \in \mathbf{B}\} \tag{85}$$

B is called a *base* for Ψ. A base is not unique: different bases can produce the same filter, however, our desire is to use a base with a small number of primitives. Design of a τ-opening requires finding an appropriate base.

Key to the design of any filter Ψ is its *invariant class*,

$$\mathrm{Inv}[\Psi] = \{A : \Psi(A) = A\} \tag{86}$$

If possible, we would like a filter to pass unchanged those images considered to be uncorrupted while removing noise from corrupted images. Invariant classes of τ-openings are easily characterized: the invariant class of a τ-opening consists of all images that are formed as unions of translates of the base primitives. This is in complete accord with the situation for a single opening, for we have already noted that the invariants of a single opening are those images that are unions of translates of the opening structuring element.

For design purposes in the presence of union noise, if we can express an image as a union of desirable primitives, we can construct a τ-opening that passes the image in its entirety. Of course, should the noise also be made up in part of some of the same primitives, some of it will also be passed. Just as in the case of linear filtering, τ-opening filtering requires a trade-off: it may be necessary to filter out some of the image and to pass some of the noise in order to obtain an optimally filtered image. Design of optimal τ-openings has been considered [11] and design of optimal alternating sequential filters has also been considered [35].

Openings satisfy four filter properties: translation invariance, antiextensivity, increasing monotonicity, and idempotence. Because τ-openings are formed as unions of openings, it might be conjectured that they, too, satisfy the four properties. In fact, as shown by Matheron [31], a filter satisfies the four properties if and only if it is a τ-opening. Because closing is dual to opening, the preceding discussion applies almost without change. Specifically, an intersection of closings is called a τ-*closing*, and a filter is a τ-closing if and only if it is translation invariant, increasing, extensive, and idempotent.

Dilation, erosion, opening, and closing are both increasing and translation invariant. So too are many other filters. For instance, for a digital domain, the binary median, as well as the weighted median, are both increasing and translation invariant. With every translation-invariant filter Ψ there is associated a set of images called the *kernel*, written Ker[Ψ]. An image lies in the kernel if and only its filtered version contains the origin:

$$\mathrm{Ker}[\Psi] = \{A : 0 \in \Psi(A)\} \tag{87}$$

One of the central propositions of morphological filtering, the *Matheron representation theorem* [31], states that every increasing, translation-invariant filter Ψ can be expressed as the union of erosions by all elements in its kernel:

$$\Psi(A) = \cup \{A \ominus B : B \in \text{Ker}[\Psi]\} \tag{88}$$

The representation of Eq. (88) is typically redundant, with many erosions being unnecessary. The reason is straightforward: if B_1 and B_2 are kernel elements for which $B_1 \subseteq B_2$, then $A \ominus B_1 \supseteq A \ominus B_2$, so when taking the union in Eq. (88), erosion by B_2 is unnecessary. Redundancy can usually be eliminated. A subclass of Ker[Ψ] is a *basis* for Ψ if (1) no image in the basis is a proper subset of another image in the basis; and (2) for any image in the kernel, there exists an image in the basis that is a subimage of the kernel image. If a basis exists, it is unique and is denoted by Bas[Ψ]. It is not our intention here to go into theoretical questions concerning when a basis exists and when it does not. Let it suffice to say that in the digital setting, unless an operator is somewhat pathological, it possesses a basis. When a basis exists, the representation of Eq. (88) can be reduced to an expansion over Bas[Ψ] instead of Ker[Ψ]. For more complete, theoretical discussions of the basis representation, see Giardina and Dougherty [16] or Maragos and Schafer [29,30].

Example. If $\Psi(A) = A \circ E$, it is straightforward to show that Bas[Ψ] consists of all translates of E containing the origin. For instance, if E is the strong-neighbor mask at the origin (which consists of the origin together with its horizontal and vertical neighbors), the basis consists of five images:

$$\begin{pmatrix} 0 & 1 & 0 \\ 1 & 1 & 1 \\ 0 & 1 & 0 \end{pmatrix} \quad \begin{pmatrix} 0 & 1 & 0 \\ \mathbf{1} & 1 & 1 \\ 0 & 1 & 0 \end{pmatrix} \quad \begin{pmatrix} 0 & 1 & 0 \\ 1 & \mathbf{1} & 1 \\ 0 & 1 & 0 \end{pmatrix} \quad \begin{pmatrix} 0 & 1 & 0 \\ 1 & 1 & \mathbf{1} \\ 0 & 1 & 0 \end{pmatrix} \quad \begin{pmatrix} 0 & 1 & 0 \\ 1 & 1 & 1 \\ 0 & \mathbf{1} & 0 \end{pmatrix}$$

$A \circ E$ can be evaluated by eroding A by the five structuring elements and then taking the union.

Example. If Ψ is the binary median over the strong-neighbor mask, its basis is composed of the following 10 elements:

$$\begin{pmatrix} 0 & 1 & 0 \\ 1 & \mathbf{1} & 0 \\ 0 & 0 & 0 \end{pmatrix} \quad \begin{pmatrix} 0 & 1 & 0 \\ 1 & \mathbf{0} & 1 \\ 0 & 0 & 0 \end{pmatrix} \quad \begin{pmatrix} 0 & 1 & 0 \\ 0 & 1 & 1 \\ 0 & 0 & 0 \end{pmatrix} \quad \begin{pmatrix} 0 & 1 & 0 \\ 1 & \mathbf{0} & 0 \\ 0 & 1 & 0 \end{pmatrix} \quad \begin{pmatrix} 0 & 1 & 0 \\ 0 & 1 & 0 \\ 0 & 1 & 0 \end{pmatrix}$$

$$\begin{pmatrix} 0 & 1 & 0 \\ 0 & \mathbf{0} & 1 \\ 0 & 1 & 0 \end{pmatrix} \quad \begin{pmatrix} 0 & 0 & 0 \\ 1 & 1 & 1 \\ 0 & 0 & 0 \end{pmatrix} \quad \begin{pmatrix} 0 & 0 & 0 \\ 1 & 1 & 0 \\ 0 & 1 & 0 \end{pmatrix} \quad \begin{pmatrix} 0 & 0 & 0 \\ 1 & \mathbf{0} & 1 \\ 0 & 1 & 0 \end{pmatrix} \quad \begin{pmatrix} 0 & 0 & 0 \\ 0 & 1 & 1 \\ 0 & 1 & 0 \end{pmatrix}$$

Direct observation shows that none of the 10 structuring elements is a proper subset of another. Second, suppose that $A \in \text{Ker}[\Psi]$. Then $\Psi(A)$ contains the origin. Hence when the strong-neighbor mask is placed at the origin, at least three of A's pixels in the mask must be 1, so that at least one of the 10 structuring elements must be a subimage of A. According to the Matheron representation, the strong-neighbor median can be evaluated by eroding by each of the 10 images and then forming the union.

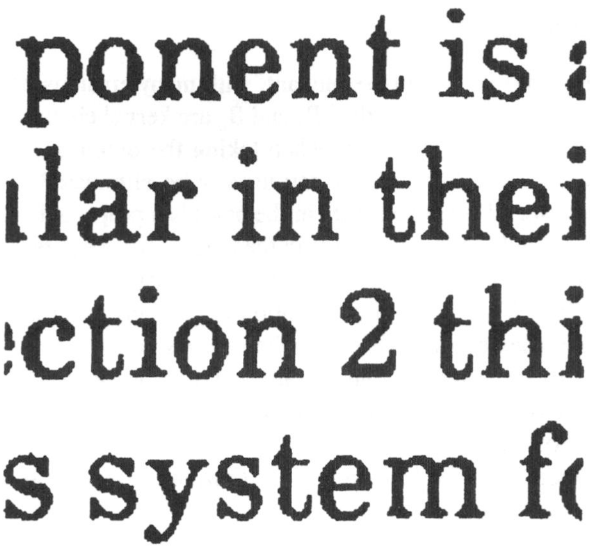

ponent is a
lar in thei
ction 2 thi
s system fo

Figure 9 Text image.

The preceding example shows a strong relation between the median and erosions, the former being a union of the latter. Although the example only treats the binary case, the fact that gray-scale medians can be evaluated using binary medians via threshold decomposition at once implies that all medians can be evaluated by means of erosion. But this reasoning goes much further. Every stack filter can be evaluated via threshold decomposition with the binary operator being a positive Boolean function, which is itself an increasing, translation-invariant binary filter. At once we see that stack filters possess representations in terms of binary erosions [30].

All increasing, translation-invariant filters are unions of erosions. Thus design of good increasing, translation-invariant filters involves searching for a basis that produces the desired filter effects. The next example, reproduced from [7], illustrates restoration of an image degraded by salt noise.

Example. Consider the 256×256 text image of Fig. 9, area coverage being 18.17%. A degraded version of the image is shown in Fig. 10, degradation resulting from subtractive noise. Noise coverage is 10% and each noise component is a randomly generated subset of 3×3 square. Mean-square error for the noisy image is 1.93%. The restored image, shown in Fig. 11, is obtained by way of the Matheron expansion using the eight structuring elements of Fig. 12, and its mean-square error is 0.75%.

Obtaining a basis of structuring elements that provide good restoration is a highly nontrivial problem and is currently under investigation. The basis of the

ponent is :

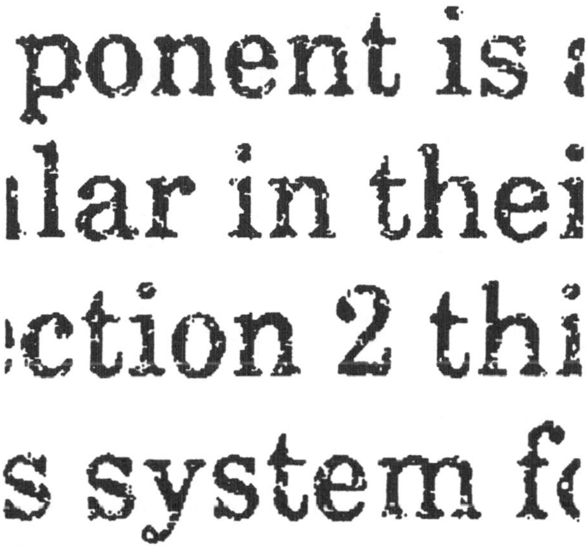

lar in thei

ction 2 thi

s system f(

Figure 10 Noisy text image.

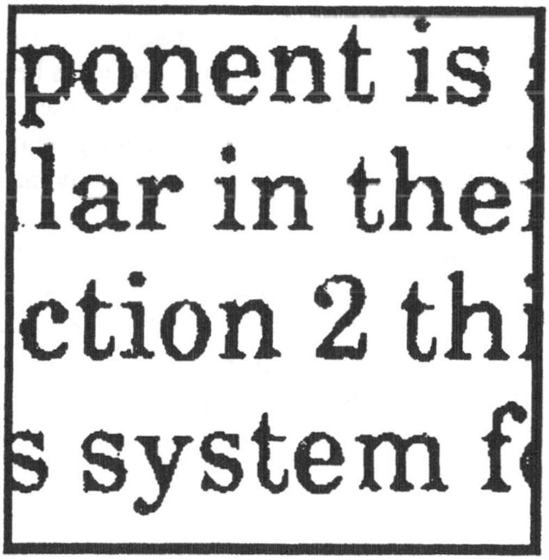

ponent is

lar in the

ction 2 th

s system f

Figure 11 Restored text image.

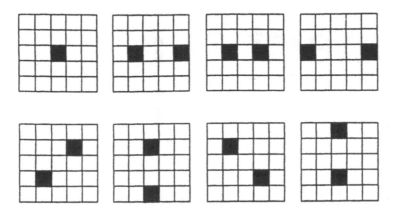

Figure 12 Structuring elements for restoring text image.

preceding example was found by applying a constrained optimization approach to produce a filter minimizing the mean-square error between the restored image and the uncorrupted original image [9, 26,27].

B. Gray-Scale Filters

Gray-scale mathematical morphology is of more recent vintage than binary. For theoretical reasons we assume that the morphological operators act on functions taking values inclusively between minus and plus infinity, these functions being defined on n-dimensional Euclidean space or the n-dimensional Cartesian grid. For signals, $n = 1$, and for images, $n = 2$. For notational and geometric simplicity we consider only signals. Once the underlying gray-scale theory has been presented for signals, one need only recognize that by treating points on the line as spatial points in the plane, the theory at once goes over into the imaging domain, the key point being that the theory itself is independent of domain dimensionality.

There are some basic mathematical building blocks. The graph of a signal can be translated in two ways, spatially or vertically. A spatial translation by x of a signal f is defined by $f_x(z) = f(z - x)$. A vertical translation of f by y, called an *offset*, is defined by $(f + y)(z) = f(z) + y$. These are typically used in conjunction to obtain a translation in the plane of the function's graph; that is,

$$(f_x + y)(z) = f(z - x) + y \tag{89}$$

The order relation in the gray scale is pointwise function ordering, where it must be kept in mind that $f(z) = -\infty$ at any point z outside the signal (image) frame. Since digital signals are typically defined on a finite domain, it is best to give the

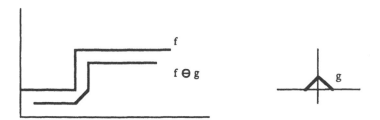

Figure 13   Erosion by a triangular structuring element.

definition in terms of $D[f] = \{z : f(z) > -\infty\}$. To that end, g is *beneath* f if $D[g] \subseteq D[f]$, and for any $x \in D[g]$, $g(x) \leq f(x)$. We write $g \ll f$. Since maximum and minimum play key roles in gray-scale morphology, as opposed to union and intersection in binary morphology, the convention regarding $-\infty$ must be kept in mind. Specifically, $\max\{y, -\infty\} = y$ and $\min\{y, -\infty\} = -\infty$. Consequently, letting $f \vee g$ and $f \wedge g$ denote the maximum and minimum, respectively, between signals f and g,

$$(f \wedge g)(z) = \begin{cases} \min\{f(z), g(z)\}, & \text{if } z \in D[f] \cap D[g] \\ -\infty & \text{otherwise} \end{cases} \tag{90}$$

$$(f \vee g)(z) = \begin{cases} \max\{f(z), g(z)\} & \text{if } z \in D[f] \cap D[g] \\ f(z) & \text{if } z \in D[f] - D[g] \\ g(z) & \text{if } z \in D[g] - D[f] \\ -\infty & \text{otherwise} \end{cases} \tag{91}$$

Finally, we define signal reflection by $h^{\wedge}(x) = -h(-x)$, which is accomplished by first reflecting the signal through the vertical axis and then through the horizontal axis.

Because erosion and dilation satisfy a number of algebraic identities, there are a number of equivalent ways of defining them. Since the genesis of morphology is fitting and gray-scale morphological processing is concerned with the topography of a signal's (image's) graph, we define the gray-scale operations directly in terms of fitting. The *erosion* of signal f by structuring element g (also a signal) is defined pointwise by

$$(f \ominus g)(x) = \max\{y : g_x + y \ll f\} \tag{92}$$

Geometrically, to find the erosion of a signal by a structuruing element at a point x, slide the structuring element spatially so that its origin is located at x and then find the maximum amount the structuring element can be "pushed up" and still be beneath the signal. Erosion is illustrated in Fig. 13, the effect there being as if the triangular structuring element were slid under the signal and the origin

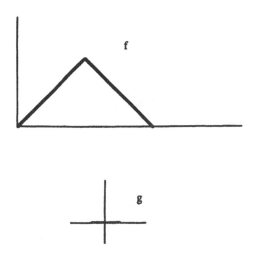

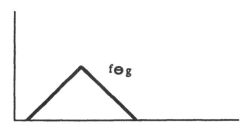

Figure 14   Erosion by a flat structuring element.

traced, there being the restriction that the element can never be translated so that it is not beneath the signal. Erosion by a flat structuring element is illustrated in Fig. 14.

The maximum of Eq. (92) is computed by taking the minimum difference between the signal and translated structuring element. Thus

$$(f \ominus g)(x) = \min\{f(z) - g_x(z)\} \tag{93}$$

Example.   Consider the signal

$$f = (* \quad * \quad 0 \quad 2 \quad 1 \quad 5 \quad 9 \quad 6 \quad 1 \quad 0)$$

where the asterisk indicates that the signal is undefined (negative infinity) at the point, and the bold character indicates the origin position relative to the signal. We wish to erode f by the structuring element $g = (5 \quad 5 \quad 4)$. Translating g to the right, the first time it lies beneath f is when it is translated 2 units, so that we are considering the translate

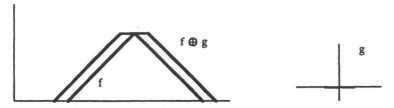

Figure 15 Dilation by a flat structuring element.

$$g_2 = (* \quad * \quad 5 \quad 5 \quad 4)$$

According to Eq. (93),

$$(f \ominus g)(2) = \min\{0 - 5, 2 - 5, 1 - 4\} = -5$$

Successively translating g rightward and taking the minima yields the eroded signal

$$f \ominus g = (* \quad * \quad -5 \quad -4 \quad -4 \quad 0 \quad -3 \quad -4)$$

Dilation is defined in a dual manner to erosion. Instead of translating the structuring element and finding the maximum the translated element can be pushed up and still be beneath the signal, we take the reflection of the structuring element and find the minimum it needs to be pushed up to be above the signal, when the signal is restricted to the domain of the translated structuring element. This last proviso is necessary because otherwise the domain of the signal would probably extend outside the domain of the translated reflected structuring element and the signal would never lie beneath the translation of the reflected structuring element. The *dilation* of f by g is defined pointwise by

$$(f \oplus g)(x) = \min\{y: (g^{\wedge})_x + y \geqslant f \mid _{D[g^{\wedge}]}\} \tag{94}$$

where $f \mid _{D[g^{\wedge}]}$ denotes f restricted to the domain of $g^{\wedge}$. Figure 15 illustrates dilation for a flat structuring element. Dilation is both commutative and associative. Duality is given by

$$f \oplus g = -[(-f) \ominus (-g^{\wedge})] \tag{95}$$

Discussion of filter properties for gray-scale morphology entails a reformulation of the basic filter properties. When treating gray-scale signals (images) morphologically, it is the topography of the graph as a subset of the plane (space) that plays the central role. Consequently, translation invariance is defined relative to both spatial and vertical translation. Filter Ψ is *translation invariant* if

$$\Psi(f_x + y) = \Psi(f)_x + y \tag{96}$$

for any signal f and any x and y. Translating the graph and then filtering is equivalent to filtering and then translating. If Ψ is translation invariant, then ipso facto it is both spatially translation invariant and offset invariant: $\Psi(f_x) = \Psi(f)_x$ and $\Psi(f + y) = \Psi(f) + y$. Both gray-scale erosion and dilation are translation invariant.

In the gray-scale, a filter Ψ is *monotonically increasing* if $f \ll g$ implies that $\Psi(f) \ll \Psi(h)$, so that order is preserved. Both erosion and dilation are increasing filters. From the perspective of structuring elements, the order relation is preserved for dilation and inverted for erosion: if $g \ll k$, then $f \oplus g \ll f \oplus k$ and $f \ominus g \gg f \ominus k$. The latter relation is important to morphological filter theory. It is straightforward because if g is beneath k, then g can be pushed up at least as much as k and still lie beneath f.

In the gray scale, the *dual* of filter Ψ is defined by $\Psi^*(f) = -\Psi(-f)$. Gray-scale erosion and dilation are dual. Gray-scale *opening* is defined by

$$f \circ g = (f \ominus g) \oplus g \tag{97}$$

Closing is defined by duality:

$$f \cdot g = -[(-f) \circ (-g)] \tag{98}$$

As in the binary setting it is usually better to view opening in terms of fitting:

$$f \circ g = \vee \{g_x + y : g_x + y \ll f\} \tag{99}$$

Accordingly, the opening is found by taking the maximum over all translations of the structuring element that fit beneath the input signal. The fitting fomulation gives the geometric intuition for opening: slide the structuring element along beneath the signal and at each point record the point on the structuring element translate that is highest at that point. The position of the origin relative to the structuring element is irrelevant. Opening is illustrated in Fig. 16.

As a filter, opening is increasing and translation invariant. It is also *antiextensive* $[f \circ g \ll f]$ and *idempotent*

$$(f \circ g) \circ g = f \circ g \tag{100}$$

A gray-scale filter that is translation invariant, monotonically increasing, antiextensive, and idempotent is called a τ-*opening*. Any gray-scale τ-opening is of the form

$$\Psi(f) = \vee \{f \circ g : g \in \mathbf{B}\} \tag{101}$$

where $\mathbf{B}$ is a *base* for Ψ [16]. In the gray-scale case, Inv[Ψ] consists of all maxima of translations of signals in $\mathbf{B}$. From a filtering perspective, openings and closings can be employed to attenuate spikes jutting above and below the signal, respectively. As in the binary setting, they are often used iteratively to form close-open, open-close, and alternating sequential filters.

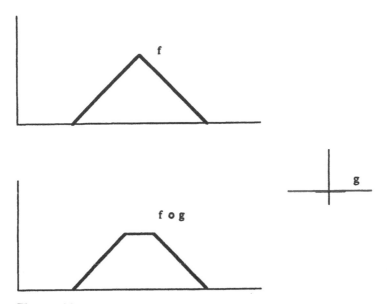

Figure 16 Opening by a flat structuring element.

Historically, flat structuring elements have played a key role in gray-scale morphology, where by a "flat element" we mean one that is constant over its domain. Owing to translation invariance, when dealing with a flat structuring element we might as well assume that it is zero on its domain; if it is not, we can always offset it, operate, and then reoffset in the opposite direction. Thus the class of flat structuring elements can be viewed as the class of structuring elements defined by their domains and zero thereon. This class is identical with the set of subsets of the line (plane for images). Consequently, we may consider erosion or dilation by a flat structuring element as erosion or dilation of a signal by a set. Hence if g is zero on its domain D, it is common to write $f \ominus D$ and $f \oplus D$ to denote the erosion and dilation of f by g, respectively. According to Eq. (93),

$$(f \ominus D)(x) = \min\{f(z) : z \in D + x\} \tag{102}$$

so that if $f \ominus D$ is simply a moving-minimum filter over the window D. Correspondingly,

$$(f \oplus g)(x) = \max\{f(z) : z \in D + x\} \tag{103}$$

so that $f \oplus D$ is a moving-maximum filter over the window D. Thus erosion and dilation by flat structuring elements are special cases of order-statistic filters.

As stack filters, erosion and dilation by flat structuring elements can be evaluated via threshold decomposition. Since erosion by a set D is a minimum operation, each threshold operation is a minimum over D, and a binary minimum over D is erosion by D. Consequently,

$$(f \ominus D)(x) = \max\{y : x \in A^y \ominus D\} \tag{104}$$

where A^y is the threshold set $\{z : f(z) \geq y\}$. Analogously,

$$(f \oplus D)(x) = \max\{y : x \in A^y \oplus D\} \tag{105}$$

The entire matter can be made transparent if we stick to the digital-signal setting and consider M-valued stack filters as defined in Eq. (37). The positive Boolean function of Eq. (36) can be expressed as a minimal sum of products,

$$y^{(i)}(n) = \sum_r \prod_{j=-k}^{k} [x^{(i)}(n + j)]^{p[r,j]} \tag{106}$$

where the sum (over r) is over all possible products and where $p[r; j]$, the power of the binary variable $x^{(i)}(n + j)$, is $+1$ if the variable appears in the product

$$q[r; i, n, j] = \prod_{j=-k}^{k} [x^{(i)}(n + j)]^{p[r; j]} \tag{107}$$

and is 0 if the variable does not appear in the product. If D_r is a subset of the window $W = \{-k, -k + 1, \ldots, k\}$, then D_r defines a vector

$$\mathbf{p}[r] = (p[r; -k], p[r; -k + 1], \ldots, p[r; k]) \tag{108}$$

by $p[r; j] = 1$ if and only if $j \in D_r$. Conversely, every vector $\mathbf{p}[r]$ defines a subset D_r of the window, so that there is a one-to-one correspondence between D_r and $\mathbf{p}[r]$. If there exists only a single product term (single r) in Eq. (106), that product defines a binary maximum over the variables corresponding to D_r and is therefore equivalent to binary erosion by D_r. Replacing the sum in Eq. (37) by maximum yields the stack filter definition

$$y(n) = \max\{i : y^{(i)}(n) = 1\} \tag{109}$$

and when $y^{(i)}(n)$ consists of the single product corresponding to D_r, this is precisely the maximum of Eq. (104). Thus we see that a single-product-term stack filter is an erosion.

More generally, each product term in Eq. (106) corresponds to an erosion (r corresponding to some subset D_r of the window), so that Eq. (109) shows that every stack filter can be expressed in terms of unions of binary erosions by flat structuring elements,

$$y(n) = \max\{i : (\cup_r x^{(i)} \ominus D_r)(n) = 1\} \tag{110}$$

where it must be kept in mind that the union evaluated at n is 1 if n is an element of the union and 0 otherwise. Finding the maximum i such that the union equals

1 is equivalent to finding the maximum i for which each erosion is 1 and then taking a second maximum. We obtain

$$y(n) = \max_r\{\max\{i : (x^{(i)} \ominus D_r)(n) = 1\}\}$$
$$= \max_r\{\max\{i : n \in x^{(i)} \ominus D_r\}\} \tag{111}$$

Finally, applying Eq. (104) yields

$$y = \bigvee_r \{x \ominus D_r\} \tag{112}$$

which shows that a stack filter is a maximum of erosions by flat structuring elements [30]. The latter are sometimes termed *flat morphological filters*, so that stack filters and flat morphological filters compose the same class of filters.

The entire analysis leading up to Eq. (112) was digital; however, it could have been carried out for continuous stack filters. Furthermore, a maximum of erosions by flat structuring elements is a special case of a more general expansion, a maximum of erosions by arbitrary structuring elements:

$$\Psi(f) = \bigvee_r \{f \ominus g_r\} \tag{113}$$

Stack filters form a key subclass of the class of all such filters. What is this expanded class? In fact, filters of the form given in Eq. (113) compose the class of increasing, translation-invariant gray-scale filters, and Eq. (113) is the gray-scale form of the Matheron representation theorem. More specifically, the *kernel* of a translation-invariant gray-scale filter Ψ is defined by

$$\text{Ker}[\Psi] = \{h : h(0) \geq 0\} \tag{114}$$

and a filter is increasing and translation invariant if and only if it possesses the representation

$$\Psi(f) = \bigvee \{f \ominus g : g \in \text{Ker}[\Psi]\} \tag{115}$$

As in the binary case, redundancy is typical. A subset $\text{Bas}[\Psi]$ of $\text{Ker}[\Psi]$ is a *basis* for Ψ if (1) there exist no g_1 and g_2 in $\text{Bas}[\Psi]$ such that $g_1 \ll g_2$, and (2) for any $h \in \text{Ker}[\Psi]$ there exists $g \in \text{Bas}[\Psi]$ such that $g \ll h$. If a basis exists, it is unique. Moreover, if a basis exists, it is obvious that the kernel expansion can be replaced by a basis expansion since $g \ll h$ implies that $f \ominus h \ll f \ominus g$.

The algebraic structure of nonlinear filtering is now complete, insofar as nonlinear filtering is taken to mean filtering by increasing, translation-invariant filters. Median filters from a subset of order statistic filters, which in turn form a subset of stack filters, which in turn are equivalent to flat morphological filters, which in turn form a subclass of increasing, translation-invariant filters, which themselves are sometimes called *morphological filters* (For it to be termed "morphological," Serra [14] also requires an increasing, translation-invariant filter to be idempotent). What about filters that are simply translation invariant? These, too, can be expressed in terms of morphological operations; however, we leave this to the literature, mentioning only that the Matheron representation possesses a generalization to translation-invariant filters that are not necessarily increasing. Regarding the literature, see Giardina and Dougherty [16], Maragos and Schafer [29,30], and Maragos [28] for the theory of increasing, translation-invariant filters as applied to functions; see Banon and Barrera [4,5] for the representation of translation-invariant filters; see Serra [38], Heijmans [18], and Heijmans and Ronse [19,20] for the mathematical characterization of filters in the context of lattices; and [13,14] for a strictly logical-calculus-based interpretation of nonlinear filter theory.

At this point, as we conclude, one might ask whether all the algebraic theory we have summarized is important to practical filtering tasks. At first glance it might appear that we have gained only the ability to find erosion representations, flat or otherwise, of filters that in many cases are much more readily understood in their original forms. In fact, much more has been gained. Because we have canonical representations of various filter classes, we have a paradigm for filter construction: given that we desire a filter of a certain kind, find the erosion basis that produces the "best" such filter. We are led to conceive of optimal performance in terms of the selection of appropriate bases; indeed, as mentioned at the time, the filter basis of Fig. 12 was produced via an optimality paradigm. For the general theory of mean-square-error (also mean-absolute-error) optimization applied to digital increasing, translation-invariant filters, see Dougherty [9,10]; for optimization of binary translation-invariant filters via mathematical morphology, see Dougherty and Loce [12]; and for an adaptive approach to morphological-filter optimization, see Salembier [34]. For mean-absolute-error optimization applied to the class of stack filters, see Coyle and Lin [6], Lin et al. [25], and Gabouj and Coyle [15].

REFERENCES

1. J. Astola and Y. Neuvo, Optimal median type filters for exponential noise distributions, *Signal Process.*, 17:95–104 (1989).
2. J. Astola, P. Haavisto, and Y. Neuvo, Vector median filters, *Proc. IEEE*, 78(4):678–689 (1990).
3. J. Astola and Y. Neuvo, Matched median filtering, *IEEE Trans. Comm.* COM-40(4):722–729 (1992).
4. G. J. Banon and J. Barrera, Minimal representation of translation-invariant set mappings with mathematical morphology, *SIAM J. Appl. Math.*, 51(6):1782–1798 (1991).
5. G. J. Banon and J. Barrera, Decomposition of mappings between complete lattices by mathematical morphology, *J. Signal Processing*, 30 (1993).
6. E. J. Coyle and J. -H. Lin, Stack filters and the mean absolute error criterion, *IEEE Trans. Acoust. Speech Signal Process.* ACCS-36(8):1244–1254 (1988).

7. E. R. Dougherty, *An Introduction to Morphological Image Processing*, SPIE Press, Bellingham, Wash., 1992.
8. E. R. Dougherty, ed., *Mathematical Morphology in Image Processing*, Marcel Dekker, New York, 1993.
9. E. R. Dougherty, Optimal mean-square N-observation digital morphological filters, Part I: Optimal binary filters, *Comput. Vision Graphics Image Process. Image Understand.*, 55(1):36–54 (1992).
10. E. R. Dougherty Optimal mean-square N-observation digital morphological filters Part II: Optimal gray-scale filters, *Comput. Vision Graphics Image Process. Image Understand.*, 55(1):55–72 (1992).
11. E. R. Dougherty, R. M. Haralick, Y. Chen, C. Agerskov, U. Jacobi, and P. H. Sloth, Estimation of optimal τ-opening parameters based on independent observation of signal and noise pattern spectra, *Signal Process.*, 29 (Dec. 1992).
12. E. R. Dougherty and R. P. Loce, Optimal mean-absolute-error hit-or-miss filters: Morphological representation and estimation of the binary conditional expectation, *Opt. Engrg.* 32 (4) (1993).
13. E. R. Dougherty and R. M. Haralick, Unification of nonlinear filtering in the context of binary logical calculus, Part I: Binary filters, *Math. Imag. Vision*, 2(2):173–183 (1992).
14. E. R. Dougherty, Unification of nonlinear filtering in the context of binary logic calculus, Part II: Gray-scale filters, *Math. Imag. Vision*, 2(2) (1992).
15. M. Gabbouj and E. J. Coyle, Minimum mean absolute error stack filtering with structural constraints and goals, *IEE Trans. Acoust. Speech Signal Process.* ASSP-38(6):955–967 (1990).
16. C. R. Giardina and E. R. Dougherty, *Morphological Methods in Image and Signal Processing*, Prentice Hall, Englewood Cliffs, N.J., 1988.
17. P. Haavisto, M. Gabbouj, and Y. Neuvo, Median-based idempotent filters, *Circuits Systems Comput.* 1(1):125–148 (1991).
18. H. Heijmans, Theoretical aspects of gray-level morphology, *IEEE Trans. Pattern Anal. Mach. Intelligence*, PAMI-13 (1991).
19. H. Heijmans and C. Ronse, The algebraic basis of mathematical morphology, I: Dilations and erosions, *Comput. Vision Graphics Image Process.* 50(3) (1990).
20. H. Heijmans and C. Ronse, The algebraic basis of mathematical morphology, II: Openings and closings, *Comput. Vision Graphics Image Process. Image Understand.*, 54 (1991).
21. P. J. Huber, *Robust Statistics*, Wiley, New York, 1981.
22. S. A. Kassam and H. V. Poor, Robust techniques for signal processing: a survey, *Proc. IEEE*, 73:433–481 (Mar. 1985).
23. D. Kazakos and P. Papantoni-Kazakos, *Detection and Estimation*, Computer Science Press, New York, 1990.
24. P. M. Lewis and C. L. Coates, *Threshold Logic*, Wiley, New York, 1967.
25. J.-H. Lin, T. M. Sellke, and E. J. Coyle, Adaptive stack filtering under the mean absolute error criterion, *IEEE Trans. Acoustics Speech Signal Process.*, ASSP-38(6):938–954 (1990).

26. R. P. Loce and E. R. Dougherty, Facilitation of optimal binary morphological filter design via structuring-element libraries and observation constraints, *Opt. Engrg.*, 31(5):1008–1025 (1992).

27. R. P. Loce and E. R. Dougherty, Optimal morphological restoration: the morphological filter mean-absolute-error theorem, *J. Visual Comm. Image Representation*, 3(4) (1992).

28. P. Maragos, A representation theory for morphological image and signal processing, *IEEE Trans. Pattern Anal. Mach. Intelligence*, PAMI-11(6) (1989).

29. P. Maragos and R Schafer, Morphological filters, Part I: Their set-theoretic analysis and relations to linear shift-invariant filters, *IEEE Trans. Acoustics Speech Signal Process.* ASSP-35 (Aug. 1987).

30. P. Maragos and R. Schafer, Morphological filters, Part II: Their relations to median, order-statistic, and stack filters, *IEEE Trans. Acoustics Speech Signal Process.*, ASSP-35 (Aug. 1987).

31. G. Matheron, *Random Sets and Integral Geometry*, Wiley, New York, 1975.

32. S. Muroga, *Threshold Logic and Its Applications*, Wiley-Interscience, New York, 1971.

33. I. Pitas and A. N. Venetsanopoulos, *Nonlinear Digital Filters: Principles and Applications*, Kluwer Academic Publishers, Norwell, Mass., 1990.

34. P. Salembier, Structuring element adaptation for morphological filters, *J. Visual Comm. Image Representation*, 3(2) (1992).

35. D. Schonfeld and J. Goutsias, Optimal morphological pattern restoration from noisy binary images, *IEEE Trans. Pattern Anal. Mach. Intelligence*, PAMI-13(1) (1991).

36. J. Serra, *Image Analysis and Mathematical Morphology*, Academic Press, New York, 1983.

37. J. Serra, ed. *Image Analysis and Mathematical Morphology*, Vol. 2, Academic Press, New York, 1988.

38. J. Serra, "Introduction to morphological filters," in *Image Analysis and Mathematical Morphology*, Vol. 2, J. Serra, ed., Academic Press, New York, 1988.

39. J. Serra and L. Vincent, *Lecture Notes on Morphological Filtering*, École Nationale Supérieure des Mines de Paris, Fontainebleau, France, 1989.

40. J. W. Tukey, Nonlinear (nonsuperposable) methods for smoothing data, *Congr. Rec. EASCON'74*, 1974, p. 673.

41. P. D. Wendt, E. J. Coyle, and N. C. Gallagher, Stack filters, *IEEE Trans. Acoustics Speech Signal Process.*, ASSP-34(4):898–911 (1986).

42. L. Yin, J. Astola, and Y. Neuvo, Adaptive stack filtering with application to image processing, *IEEE Trans. Signal Process.* SP41: 162–184 (1993).

43. O. Yli-Harja, J. Astola, and Y. Neuvo, Analysis of the properties of median and weighted median filters using threshold logic and stack decomposition, *IEEE Trans. Signal Process.*, SP.39:395–410 (Feb. 1991).

2

Morphological Segmentation for Textures and Particles

Luc Vincent

Xerox Imaging Systems
Peabody, Massachusetts

Edward R. Dougherty

Rochester Institute of Technology
Rochester, New York

I. INTRODUCTION

The present chapter concerns image segmentation via the methods of morphological image processing. A generally accepted meaning of the word *segmentation* in the image-processing community is the decomposition of the image under study into its different areas of interest. Here we take the perspective that there are essentially two kinds of segmentation: segmentation of images of *texture* and segmentation of images of *particles*.

In texture segmentation, an image is partitioned into regions, each of which is defined by a set of features characteristic to the microimage structure within it, this structure typically being viewed in terms of the small *texture primitives* composing it. Typical applications include segmentation of vegetation types in aerial photographs, segmentation of text and halftones in document pages, and medical imaging applications (like the extraction of bone tissue according to trabecular structure in magnetic resonance images [15]; see Fig. 10).

The second kind of segmentation is concerned with images of particles (or objects), where textural information is either not present or cannot be used simply as a discriminating factor. The segmentation task consists of extracting the particles from the image(s) under study. In other words, the goal is to partition the image in as many connected components as there are objects or regions to extract, plus some background regions. We distinguish between binary and gray-scale particle segmentation: in the binary case (i.e., when the images under study are binary), the segmentation task consists of separating the overlapping particles (e.g., see the coffee-beans example of Fig. 11a). In the gray-scale case, the segmentation task is equivalent to a contour extraction problem (e.g., in Fig. 11b, the contours of the electrophoresis spots have to be extracted as precisely as possible).

The chapter is divided into two sections, the first concerning regional texture-oriented segmentation and the second concerning particle segmentation. The literature on image processing of textures is too abundant to be reviewed thoroughly in this chapter, even if we restricted ourselves to purely morphological methods. Instead, we focus on the approach based on *local granulometries*, which offers an intuitive formulation, can be applied to a wide range of images, and can be used for both texture segmentation and classification. In this section we deal only with binary images, but the method can be used for gray-scale images equally well (see the example of Fig. 10). As for particle segmentation, we concentrate primarily on *watershed* segmentation, which has the advantages of being very general, usually accurate and fast, and applicable to both binary and gray-scale images.

Both segmentation paradigms involve a fair amount of morphological machinery and, subject to the space constraint of a single chapter, we introduce the necessary machinery, leaving detailed theoretical descriptions to the literature. We refer to Chapter 1 for definitions and properties of the basic morphological operations, together with the notations used to represent them, and to [20] for an introductory account of the fundamentals of morphological image processing. For a more advanced coverage of mathematical morphology, the interested reader can consult the books by Serra [45,46] and the review paper by Serra and Vincent [47].

II. GRANULOMETRIC SEGMENTATION OF TEXTURES

A. Granulometries Generated by a Single Structuring Element

As originally conceived for binary images [34], granulometries are employed to characterize size and shape information within granular binary images via the manner in which they are sieved through sieves of various sizes and shapes. If an image is considered as a collection of grains, then whether or not an individual grain will pass through a sieve depends on its size and shape relative to the mesh of the sieve. By increasing mesh size, ever more grains within the image will pass through, the eventual result being that no grains will remain. Of course, this sieving model does not fully describe even a granular image, for in a real image the grains will probably overlap; nevertheless, it does serve as a means to approach the removal of nonconforming image structure and can be developed further to obtain image signatures based on the rate of sieving.

We begin by considering the most basic type of granulometry for Euclidean images, which by definition are considered to be closed subsets of the Euclidean plane. A fundamental result of mathematical morphology [34] states that a compact set B is convex if and only if whenever $r > s > 0$, rB is sB-open, the latter meaning that $rB \circ sB = rB$. (Recall that $rB = \{rb \mid b \in B\}$ is the homothetic or scalar multiple of the set B by the real number r and that $\circ$ denotes the morphological opening.) A basic property of opening is that if E is B-open, then for any set A, $A \circ E$ is a subset of $A \circ B$. Thus if B is convex, $r > s > 0$ implies that $A \circ rB$ is a subset of $A \circ sB$. If we think of grains in the image A falling through the holes sB and rB, more will fall through the hole rB, thereby yielding a more diminished filtered image. Indeed, since rB is sB-open, it can be shown not only that $A \circ rB$ is a subset of $A \circ sB$ but also that

$$(A \circ rB) \circ sB = (A \circ sB) \circ rB = A \circ rB \tag{1}$$

so that iteratively opening, in either order, is equivalent to opening only by rB.

Definition 1: Granulometry. If we consider $t > 0$ as a variable, the family $\{A \circ tB\}$ of opened images, B convex, is called a granulometry.

(*Note*: In section II.D we give a more general definition of granulometries.) If $\Omega(t)$ is the area of $A \circ tB$, with $\Omega(0)$ being the area of A itself, then $\Omega(t)$ is a decreasing function of t, known as a *size distribution*. Under the assumption that A is bounded, $\Omega(t) = 0$ for sufficiently large t.

One can now define the *normalized size distribution* as $\Phi(t) = 1 - \Omega(t)/\Omega(0)$. It can be shown that $\Phi(t)$ increases from 0 to 1 and is continuous from the left [34], and we can therefore propose the following definition:

Definition 2: Normalized Size Distribution. The normalized size distribution

$$\Phi(t) = 1 - \frac{\Omega(t)}{\Omega(0)} \tag{2}$$

is a probability distribution function known as the granulometric size distribution of A with respect to the generator B.

As a result, the derivative of $\Phi(t)$, $\Phi'(t) = d\Phi(t)/dt$ is a probability density, sometimes also called the granulometric size distribution, or, of more recent vintage, the *pattern spectrum* of the image A relative to the *generator B*. Since $\Phi'(t)$ is a probability density, it possesses moments. These are employed as image signatures.

To apply the granulometric method practically, we first have to adapt it to digital images, which in the present section are assumed to be subsets of the square Cartesian grid $\mathbf{Z}^2$. This cannot be done directly, owing to two difficulties regarding the Cartesian grid: first, the lack of an appropriate notion of convexity; and second, the inability to apply scalar multiplication by arbitrary real numbers. The method of granulometric generation we now discuss is applicable to both Euclidean images and discrete images, its main purpose being application to the latter.

Consider a sequence $E_1, E_2, \ldots$ of structuring elements of increasing size, where E_{k+1} is E_k-open for all k. Owing to the latter requirement, if S is any image, then $S \circ E_{k+1}$ is a subimage of $S \circ E_k$. Consequently, opening in turn by the structuring elements yields a decreasing sequence of images:

$$S \circ E_1 \supseteq S \circ E_2 \supseteq S \circ E_3 \supseteq \cdots \tag{3}$$

For each k, let $\Omega(k)$ be the number of pixels of $S \circ E_k$. Then $\Omega(k)$ is a decreasing function of k. Under the assumptions that E_1 consists of a single pixel and that S is finite, $\Omega(1)$ gives the original pixel count in S and $\Omega(k) = 0$ for sufficiently large k. Applying the normalization of Eq. (2) with k in place of t and 1 in place of 0 yields a normalized size distribution $\Phi(k)$. It is a discrete probability distribution function and possesses a discrete derivative

$$\Phi'(k) = d\Phi(k) = \Phi(k + 1) - \Phi(k) \tag{4}$$

which is a discrete density (probability mass function). Again, the density is called a *granulometric size distribution* or *pattern spectrum*, and its moments can be employed as image signatures.

There exists a straightforward approach to forming sequences $\{E_k\}$ such that E_{k+1} is E_k-open: choose a primitive E and let E_1 be a single pixel, $E_2 = E$, $E_3 = E \oplus E$, $E_4 = E \oplus E \oplus E$, $\ldots$. We are assured that E_{k+1} is E_k-open since $E_{k+1} = E_k \oplus E$ and the dilation of two sets is open with respect to both. Another, less elegant way to form a satisfactory sequence of structuring elements is simply to construct the desired sequence "by hand."

As an illustration, consider Fig. 1a, in which digital "balls" of four sizes are randomly dispersed about the image. The generating sequence $\{E_k\}$, from which the four balls generating the image are drawn, consists of digital balls of increasing size (the first being a single pixel). As ever-larger balls are employed for the opening structuring elements, the grains (balls) in the image are sieved from the image. As the structuring element sequence passes each of the four balls that generate the image, translates of the specific structuring element are sieved from the image, the result being the unnormalized size distribution in Fig. 2a. Also shown in Fig. 2 are the normalized size distribution and the pattern spectrum (derivative). As opposed to the situation represented by the simulated image of Fig. 1, in many real-world images, grains overlap to create larger, irregular compound grains that are not so regularly sieved by the granulometry.

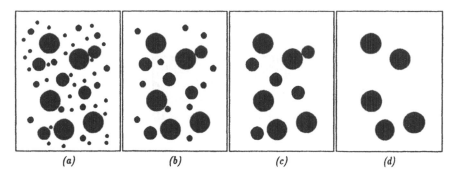

(a)   *(b)*   *(c)*   *(d)*

Figure 1 Binary image of disks (a) and its openings with respect to disks of increasing size (b)–(d).

While granulometries can be employed to describe shape [32,38], more directly related to texture segmentation is their ability to measure changes in surface condition [8] and in particle distribution processes, for instance electrophotographic images [18]. Figure 3 shows different examples of binarized particle distributions resulting from an electrophotographic process. In Fig. 3a the toner particles are fairly uniformly spread across the image, whereas in Fig. 3b the particles suffer from agglomeration, a typical problem with electrophotographic processes. Yet a different case is shown in Fig. 3c. Granulometries have been applied to these three images using a digital ball generating sequence, and the resulting pattern spectra are shown in Fig. 4. Notice how the agglomeration has resulted in a shift of the pattern spectra to the right, especially with regard to skewing to the right. We might expect this to result in significant changes in the mean, variance, and skewness of the pattern spectrum. In fact, hypothesis tests can be based on these granulometric moments to determine whether, owing to agglomeration (or some other problem), the electrophotographic process is out of control [18]. Indeed, from a statistical perspective, a captured toner particle image represents only a single selection from the population of images being generated by the electrophotographic process. Each captured image is only a realization of the process, and it is the overall random image process that is of concern.

B. Granulometries on Random Binary Images

Owing to the randomness of the image process, the pattern spectrum is actually a random function (stochastic process). Each realization of the image process yields its own particular pattern spectrum, which is a realization of the spectrum process, and each spectrum realization has its own particular moments. Thus the moments of the pattern spectrum (its mean, variance, skewness, etc.) are them-

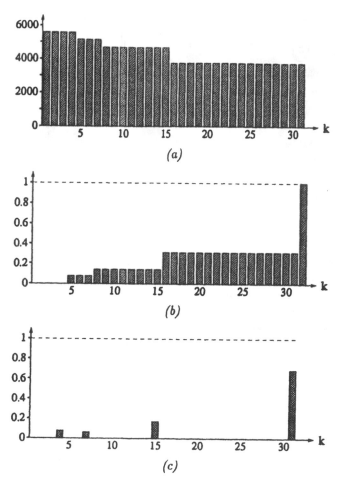

Figure 2 (a) Size distribution $\Omega(k)$, (b) normalized size distribution $\Phi(k)$, and (c) pattern spectrum $d\Phi$ corresponding to the binary image of Fig. 1a.

selves random variables. Since these pattern spectrum moments are random variables, they possess their own statistical distributions, and these in turn possess their own moments. Thus we arrive at the moments of the moments. Letting PSM, PSV, and PSS denote the mean, variance, and skewness of the pattern spectrum, PSM, PSV, and PSS are random variables possessing their own distributions. They have their own means (expectations) $E[\text{PSM}]$, $E[\text{PSV}]$, and $E[\text{PSS}]$, and their own variances, Var[PSM], Var[PSV], and Var[PSS]. Granulometric classification depends on the distributions of the pattern spectrum moments.

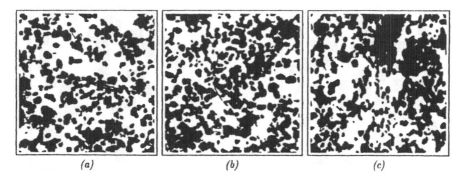

(a)   (b)   (c)

Figure 3 Different examples of particle distributions resulting from an electrophotographic process.

To illustrate the random model, we consider randomly placing $N = 50$ squares whose edge lengths are randomly chosen from a normal distribution with mean $\mu = 20$ and variance $\sigma^2 = 5$. In one case randomness of location is constrained so that the squares do not touch (nonoverlapping case) and in the other case there is no constraint on location (overlapping case). Realizations of the processes are shown in Fig. 5 and the pattern spectra for the depicted realizations are shown in Fig. 6. In all, 30 realizations have been performed for each case. For the nonoverlapping realizations the PSMs ranged from 20.0906 to 21.1104, with an average *PSM* of 20.4828. Taking the usual statistical approach, we take 20.4828 as an estimate of $E[\text{PSM}]$. The PSVs for the 30 nonoverlapping

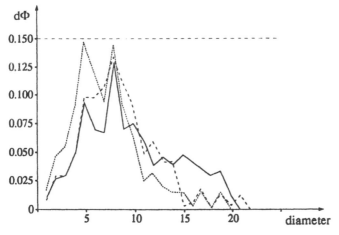

Figure 4 Pattern spectra corresponding to Fig. 3a–c: (a) dotted, (b) dashed, (c) continuous.

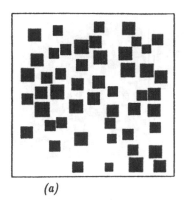 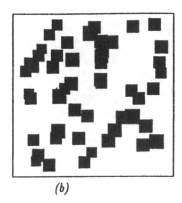

(a)   (b)

Figure 5   Randomly generated images of squares: (a) nonoverlapping; (b) overlapping.

realizations ranged from 3.0565 to 7.6976, the average PSV being 5.0759. For the overlapping realizations, average PSM was 20.7614 and average PSV was 5.3515, these being taken as estimates of E[PSM] and E[PSV], respectively. Note that overlapping has slightly shifted the mean to the right.

The preceding example brings to focus a fundamental task of granulometric analysis: Given an image model, one would like to:

1. Find expressions for the pattern spectrum moments.
2. Find expressions for the moments of the pattern spectrum moments.
3. If possible, describe the statistical distributions of the pattern spectrum moments.

Generally, these problems are quite difficult and they have been solved only in some special cases [1,2,19,42,43]. We shall consider briefly one of these and mention another.

Consider a random Euclidean image process S whose realizations are disjoint unions of N scalar multiples of a single convex, compact primitive B:

$$S = (r_1 B + x_1) \cup (r_2 B + x_2) \cdots \cup (r_N B + x_N) \tag{5}$$

where $r_i B + x_i$ is the translate of $r_i B$ by x_i. S is random, owing to both the locations (x_i) of the grains and the size (r_i) of the grains. Key to granulometric analysis is the sizing distribution governing $r_1, r_2, \ldots, r_N$. Under the model of Eq. (5), the kth central moment of the pattern spectrum for S generated by the primitive B is given by the random variable Z_{k+2}, where

$$Z_p = \frac{\sum_{i=1}^{N} r_i^p}{\sum_{i=1}^{N} r_i^2} \tag{6}$$

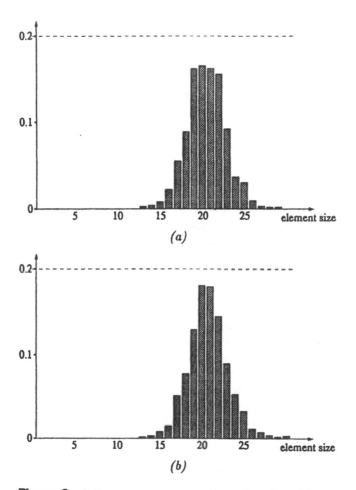

Figure 6 Pattern spectra corresponding to Fig. 5a and b.

[19,42,43]. For instance, PSM $= Z_3$ and PSV $= Z_4 - Z_3^2$. In fact, other image models besides that of Eq. (5) possess granulometric moments given by Z_{k+2}, and any grain model whose granulometric moments are given by Z_{k+2} is said to be *perfect* [42,43].

If an image process is perfect, it is possible to ascertain exactly the moments of its pattern spectrum moments [43]. More practically beneficial, under the assumption that the grain sizes are independent, it can be shown that the distributions of the pattern spectrum moments are asymptotically normal (with N) and that asymptotic expressions for the means and variances of the pattern spectrum moments can be obtained [42], so that the pattern spectrum moments are fully (asymptotically) described. For PSM and PSV, asymptotic convergence is $O(N^{-1})$ and $O(N^{-3/2})$, respectively.

For instance, if the sizing distribution is normal with mean μ and variance σ^2, then PSM is asymptotically normal with asymptotic mean

$$E[\text{PSM}] = \mu[1 + 2(1 + \mu^2\sigma^{-2})^{-1}] + O(N^{-1}) \qquad (7)$$

and asymptotic variance

$$\text{Var}[\text{PSM}] = \frac{\sigma^2(\mu^8 + 8\mu^6\sigma^2 + 12\mu^4\sigma^4 + 12\mu^2\sigma^6 + 15\sigma^8)}{N(\sigma^2 + \mu^2)^4}$$
$$+ O(N^{-3/2}) \qquad (8)$$

For the nonoverlapping process whose realization is depicted in Fig. 5a, the asymptotic value for $E[\text{PSM}]$ is 20.5000, which is very closed to the estimated value of 20.4828. For PSV, the asymptotic mean is

$$E[\text{PSV}] = \frac{\sigma^2(3\sigma^4 + \mu^4) + O(N^{-1})}{(\sigma^2 + \mu^2)^2} \qquad (9)$$

(see [42] for the asymptotic expression for Var[PSV] with a normal sizing distribution and for asymptotic expressions for a gamma sizing distribution).

In addition to the perfect grain model, which applies to noise-free random images, granulometric moments have been studied for arbitrary deterministic, digital binary images corrupted by either union or subtractive independent noise. Because the noise is random the pattern spectrum moments are random. Expressions have been found for the means of the first (PSM) and second pattern spectrum moments in the case of granulometries generated by linear structuring elements [1,2].

C. Local Granulometric Size Distributions and Texture Analysis

The image process of Eq. (5) is homogeneous in that its statistical description is invariant across the image frame. As a texture process, its realizations represent the same texture, and this texture occupies the entire image frame. More generally, we can consider an image process S defined on a frame F in such a manner that the frame is partitioned into n subframes,

$$F = F_1 \cup F_2 \cup \cdots \cup F_n \qquad (10)$$

with F_i and F_j disjoint if $i \neq j$. Furthermore, suppose that when restricted to F_i, S is defined by the image (texture) process T_i and that for $i \neq j$, T_i and T_j are not identically distributed. Then S is composed of textures from the class $\mathcal{T} = \{T_1, T_2, \ldots, T_n\}$ and the domain of T_i is F_i. The segmentation problem is to observe a realization (or realizations) of S and estimate the regions $F_1, F_2, \ldots,$

F_n. Using a pixel-based approach, the task is to associate a feature descriptor with each pixel and then classify the pixel into one of the regions according to the value of the descriptor observed. Since classification is to be texture-based and since texture is regional, the feature descriptor at a pixel x must be a function of some neighborhood (window) containing x.

In the classical global approach to granulometries, the entire image is successively opened and at each stage an image pixel count is taken. To measure image texture local to a given pixel, rather than take the pixel count across the whole image we take the count in a window W_x about the pixel x.

Definition 3: Local Granulometric Size Distribution. Consider a granulometry $\{A \circ E_k\}$ and an origin-centered window W. The local size distribution $\Omega_x(k)$ at pixel x generated by the granulometry is obtained by counting the number of pixels of $A \circ E_k$ in the translated window $W_x = W + x$ for each k. Normalization according to Eq. (2) yields a local granulometric size distribution $\Phi_x(k)$ and differencing yields a local pattern spectrum density $\Phi'_x(k)$.

Each local pattern spectrum possesses moments. If a subregion of the image possesses homogeneous texture, it is likely that the moments remain somewhat stable across the subregion and different subregions characterized by different textures can be differentiated based on the local pattern spectrum moments.

If, for the moment, we confine ourselves to the mean, then for each pixel x we have a mean PSM_x and pixel regions are segmented based on differing PSM_x values. According to the random process model, each PSM_x is a random variable and segmentation accuracy depends on the probability distributions of the PSMs across the various texture regions.

The partition of Eq. (10) induces a secondary partition based on the window W, which can be treated as a binary structuring element, namely,

$$F = (\bigcup_{i=1}^{n} (F_i \ominus W)) \cup (F \setminus \bigcup_{i=1}^{n} (F_i \ominus W)) \tag{11}$$

We write this new partition as

$$F = F_h \cup F_b \tag{12}$$

If $x \in F_h$, there exists i such that $W_x \subseteq F_i$; if $x \in F_b$, there does not exist such an i. Thus for $x, y \in F_h$, if $x, y \in F_i \ominus W$, then PSM_x is identically distributed to PSM_y; if x and y do not lie in the same region, PSM_x and PSM_y are not identically distributed. Segmentation of the subregions of F_h results from observing the local PSM values. As for the residual region F_b, pixels within it have windows lying in more than a single texture region, and therefore the local PSMs at these pixels result from mixed processes. Typically, these PSMs are also used for classification; however, estimation accuracy suffers. F_b is composed of border regions and the size of these regions is a function of window

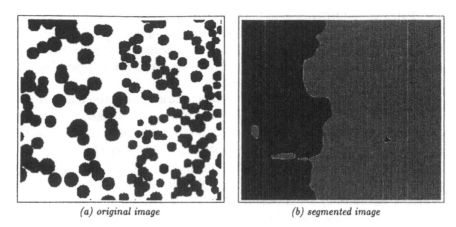

(a) original image   *(b) segmented image*

Figure 7   Texture segmentation via local pattern spectra means.

size. A small window size decreases the size of the border regions, but at the same time a small window tends to decrease the accuracy of estimation within regions because it makes the PSMs more susceptible to realization fluctuation [19].

As an illustration, consider the realization of Fig. 7a, in which each side of the image consists of randomly dispersed balls possessing random radii, the difference being that on the right side the mean radius is smaller than that on the left side. Using local granulometric size distributions generated by a ball and then segmenting based on the local pattern spectra means yields the segmented image in Fig. 7b. Here the method of segmentation is quite straightforward: since there are only two regions, the local PSMs are computed and those above a certain threshold are considered to arise from pixels on the left, while the others are considered to arise from pixels on the right. Two points should be noted. First, a single grain approach will not work because, owing to size randomness, grains resulting from the larger sizing distribution can be smaller than grains resulting from the smaller sizing distribution. Second, owing to overlapping, grains in the small size region can cluster and thus be seen as forming a large grain region, and owing to random location, there can be gaps in the large grain region that result in pixel misclassification into the small grain region. At least insofar as the classification of Fig. 7a is concerned, these misclassifications can be mitigated by performing an opening followed by a closing with a relatively small ball, the only cost being a slight smoothing of the classification boundary.

Generally, one requires more features than merely the local PSM to classify a pixel; variances and other higher-order pattern spectrum moments might be

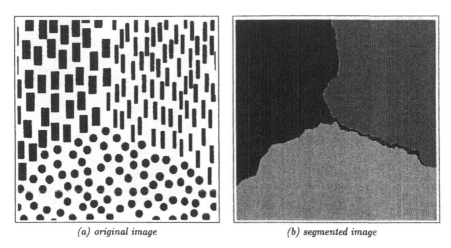

(a) original image   *(b) segmented image*

Figure 8   Segmentation of triphased binary texture.

necessary. The choice of structuring element sequence is also important; in fact, good segmentation will probably require a number of local size distributions generated by various structuring element sequences. An example of such a situation occurs in Fig. 8a. A ball sequence can separate the thick rectangles and balls from the thin rectangles, and a vertical linear sequence one pixel wide can separate the rectangles from the balls. Using the local pattern spectra resulting from each of these sequences in conjunction yields the segmented image of Fig. 8b. Segmentation has been accomplished by forming two partially segmented images, the balls and thick rectangles from the thin rectangles and the rectangles from the balls, and then logically operating to obtain the final segmentation.

The image of Fig. 9a illustrates a different problem. We wish to employ a ball-generated sequence to segment the left side from the right. The local pattern spectra on the left tend to be spikelike and centered at the common ball radius; the local pattern spectra on the right tend to possess two spikelike concentrations centered at the two radii of the balls composing that side of the image. In both regions, the local PSMs cluster around the common ball radius of the left side. Hence local PSMs cannot achieve segmentation. However, segmentation can be achieved via local PSVs, since the local PSVs on the left are close to zero, whereas the local PSVs on the right reflect two-spiked local pattern spectra. The segmentation shown in Fig. 9b results from classifying together all pixels with PSV above a certain threshold, these composing the right side of the segmentation.

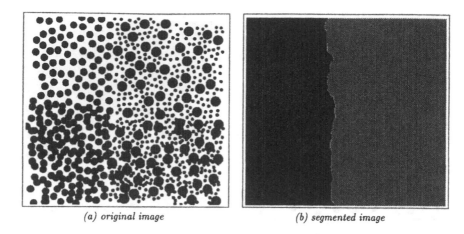

(a) original image   (b) segmented image

Figure 9   Texture segmentation via local pattern spectra variances.

More generally, segmentation of real-world images requires application of statistical classification methods to local pattern spectrum moments. The general approach can be described via the texture family $\mathcal{T} = \{T_1, T_2, \ldots, T_n\}$. If $x[i]$ is an arbitrary pixel in region $F_i \ominus W$, so that $W_{x[i]} \subseteq F_i$, for a given granulometry there is a local pattern spectrum associated with $x[i]$, which, as a random function, is identically distributed with the local pattern spectrum for any other pixel in $F_i \ominus W$. This local pattern spectrum possesses moments, in particular, PSM_i, PSV_i, and PSS_i, so that we can associate the random feature vector $\mathbf{F}_i = (\text{PSM}_i, \text{PSV}_i, \text{PSS}_i)$ with $x[i]$, or, more generally, with F_i. In fact, as evidenced by a preceding example, we might need to employ more than a single granulometry. If we employ m different granulometries, then $\mathbf{F}_i$ will have $3m$ component features.

Pixel classification is accomplished in the following manner. Assume that we know the multivariate probability distribution of $\mathbf{F}_i$ for $i = 1, 2, \ldots, n$. If we then observe a realization of the image process and select a pixel x, the same granulometric features that compose the feature vectors $\mathbf{F}_i$ are computed locally for x and this results in a feature vector $\mathbf{F}_x$. By employing some statistical measure of closeness, a determination is made as to which feature vector, $\mathbf{F}_1, \mathbf{F}_2, \ldots,$ or $\mathbf{F}_n$, the observed vector $\mathbf{F}_x$ is closest. If $\mathbf{F}_x$ is closest to $\mathbf{F}_i$, then x is classified into region F_i (classified as belonging to a texture region of the type T_i). Application of the method depends on a choice of classification methodology. The method thus far employed has been Gaussian maximum-likelihood classification [22].

A number of questions arise regarding granulometric texture classification. Two important ones having been addressed [22] concern robustness and classifier minimization. First, the method is quite robust with respect to certain types of union noise. This should be expected since openings filter out union noise, and as long as the noise components are sufficiently small, we should expect them to be filtered out early in the granulometric process. In addition, training in noise has been seen to be quite beneficial. Regarding classifier minimization, although we might initially employ a large number of granulometric features, many might be redundant, and worse yet, some might actually confuse the classification process. Moreover, large feature sets require large computation times. In [22], feature sets have been greatly reduced via a minimization procedure; indeed, very high accuracy has been achieved for the texture class under study there by using only six features.

D. General Granulometries

Thus far we have considered only the most basic granulometries, those generated by a single primitive generator. Next we describe Matheron's general theory of binary granulometries [34].

Definition 4: Granulometry. A granulometry is a collection of binary image operators $\{\Psi_t\}$, $t > 0$, such that (1) Ψ_t is antiextensive for all t, (2) Ψ_t is increasing for all t, and (3)

$$\Psi_t \Psi_s = \Psi_s \Psi_t = \Psi_{\max\{t,s\}} \tag{13}$$

If we view t as a sieving parameter for sieves of increasing mesh size, we see the genesis of the three properties:

1. The image remaining after any sieving operation is a subset of the original.
2. If one image is a subset of another, the sieved images maintain the same subset relation.
3. If Ψ_t and Ψ_s are two sieves in the process, the order of sieving does not matter, the remaining image being the same as if one were only to sieve through the largest of the mesh sizes.

Two additional properties are deduced from the three basic postulates [34]: if $r \geq s$, then $\Psi_r(A) \subseteq \Psi_s(A)$, and the invariant class of Ψ_r is a subclass of the invariant class of Ψ_s. In line with the sieving model, there will be less residue remaining after sieving with the larger sieve, and those images that are invariant under the larger sieve must also be invariant under the smaller sieve.

The elementary opening granulometries $\{A \circ tB)\}$, B compact and convex, satisfy the three granulometric postulates. They also satisfy two other fundamental properties:

4. They are translation invariant.
5. The following equation is satisfied:

$$\Psi_t(A) = t\Psi_1(t^{-1}A) \tag{14}$$

for any $t > 0$ and any binary Euclidean image A.

Property 5 is the most interesting: it says that there is a *unit* sieve, Ψ_1, and that any other sieve in the process can be evaluated by first scaling the image by the reciprocal of the parameter, filtering by the unit sieve, and then rescaling. If one thinks of sieving particles through a mesh, the property appears quite intuitive.

Definition 5: Euclidean Granulometry.   If a granulometry $\{\Psi_t\}$ satisfies properties (4) and (5), it is called a Euclidean granulometry.

Every Euclidean granulometry can be expressed in terms of openings [34], an elementary opening generated by a convex, compact primitive being the simplest Euclidean granulometry. While we leave a full discussion to more complete texts [17,23], we note that the most important example of a Euclidean granulometry is a finite union of openings, each by a parameterized convex, compact primitive:

$$\Psi_t(A) = (A \circ tB_1) \cup (A \circ tB_2) \cup \cdots \cup (A \circ tB_n) \tag{15}$$

The set of primitives $\mathscr{B} = \{B_1, B_2, \ldots, B_n\}$ is called the *generator* of the granulometry and the invariant class of Ψ_t is composed of all binary images that can be formed as unions of translates of scalar multiples tB_i of B_i where $t \geq 1$. The key to the relative simplicity of Eq. (15), as opposed to more complicated Euclidean granulometries, is that the generator is composed of convex shape primitives. In effect, rather than construct granulometric size distributions by utilizing a single convex shape primitive, one examines the sieving effect by a generator consisting of several convex shape primitives. Like single-primitive granulometries, more general granulometries induce size distributions.

As in the case of a granulometry generated by a single primitive, some care must be exercised when going to the digital setting. Since the Euclidean property, Eq. (14), is not applicable in the digital setting, we content ourselves with satisfying the first four properties. Consequently, for digital images we employ granulometries of the form

$$\Psi_k(S) = (S \circ E_{1k}) \cup (S \circ E_{2k}) \cup \cdots \cup (S \circ E_{nk}) \tag{16}$$

where, for $j = 1, 2, \ldots, n$, $E_{j,k+1}$ is $E_{j,k}$-open.

It is interesting to note that the asymptotic methods of [42] apply to multiply generated Euclidean granulometries of the form given in Eq. (15) if the primitives composing the generator possess the *orthogonality* property defined therein; that is, under the condition of generator orthogonality, the granulometric moments are asymptotically normal and asymtotic expressions for the means and variances of the moments can be derived.

E. Extensions of Granulometric Texture Classification

Two extensions of the granulometric approach to texture classification deserve mention: (1) granulometries employing the image complement and (2) gray-scale texture analysis. If we consider the image frame F as a binary image, then given an image A, we can treat the set subtraction $F\backslash A$ as a complementary image (also denoted A^C). If A represents the foreground, $F\backslash A$ represents the background. Granulometries of the type $\{A \circ tB\}$ provide information via filtering from the inside of A; we can also consider granulometries of the type $\{(F\backslash A) \circ tB\}$. These complemenatary granulometries provide information via filtering the complement of A. Pattern spectra moments formed from granulometries on $F\backslash A$ can be used to augment those formed from granulometries on A.

Binary and gray-scale openings share the same definition in terms of erosion and dilation, and whereas the binary opening of set A by structuring element B is given by the union of all translates of B that are subsets of A, the gray-scale opening of a function f by a structuring element (function) g is given by the supremum of all gray-scale translates of g that lie beneath f. Like binary openings, gray-scale openings possess four fundamental filter properties: translation invariance, increasing monotonicity, antiextensivity, and idempotence. Furthermore, Matheron's Euclidean granulometric theory extends to gray-scale signals [21] and granulometric size distributions can be constructed for texture analysis.

These gray-scale granulometries have been used, for example, to detect the presence of osteoporosis in magnetic resonance (MR) images [15], as illustrated in Fig. 10. Figure 10a shows an MR image of a wrist bone, the bone itself horizontally traversing the upper half of the image. The trabecular structure of the

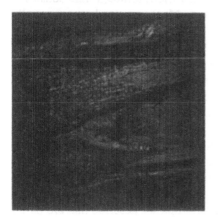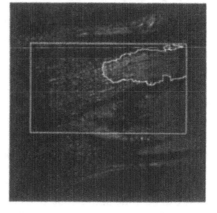

Figure 10 Granulometric segmentation of gray-scale MR images of osteoporotic bone regions: (a) original image; (b) segmented image.

bone appears as a grainy texture and osteoporosis is evidenced by diminished trabecular structure. Gray-scale granulometric features have been employed to classify bone pixels: the osteoporotic region of the bone extracted by the classifier is outlined in Fig. 10b. Note that even though the extracted region extends beyond the bone, only pixels within the bone region have been classified. Refer to [15] for details.

In analogy to [22], texture classification via gray-scale granulometric moments has been studied in the context of Gaussian maximum-likelihood classification [13,14]. The effects of employing complementary granulometries, recognition in noise, training in noise, and Karhunen–Loeve feature reduction have been investigated.

III. BINARY AND GRAY-SCALE PARTICLE SEGMENTATION

As mentioned previously, the purpose of this section is to show how morphology can be applied successfully to a wide range of object segmentation problems. In this section, *segmentation* of an image refers to the task consisting of extracting from it the objects or particles of interest as precisely as possible. By region–contour duality, this is equivalent to dividing the image into object regions and background regions.

Segmenting an image *I* is more than simply counting its objects and pointing at them in the image: it also encompasses the extraction of the objects' contours. However, as we shall repeat throughout this section, the contour extraction step usually requires prior *marking* of the objects to extract. By a *marker* of an object or set *X*, we simply mean a set *M* that is included in *X*. We also generally assume that markers have the same *homotopy* as the set they are marking. For example, a marker of a connected set is usually connected itself, although this constraint can in fact be loosened in many practical cases. Moreover, markers are usually located toward the central part of the objects they mark.

This marker extraction step is shown to be the most important step in many complex segmentation applications that are undertaken using morphology. Intuitively, this marking is an algorithmic simulation of human behavior: when asked to show the objects present in an image, a human being does not start right away by outlining these objects precisely. On the contrary, he or she will first point at the objects, one after the other, thus marking them. This marking can be seen as a first estimate of the objects; it is then refined by outlining the objects' contours. From an image analysis point of view, it is generally admitted that finding object markers is a less difficult task than directly extracting accurate contours. In this section we show that going from markers to actual segmentation can be done automatically using procedures that we describe that are based primarily on the watershed transformation.

In the binary case, segmentation refers to the extraction of the connected components representing objects of interest, as well as the separation of the overlapping objects. For example, the binary image shown in Fig. 11a represents coffee beans. This image will be used throughout this section to illustrate the effect of the morphological transformations we will be dealing with. We can count the beans present in the image simply by extracting a connected marker for each bean and counting the number of markers. However, this is not sufficient if one wishes to perform measurements on each individual bean (area, perimeter, elongation, etc). For this purpose, the beans need to be segmented (i.e., separated from one another). In the following, binary segmentation means "separation of overlapping objects in a binary image."

This marker-based methodology will also provide a framework for gray-scale segmentation. We will illustrate our approach on the ultraclassic image of two-dimensional electrophoresis gels shown in Fig. 11b. On this image, we not only want to find all the dark spots, but also correctly outline them in order to measure, for example, their respective areas, the density of black under them, and so on. In both the binary and the gray-scale segmentation cases, the *watershed transformation* (see Section III.C) will be used to extract precise object outlines from markers. We will talk about *marker-driven watershed segmentation*.

Note that in many simpler cases, there is no need to apply this approach. For example, the *tophat* transformation originally proposed in [36] provides an excellent tool for extracting light (respectively, dark) objects from an uneven background. Its relies on the fact that by gray-scale opening, one removes from an image the light areas that cannot hold the structuring element. Subtracting the opened image from the original one yields an image where the objects that have been removed by opening clearly stand out, and that image can then easily be thresholded (see Fig. 12).

Archetypically, the structuring element used in the opening step is a disk or a discrete approximation of a disk. However, infinite variations of this transformation can be derived. We may cite the following:

1. Using linear elements, one can specifically extract objects that are elongated in one direction,
2. Using a closing instead of an opening and subtracting the original image from the closed one allows us to extract dark objects on a lighter background. In this case we talk about "black tophat" as opposed to the "white tophat" (by opening).
3. If the background contains a lot of thin and elongated objects, one can use maxima of openings (respectively, minima of closings) with linear elements, followed or not by gray-scale reconstruction (see Section III.B.4) or area openings (respectively, closings) [58].

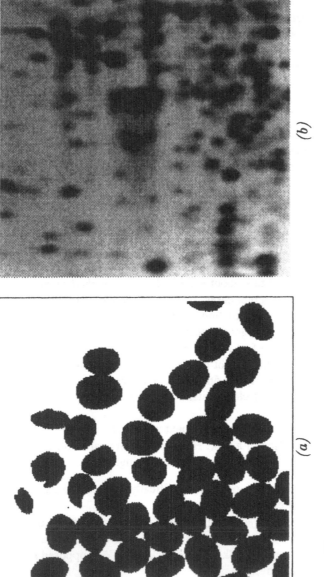

Figure 11 (a) Binary image of coffee beans that need to be separated; (b) gray-scale image of two-dimensional electrophoresis gel whose spots have to be extracted.

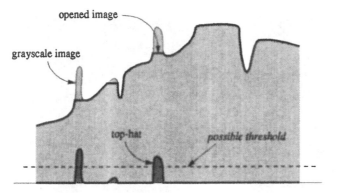

grayscale image

Figure 12 Peaks extracted via tophat followed by thresholding.

An application is shown in Fig. 13: Fig. 13a is a scanning electron micros-copy image where the balls in the lower right corner are to be extracted. These being compact and light compared to the background around them, they are re-moved by an opening of size 2 (see Fig. 13b). After subtraction of (b) from (a) [i.e., tophat (see Fig. 13c)] these small balls stand out and this image can easily be thresholded into Fig. 13d. The desired balls (right side) can now be extracted as the balls contained in the largest connected component of the dilation of Fig. 13d. The dilated image is shown in Fig. 13e, and the resulting segmentation is shown in Fig. 13f.

In the following, we concentrate on more complex segmentation problems, where tophats cannot provide satisfactory solutions. We first deal with binary segmentation, and the tools and techniques we are lead to use turn out to be even more useful later, for gray-scale segmentation. In Section III.A we present the concepts of maximal balls and skeletons and show how they can be used to mark robustly the centroid of overlapping objects in binary images. To go further and, from these markers, derive the desired binary segmentation, we need to make use of *geodesic operators*, which are defined in Section III.B. The gray-scale version of these operators is also very useful for gray-scale segmentation. Fi-nally, in Section III.C we describe the watershed transformation in detail and show how it unifies both binary and gray-scale segmentations. Its use is illus-trated on the segmentation of images of two-dimensional electrophoresis gels.

Some Reminders and Notations. From this point on we are concerned only with the discrete case (i.e., our workspace is the discrete plane Z^2). In this plane, a *grid G* provides the neighborhood relationships between pixels. Com-monly used grids are the square grid, for which a pixel p has either four (in 4-connectivity) or eight neighbors (in 8-connectivity), as well as the hexagonal grid (6-connectivity). Two neighboring pixels p and q form an edge of G. The

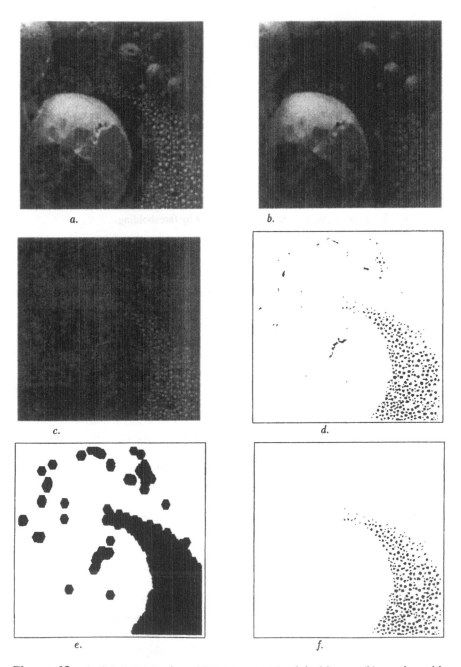

Figure 13   Tophat segmentation of SEM image: (a) original image; (b) opening with disk of radius 2; (c) corresponding top hat; (d) threshold; (e) dilation of (d); (f) final result.

grid G induces a discrete distance in $\mathbb{Z}^2$, the distance between two pixels being the minimal number of edges required to join them.

Discrete images are considered as mappings from $\mathbb{Z}^2$ onto $\mathbb{Z}$; gray-scale images take their values in a range $\{0, 1, \ldots, N\}$, whereas binary images can only take the values 0 and 1. The information content of a binary image is contained in its pixels with value 1, and therefore, binary images are often regarded as sets. For this reason, the binary transformations described in the following are often defined as set transformations.

A. Maximal Balls, Skeletons, and Ultimate Erosions

We have seen in Section II that granulometric analyses allow one to extract size information about an image without the need to segment it. For example, consider the coffee-beans image X shown in Fig. 11. Depending on the context, X shall either refer to the image itself or to the set of its black pixels. The granulometric analysis of this image may be undertaken using squares of increasing sizes, denoted S_1, S_2, and so on; openings with the S_i's are thus performed for $i = 1$ to the first value n such that $X \circ S_n = \varnothing$. At each step i, the area of image $X \circ S_i$ is determined.

The $(X \circ S_i)_{0 \le i \le n}$ constitutes a decreasing sequence of sets, so that as shown in [52], one can synthesize all the information contained in this sequence via a single function called the *granulometry function*.

Definition 6: Granulometry Function.   The granulometry function g_X associated with a set X and the family of convex and homothetic elements $(S_i)_{i \ge 0}$ maps each pixel p of X to the first i such that $p \notin X \circ S_i$:

$$g_x : p \in X \mapsto \min\{i \in \mathbb{N}, p \notin X \circ S_i\} \tag{17}$$

Alternatively, one can say that the granulometry function maps each pixel of X to the maximal i such that there exists a translation $t(S_i)$ of S_i satisfying $p \in t(S_i) \subseteq X$. The granulometry function of our coffee-beans image is shown in Fig. 14. From these images it becomes clear that our coffee beans can be described as the areas where the largest balls (in this particular case: squares) can be included. Therefore, to extract markers of our beans, we shall start by looking at the image zones where the "largest" balls can be included.

1.   Definitions

In this section and the following, the notion of *ball* stems directly from the distance being used. For example, in the plane $\mathbb{R}^2$ equipped with the usual Euclidean distance, the balls are standard discs. In the discrete plane $\mathbb{Z}^2$, the balls are hexagons if the hexagonal grid is used (6-connectivity) or squares in square grids. The unit size ball B (ball of radius 1) corresponds to either S_1, H or S_2 depending on whether 4-, 6-, or 8-connectivity is used (see Fig. 15).

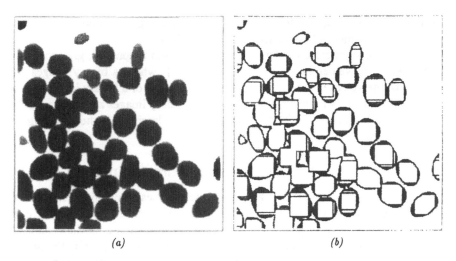

(a)   *(b)*

Figure 14   (a) Granulometry function of Fig. 11a with respect to a family of squares;
(b) level lines of this function.

Calabi's definition of the skeleton is based on the following notion of maximal ball:

Definition 7: Maximal Ball.   A ball B included in X is said to be maximal if and only if there exists no other ball included in X and containing B:

$$\forall B' \text{ ball}, \quad B \subseteq B' \subseteq X \Rightarrow B' = B \tag{18}$$

This concept is illustrated by Fig. 16, and the definition of the skeleton follows from it:

Definition 8: Skeleton by Maximal Balls.   The skeleton $S(X)$ of a set $X \subset \mathbb{Z}^2$ is the set of the centers of its maximal balls:

$$S(X) = \{p \in X \mid \exists r \geq 0, B(p, r) \text{ is a maximal ball of } X\} \tag{19}$$

The skeleton is an intuitive notion: the skeleton of a ball is reduced to its center, that of a band yields a unit thickness line, and so on. Examples of skeletons of simple shapes are shown in Fig. 17. One can see why the skeleton is often called the *medial axis transform*: it provides a description of sets in terms of thin, central lines.

Figure 15   Unit size ball B in 4-, 6-, and 8-connectivity, respectively.

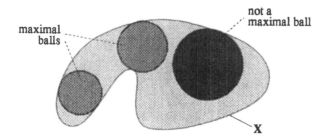

Figure 16 Concept of maximal ball in the Euclidean plane.

Unfortunately, things are not as easy as they look: in the continuous case (Euclidean plane $\mathbb{R}^2$), for example, the skeleton of two tangent disks is reduced to the two centers of these disks (see Fig. 18) instead of being a straight line joining these two points. In other words, the *homotopy* or *connectivity* of the original set is not necessarily preserved (see [46, Chaps. 11 and 12] for more details).

In the discrete case, let us denote by nB the ball of radius n in the connectivity considered:

$$nB = \underbrace{B \oplus B \oplus \cdots \oplus B}_{n \text{ times}},$$

with B being equal to either S_1, H, or S_2. Lantuéjoul proved that the skeleton by maximal balls can be obtained by the following formula:

$$S(X) = \bigcup_{n=0}^{+\infty} X \ominus nB \setminus ((X \ominus nB) \circ B) \tag{20}$$

In words, the skeleton by maximal balls can be obtained as the union of the residues of openings of X at all scales. Unfortunately, once again the skeleton does not behave as one would hope: direct application of formula (20) yields disconnected skeletons, as illustrated in Fig. 19.

  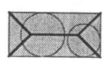 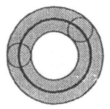

Figure 17 Skeleton of simple shapes in the Euclidean plane.

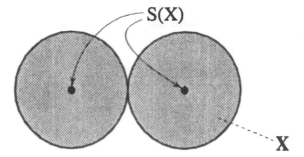

Figure 18 In the continuous Euclidean plane, the skeleton of a connected set is not necessarily connected.

Extracting correct homotopic (connected) skeletons from discrete binary images is thus not a straightforward matter. The literature on skeletons is very abundant and we certainly do not intend to cover the extraction of connected skeletons in this chapter. Let us just mention that the method recently proposed in [53] allows very efficient computation of a connected skeleton that is a min-

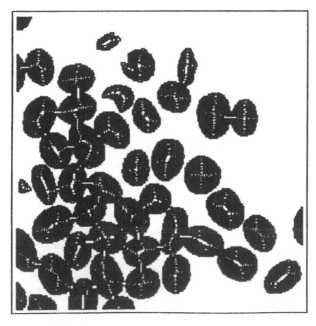

Figure 19 Example of skeleton by maximal balls on the coffee-beans image, using the balls of the 8-connected distance.

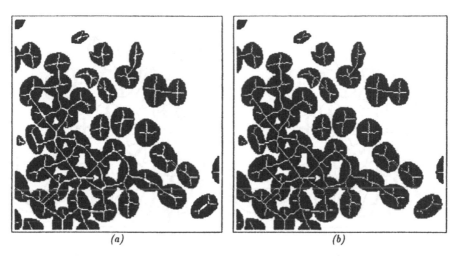

(a) *(b)*

Figure 20 (a) 8-Connected skeleton; (b) thinned 8-connected skeleton.

imal superset of the skeleton by maximal balls. An example of a connected skeleton computed in 8-connectivity using this method is shown in Fig. 20a. Since the skeleton by maximal balls is not necessarily of unit-pixel thickness, the connected skeleton of Fig. 20a is not either. For some applications it can be of interest to use thinning techniques to reduce it to a single pixel thickness, as shown in Fig. 20b.

2. Quench Function

Let us go back to our description of binary sets in terms of maximal balls and see what more can be said about these descriptions. By definition, to every pixel p in the skeleton, there corresponds a maximal ball. Let us denote by $q_X(p)$ the radius of this ball. We thus define the *quench function*.

Definition 9: Quench Function. The quench function associates with every pixel $p \in S(X)$ the radius of the corresponding maximal ball.

One of the most important results about the quench function is that its data are sufficient to reconstruct the original set completely:

Theorem 1. *A set X is equal to the union of its maximal balls*:

$$X = \bigcup_{p \in S(X)} (p + q_X(p)B) \tag{21}$$

The quench function thus allows us to do lossless encoding of binary images, and it has been studied extensively for image compression. Derived versions of this concept equal in performance the famous *CCITT group 4* encoding scheme on some types of document images [10].

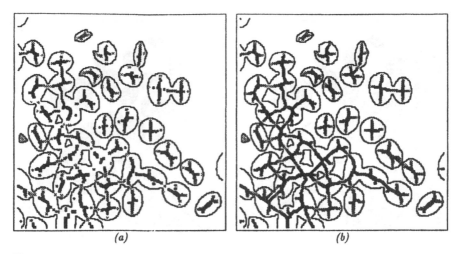

Figure 21   Quench function (a) and connected quench function (b). Their supports have been dilated for clarity.

The quench function of our image of coffee beans is presented Fig. 21a. Obviously, since this mapping is defined on the skeleton by maximal balls $S(X)$, its support is not connected. However, using the skeletonization technique mentioned briefly above [53], it is possible to reconnect the skeleton by maximal balls and to extract the radii corresponding to pixels on the connecting arcs. This produces the *connected* quench function shown in Fig. 21b.

The other major interest of the quench function is the definition of the *ultimate erosion*. We now have a way to describe a set X as the union of its maximal balls; to define markers of our coffee beans (i.e., of the convex blobs of the binary image under study) we look for the largest among these maximal balls. Clearly, for a given connected component C of X, a largest maximal ball is one of the largest balls that can be included in C, and its center marks an important object. However, if C is made of two overlapping objects, this crude method allows us to mark only one of them.

Let us consider the simple case where X is equal to the union of two overlapping disks. As shown on Fig. 22, the skeleton of X is the line segment joining the centers of these two disks. Now, upon examination of the quench function, one can notice that it exhibits two maxima, located at the exact centers a and b of our disks. (In this particular case, these maxima also happen to be located on extremities of the skeleton, which is uncommon.) These maxima therefore define *markers* of our overlapping objects, and their set constitutes the *ultimate erosion* of X:

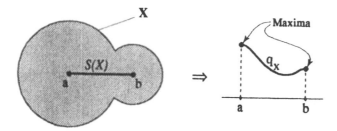

Figure 22 Skeleton of a set X, associated quench function q_X, and its maxima.

Definition 10: Ultimate Erosion. The ultimate erosion of a set X, denoted Ult(X), is the set of the (regional) maxima of the quench function q_X of X.

At this point we need to recall the definition of *maxima* (also called *regional maxima*) for gray-scale images:

Definition 11: Regional Maximum. A regional maximum M of a gray-scale image I is a connected component of pixels with a given value h (plateau at altitude h) such that every pixel in the neighborhood of M has a value strictly lower than h.

One should make a clear distinction between *regional* maxima and *local* maxima. A local maximum is defined as follows:

Definition 12: Local Maximum. A pixel p of a gray-scale image I is a local maximum if and only if for every pixel q that is a neighbor of p, $I(p) \geq I(q)$.

Obviously, if M is a regional maximum of I, then

$$p \in M \Rightarrow p \text{ is a local maximum}$$

but the converse does not hold. Similarly, one can define local and regional minima.

The problem with Definition 10 is that the discrete quench function is defined on a nonconnected support, so that its maxima are not really defined. To get around this problem, we can compute the maxima of the connected quench function presented earlier (see Fig. 21b). The ultimate erosion of the coffee-beans image thus extracted is shown in Fig. 23.

We can see that this ultimate erosion provides a reasonably good marking of our beans. We are therefore close to reaching our first goal, the extraction of one (connected) marker per object. There remain two points to address:

1. The marking is still not perfect: some beans are multiply marked. How can we reconnect some markers to end up with a single marker per bean?
2. How can we extract ultimate erosions in a more straightforward manner?

Figure 23   Ultimate erosion of the coffee-beans image.

These two issues are addressed in the next section with the description of the distance function.

3.   Ultimate Erosion and Distance Function

In this section we give a completely different interpretation of the ultimate erosion. Let X be a set made of overlapping components. When performing iterative erosions of this set with respect to the unit size ball B, its components are progressively shrunk and *separated* from the rest of the set before they are completely removed by the erosion process. This is illustrated in Fig. 24.

If, throughout this erosion process, we keep aside each connected component just before it is removed, it can be proved that the set thus obtained is exactly the ultimate erosion of X. Given two sets A and B such that $B \subseteq A$, let us denote by $\rho_A(B)$ the union of the connected components of A that have a nonempty intersection with B. This operator is called *reconstruction* and is discussed more fully in Section III.B. Our new ultimate erosion algorithm can now be expressed by the following formula:

$$\text{Ult}(X) = \bigcup_{n \in \mathsf{N}} [(X \ominus nB) \setminus \rho_{X \ominus nB}(X \ominus (n + 1)B)] \tag{22}$$

The resulting ultimate erosion of the set of Fig. 24 is shown in Fig. 25.

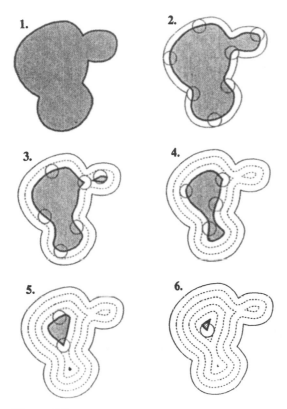

Figure 24 Successive erosions of a set. Each component is separated from the rest of the set before it is removed by the erosion process.

This method suggests yet another way to determine the ultimate erosion of a X. Indeed, there is a morphological transformation that synthesizes all the information contained in the successive erosions of a set X. This transformation is called the *distance function* and associates with each pixel p of X the size of the first erosion of X that does not contain p:

Definition 13: Distance Function. The distance function dist_X associated with a set X is given by

$$\forall p \in X, \qquad \text{dist}_X(p) = \min\{n \in \mathbb{N} \mid p \notin X \ominus nB\} \qquad (23)$$

For each pixel $p \in X$, $\text{dist}_X(p)$ is the distance between p and the background (i.e., X^C). The 8-connected distance function of the coffee-beans image is shown in Fig. 26.

By definition, the regional maxima at altitude h of the distance function are the connected components at altitude h of dist_X such that every neighboring

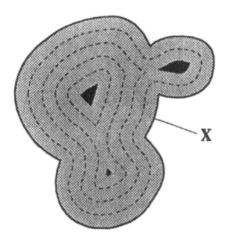

Figure 25   Ultimate erosion of a set.

pixel is of altitude strictly smaller than h. Any of these regional maxima are removed by a unit-size erosion. Indeed, if this was not true, there would exist pixels q located inside the regional maximum and verifying $\text{dist}_x(q) > h$, which is
absurd. These maxima thus belong to the ultimate erosion of X and the following
proposition can be derived:

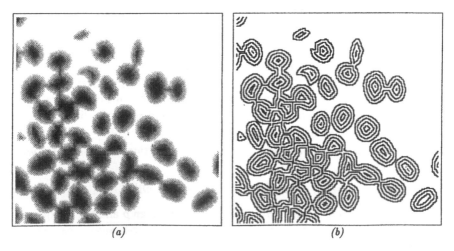

Figure 26   (a) 8-Connected distance function of the coffee-beans image; (b) level
lines of this function.

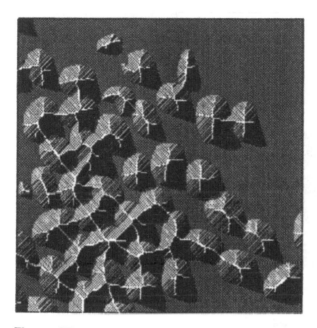

Figure 27. The skeleton follows the crest lines of the distance function.

Proposition 2. *The ultimate erosion of a set* X *is equal to the union of the regional maxima of the distance function of* X.

Since distance functions and regional maxima can be computed very efficiently in discrete images (see, e.g., [57]), this last proposition provides the best computational method for extracting ultimate erosions. As a matter of curiosity, what happens if we now extract the *local* maxima of the distance function? The following proposition holds:

Proposition 3. *The skeleton by maximal balls of a set* X *is equal to the set of local maxima of its distance function.*

The distance function is therefore at the basis of a very large number of morphological algorithms. Later in this chapter it will be used in conjunction with the watershed algorithm. For completeness in this section, we shall also mention that after local and regional maxima, the *crest lines* of the distance function are of interest: following them allows us to extract the connected skeleton of X (see Fig. 27). This forms the basis of a series of algorithms proposed by F. Meyer [46, Chap. 13].

We have now defined the morphological tools that will allow us to obtain perfect bean markers. We saw on Fig. 23 that the ultimate erosion does not quite yield perfect markers of our beans: all beans are marked, but some have multiple markers. The disconnections are caused by our discrete workspace as well as

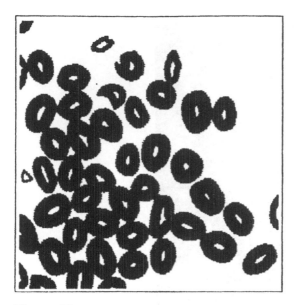

Figure 28 Final bean markers. For this image as well as for Fig. 23, the exact Euclidean distance function was used.

small contour irregularities of the beans. For this particular image, a unit-size dilation would be good enough to reconnect the markers and end up with a single marker per bean. However, this might not work in the general case: components of the ultimate erosion marking the same coffee bean may indeed be separated by arbitrarily large distances.

The method used instead relies on the fact that two components of the ultimate erosion marking the same bean are pretty much on the same "maximal zone" of the distance function. In fact, it is possible to go from one to the other on the distance function by going down no more than one level. Thus if we subtract 1 from the distance function at the location of all the components of the ultimate erosion, we obtain a modified distance function whose maxima are exactly the desired bean markers, as illustrated in Fig. 28.

4. Skeleton by Influence Zones

The last concept that we need to define in this section is that of *skeleton by influence zones*, also called *SKIZ*.

Definition 14: *Influence Zone*. Let X be a set made of n connected components $(X_i)_{1 \leq i \leq n}$. The influence zone $Z(X_i)$ of X_i is the locus of the points that are closer to it than to any other connected component of X:

$$Z(X_i) = \{p \in Z^2 \mid \forall j \neq i, \, d(p, X_i) \leq d(p, X_j)\} \qquad (24)$$

The distance d used in this equation is the discrete distance induced by the grid we are using (4-, 6-, or 8-connectivity). The SKIZ is then defined as follows:

Definition 15: SKIZ. The SKIZ of set X, denoted SKIZ(X), is the set of the boundaries of the influence zones $\{Z(X_i)\}_{1 \leq i \leq n}$.

An example of skeleton by influence zones is shown on Fig. 29a. Just as the skeleton follows the crest lines of the distance function, one can view the SKIZ as following the valley lines of the inverted distance function of the background (Fig. 29b).

B. Geodesic Transformations

1. Geodesic Distances

At this point we have achieved the first step of our segmentation as outlined at the beginning of Section III: the marker extraction. It now remains to make good use of these markers for the extraction of correct set boundaries. The idea is to define each bean as the image region centered around its marker. Our intent is therefore to "grow" these markers back in the mask of the coffee-beans image. For this purpose we now need the notion of *geodesic* operators introduced by Lantuéjoul [29,30].

Contrary to classic "Euclidean" morphological operations, geodesic operations do not function over the entire space, but on a finite set X called the *mask*. They are based on the notion of geodesic distance:

Definition 16: Geodesic Distance. The geodesic distance between two points x and y of X is the infimum of the length of the paths between x and y in X, if such paths exist:

$$d_X(x, y) = \inf\{l(C_{x,y}) \mid C_{x,y} \text{ is a path between } x \text{ and } y \text{ included in } X\} \quad (25)$$

If there are no such paths, we set $d_X(x, y) = +\infty$.

This definition is illustrated by Fig. 30. We call the *geodesic ball* of radius n and of center $p \in X$ the set $B_X(p, n)$, defined by

$$B_X(p, n) = \{p' \in X \mid d_X(p', p) \leq n\} \quad (26)$$

2. Geodesic Dilations and Erosions

Suppose now that X is equipped with its associated geodesic distance d_X. Given $n \geq 0$, we consider the *structuring function* [46] mapping each pixel $p \in X$ to the geodesic ball $B_X(p, n)$ of radius n centered at p. This leads to the definition of the *geodesic dilation* of a subset Y of X:

Definition 17: Geodesic Dilation. The geodesic dilation $\delta_X^{(n)}(Y)$ of size n of set Y inside set X is given by

$$\delta_X^{(n)}(Y) = \bigcup_{p \in Y} B_X(p, n) = [\{p' \in X \mid \exists p \in Y, d_X(p',p) \leq n\}] \quad (27)$$

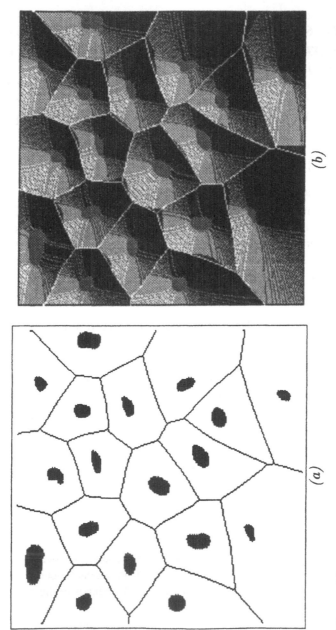

Figure 29 (a) SKIZ of a binary image; (b) the SKIZ follows the valley lines of the inverted distance function of the background. This example was computed using Euclidean distance.

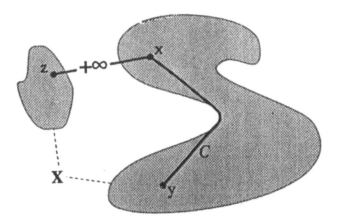

Figure 30   Geodesic distance in a set X.

The dual formulation of the *geodesic erosion* of size n of Y inside X is the following:

$$\epsilon_X^{(n)}(Y) = \{p \in Y \mid B_X(p, n) \subseteq Y\}$$
$$= \{p \in Y \mid \forall p' \in X \backslash Y, d_X(p, p') > n\} \tag{28}$$

Examples of geodesic dilation and erosion are shown in Fig. 31.

As already mentioned, the result of a geodesic operation on a set $Y \subseteq X$ is always included in X, which is our new workspace. As far as implementation is concerned, an elementary geodesic dilation (of size 1) of a set Y inside X is obtained by intersecting the result of a unit-size dilation of Y (with respect to the unit ball B) with the workspace X:

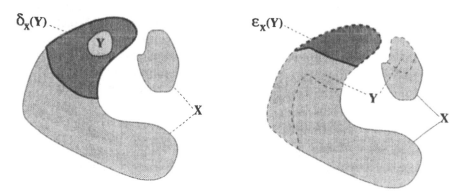

Figure 31   Examples of a geodesic dilation and of a geodesic erosion of set Y inside set X.

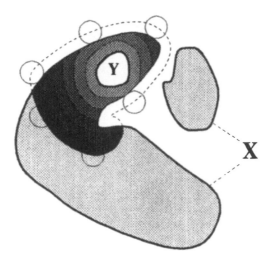

Figure 32   Successive geodesic dilations of set Y inside set X.

$$\delta_X^{(1)}(Y) = (Y \oplus B) \cap X \tag{29}$$

A geodesic dilation of size n is obtained by iterating n elementary geodesic dilations:

$$\delta_X^{(n)}(Y) = \underbrace{\delta_X^{(1)}(\delta_X^{(1)}(\cdots \delta_X^{(1)}(Y)))}_{n \text{ times}} \tag{30}$$

One can derive similar equations for geodesic erosions.

3.   Reconstruction and Applications

One can notice that by performing successive geodesic dilations of a set Y inside a set X, it is impossible to intersect a connected component of X that did not initially contain a connected component of Y. Moreover, in this successive geodesic dilation process, we progressively "reconstruct" the connected components of X that were initially *marked* by Y. This is shown in Fig. 32.

Now, the sets with which we are concerned are finite ones. Therefore, there exists n_0 such that

$$\forall n > n_0, \qquad \delta_X^{(n)}(Y) = \delta_X^{(n_0)}(Y)$$

At step n_0, we have entirely reconstructed all the connected components of X that were initially marked by Y. This operation is naturally called *reconstruction*:

Definition 18: Reconstruction.   The reconstruction $\rho_X(Y)$ of the (finite) set X from set $Y \subseteq X$ is given by the following formula:

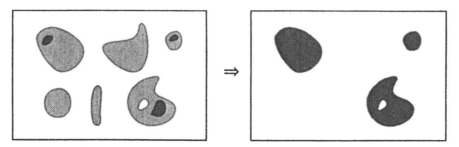

Figure 33 Reconstruction of X (light set) from Y (dark set).

$$\rho_X(Y) = \lim_{n \to +\infty} \delta_X^{(n)}(Y) \tag{31}$$

Figure 33 illustrates this transformation. Some applications require that the various markers remain unconnected (this is the case, for instance, for the binary segmentation problem with which we are concerned). In such cases, the *geodesic influence zones* of the connected components of set Y inside X are used. Indeed, the notions of influence zones and of SKIZ presented in Section III.A.4 easily extend to the geodesic case, as shown by Fig. 34.

4. Gray-Scale Reconstruction

At present, all the tools required for solving our bean segmentation problem have been defined. However, for gray-scale segmentation, we will also need to extend the concept of geodesy to gray-scale images.

It has been known for several years that—at least in the discrete case—any increasing transformation defined for binary images can be extended to gray-scale images [45,47,60]. By increasing, we mean a transformation ψ such that

$$\forall X, Y \subset Z^2, \qquad Y \subseteq X \Rightarrow \psi(Y) \subseteq \psi(X) \tag{32}$$

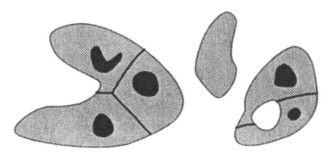

Figure 34 Example of geodesic SKIZ.

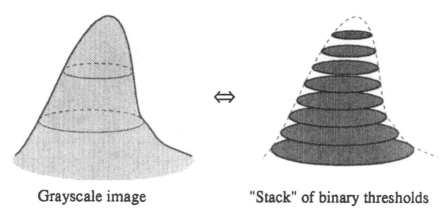

Grayscale image   "Stack" of binary thresholds

Figure 35   Threshold decomposition of a gray-scale image.

To extend such a transformation ψ to gray-scale images I taking their values in $\{0, 1, \ldots, N\}$, it suffices to consider the successive thresholds $T_k(I)$ of I, for $k = 0$ to N:

$$T_k(I) = \{p \in D_I \mid I(p) \geq k\} \tag{33}$$

where D_I is the domain of image I. They are said to constitute the *threshold decomposition* of I [33]. As illustrated by Fig. 35, these sets obviously satisfy the following inclusion relationship:

$$\forall k \in [1, N], \qquad T_k(I) \subseteq T_{k-1}(I)$$

When applying the increasing operation to each of these sets, their inclusion relationships are preserved. Thus we can now extend ψ to gray-scale images as follows:

$$\forall p \in D_I, \qquad \psi(I)(p) = \max\{k \in [0, N] \mid p \in \psi(T_k(I))\} \tag{34}$$

In the present case, binary geodesic reconstruction is an increasing transformation in that it satisfies:

$$Y_1 \subseteq Y_2, \quad X_1 \subseteq X_2, \qquad Y_1 \subseteq X_1, Y_2 \subseteq X_2 \Rightarrow \rho_{X_1}(Y_1) \subseteq \rho_{X_2}(Y_2) \tag{35}$$

Therefore, following the threshold superposition principle of equation (34), we define gray-scale reconstruction as follows [59]:

Definition 19: Gray-Scale Reconstruction.   Let J and I be two gray-scale images defined on the same domain, taking their values in the discrete set $\{0, 1, \ldots, N\}$ and such that $J \leq I$ [i.e., for each pixel $p \in D_I$, $J(p) \leq I(p)$]. The gray-scale reconstruction $\rho_I(J)$ of I from J is given by

$$\forall p \in D_I, \qquad \rho_I(J)(p) = \max\{k \in [0,N] \mid p \in \rho_{T_k(I)}(T_k(J))\}$$

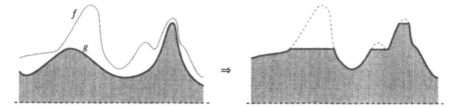

Figure 36 Gray-scale reconstruction of mask *f* from marker *g*.

Figure 36 illustrates this transformation. Just as binary reconstruction extracts those connected components of the mask that are marked, gray-scale reconstruction extracts the *peaks* of the mask which are marked by the marker image.

By duality, we are also able to define the *dual gray-scale reconstruction*, or reconstruction by erosion:

Definition 20: Dual Gray-Scale Reconstruction. Let J and I be two gray-scale images defined on the same domain, taking their values in the discrete set $\{0, 1, \ldots, N\}$ and such that $J \geq I$ [i.e., for each pixel $p \in D_I, J(p) \geq I(p)$]. The gray-scale reconstruction $\rho_I^*(J)$ of I from J is given by

$$\forall p \in D_I, \qquad \rho_I^*(J)(p) = N - \rho_{N-I}(I - J)$$

5. Binary Segmentation

Let us now use all these tools to design a powerful binary segmentation algorithm. Starting from the markers of our objects (i.e., from the ultimate erosion), our goal is to outline these objects accurately. We could consider using the geodesic SKIZ, and defining each object as the geodesic influence zone of its marker inside the initial set. Unfortunately, this is not a satisfactory algorithm. Indeed, as shown in Fig. 37, the separating lines thus defined between objects are poorly located. This is due to the fact that the *altitudes* of the various markers (i.e., the value associated with them by the quench function) is not accounted for by this method.

The way to design a good segmentation procedure—in taking the foregoing altitudes into account—is to use the geodesic SKIZ repeatedly. Let n_m be the size of the largest nonempty erosion of X:

$$X \ominus n_m B \neq \varnothing \quad \text{and} \quad X \ominus (n_m + 1)B = \varnothing$$

$X \ominus n_m B$ is obviously a subset of the ultimate erosion of X. Denote this set by X_{n_m}. Now consider the erosion of size $n_m - 1$ of X [i.e., $X \ominus (n_m - 1)B$]. Obviously, the following inclusion relation holds:

$$X_{n_m} \subseteq X \ominus (n_m - 1)B$$

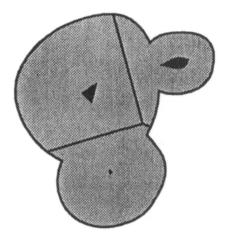

Figure 37   Bad segmentation algorithm: geodesic SKIZ of the ultimate erosion of X inside X.

Now let Y be a connected component of $X \ominus (n_m - 1)B$. There are three possible inclusion relations between Y and $Y \cap X_{n_m}$:

1. $Y \cap X_{n_m} = \emptyset$. In this case, Y is another connected component of Ult(X).
2. $Y \cap X_{n_m} \neq \emptyset$ and is connected. Here Y is used as a new marker.
3. $Y \cap X_{n_m} \neq \emptyset$ and is not connected. In this last case, the new markers are the geodesic influence zones of $Y \cap X_{n_m}$ inside Y.

These three cases are shown in Fig. 38.

Let $X_{n_m - 1}$ be the set of markers produced after this step. To summarize what we have just said, $X_{n_m - 1}$ is made of the union of:

1.   The geodesic influence zones of X_{n_m} inside $X \ominus (n_m - 1)B$.
2.   The connected components of Ult(X) whose *altitude* is $n_m - 1$.

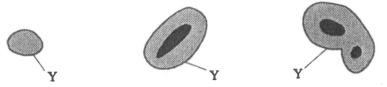

Figure 38   The three possible inclusion relations between Y and $Y \cap X_{n_m}$.

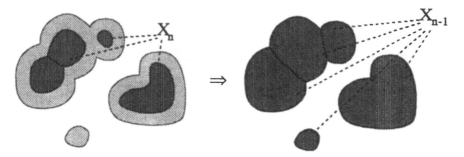

Figure 39 How to obtain X_{n-1} from X_n.

This procedure is then iterated at levels $n_m - 2$, $n_m - 3$, and so on, until level 0 is reached. In a more formal way, for every $0 < n < n_m$, let us introduce the following notations:

(a) $u_n(X)$ is the set of connected components of Ult(X) having altitude n:

$$p \in u_n(X) \Leftrightarrow p \in \text{Ult}(X) \quad \text{and} \quad \text{dist}_X(p) = n$$

(b) For every set $Y \subseteq X$, $z_X(Y)$ designates the set of geodesic influence zones of the connected components of Y inside X.

The recursion formula between levels n and $n - 1$ can now be stated:

$$X_{n-1} = z_{X \ominus (n-1)B}(X_n) \cup u_{n-1}(X) \tag{36}$$

It is illustrated in Fig. 39.

The set X_0 that is finally obtained after applying this algorithm constitutes a correct segmentation of X. Figure 40 presents an example of this binary segmentation algorithm. Applying these notions to the bean segmentation problem, we see in Fig. 41 that whereas a geodesic SKIZ of our markers results in improper separating lines, the segmentation algorithm we just described yields an accurate segmentation of the beans.

C. Watersheds and Gray-Scale Segmentation

1. Deriving a General Segmentation Approach

As presented in Section III.B.5, our morphological binary segmentation algorithm is rather complicated. In the present section we give a much more intuitive approach to it. Consider the function (gray-scale image) $-\text{dist}_X$, where dist_X is the distance function introduced in Section III.A.3, and regard it as a topographic surface. The *minima* of this topographic surface are located at the various connected components of the ultimate erosion of X. Now, if a drop of water

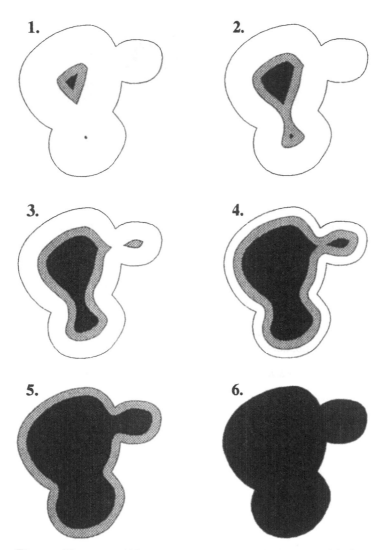

Figure 40   Correct binary segmentation algorithm presented in Section III.B.5.

falls at a point of p of $-\text{dist}_X$, it will slide along the topographic surface, following some steepest slope path, until it finally reaches one of its minima. We define the *catchment basin* $C(m)$ associated with a minimum m of our topographic surface in the following way:

Definition 21: Catchment Basin.   The catchment basin $C(m)$ associated with a (regional) minimum m of a gray-scale image regarded as a topographic surface

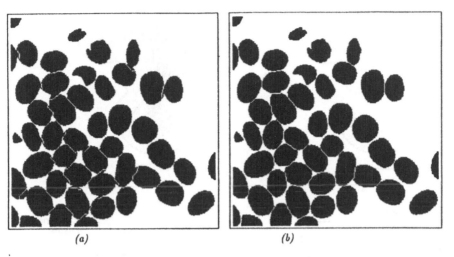

(a) *(b)*

Figure 41 Segmentation of coffee beans from their markers: (a) geodesic SKIZ; (b) correct segmentation.

is the locus of the points p such that a drop falling at p slides along the surface until it reaches m.

This definition is not very formal but has the advantage of being intuitive. In our example the catchment basins of the function $-\text{dist}_X$ exactly correspond to the regions that were extracted by the algorithm presented in Section III.B.5, as illustrated by Fig. 42. The segmentation achieved in Section III.B.5 exactly corresponds to extracting the catchment basins of the opposite of the distance function.

In fact, the notion of catchment basin can be defined for any kind of gray-scale image. Moreover, the algorithm of Section III.B.5 can easily be adapted to the determination of the basins of any gray-scale image I: it suffices to replace the successive erosions $X \ominus nB$—which correspond to the different thresholds of the distance function of X—by the successive thresholds of I (for more details, refer to [54]). The crest lines separating different basins are called *watersheds lines* or simply *watersheds*.

Definition 22: Watersheds. The watersheds (lines) of a gray-scale image I are the lines that separate the various catchment basins of I.

These notions are illustrated by Fig. 43.

Watersheds stand out as a powerful morphological crest-line extractor. It is therefore most interesting to apply the watershed transformation to gradient images: indeed, the contours of a gray-scale image can be viewed as the regions where the gray levels exhibit the fastest variations (i.e., the regions of maximal gradient). These regions are the *crest lines of the gradient*. This remark is il-

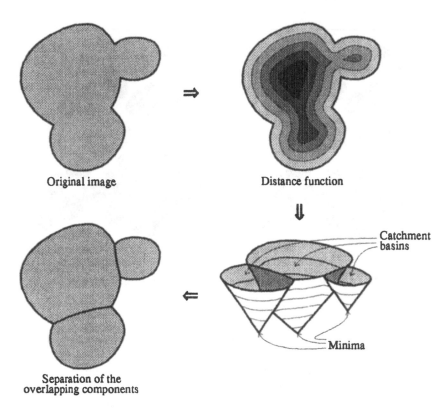

Original image   Distance function

Separation of the
overlapping components

Figure 42   Interpretation of binary segmentation in terms of catchment basins of the opposite of the distance function.

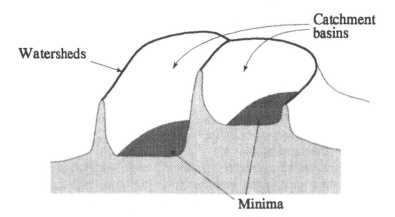

Figure 43   Regional minima, catchment basins, and watershed lines.

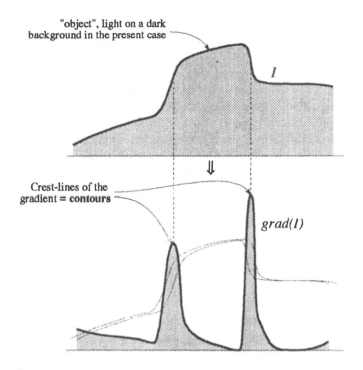

"object", light on a dark
background in the present case

I

Crest-lines of the
gradient = **contours**

grad(I)

Figure 44 Principle of gray-scale segmentation via watersheds of the gradient.

lustrated by Fig. 44 and is at the basis of the use of watersheds for gray-scale segmentation, as described and illustrated in [3,5,52,54].

Note that in morphology, the word *gradient* refers to an operation associating with each image pixel the *modulus* of its gradient—in the classical sense of the word. Most of the time, the gradient called the *morphological gradient* [45] is used, which is obtained as the algebraic difference of a unit-size dilation and a unit-size erosion of *I*:

$$\text{grad}(I) = (I \oplus B) - (I \ominus B)$$

Nonetheless, depending on the type of image contours to be extracted, other gradients may be of interest: directional gradients, asymmetric gradients, regularized gradients, and so on [39].

The watershed transformation always provides *closed* contours and constitutes a very general approach to contour detection. However, it can very rarely be used directly on gradient images without resulting in dramatic *oversegmentations*: the image gets partitioned in far too many regions (i.e., the correct contours are lost in a large number of irrelevant ones). This problem is due primarily to noise in the data: noise in the original image results in noise in its morphological gradient, in turn causing it to exhibit far too many regional minima. This translates directly into far too many catchment basins (i.e., oversegmentation).

Several approaches to overcome oversegmentation have been proposed in the literature: for example, some techniques remove arcs of the watersheds based on an integration of the gradient's gray values along them. Others take the dual point of view and merge adjacent regions (i.e., catchment basins here) when the gray level of the original image over them is comparable. None of these techniques is satisfactory in that it is very difficult to incorporate in them knowledge specific to the collection of images under study. Besides, they go against the point of view presented at the beginning of this section, where we claim that marker extraction should be the first step in every segmentation.

Therefore, the morphological approach to this problem consists of making use of image-specific knowledge (e.g., size, shape, location, or brightness of the objects to extract) to design robust object marking procedures [7,52,54]. This step in the segmentation can be completely different from one problem to another. Not only must each object be marked uniquely, but the *background* also needs its own marker(s). In a second step, this binary image of markers is used not to guide region-merging or arc-removal algorithms, but on the contrary, to *modify the gradient image* on which watersheds are computed.

More precisely, let I denote the original gray-scale image, $J = \mathrm{grad}(I)$ its morphological gradient, and let M denote the binary image of markers. The "modification" of J should result in a gray-scale image J' with the following characteristics: (1) its only regional minima are exactly located on the connected components of M (M is the set of "imposed" minima); and its only crest lines are the highest crest lines of J that are located between the minima imposed. The watersheds of J' are thus the highest crest lines of $\mathrm{grad}(I)$ that separate our markers. Hence they are the *optimal contours* corresponding to set of markers M and gradient J.

The actual computation from J and M of an image J' with these characteristics has been achieved classically using a three-step process [52]:

1. Set to $h_{\min}$ any pixel of J that is located on a marker, $h_{\min}$ being chosen such that for any $p, h_{\min} < J(p)$. This results in a new image J^*:

$$\forall p, \qquad J^*(p) = \begin{cases} h_{\min} & \text{if } M(p) = 1 \\ J(p) & \text{otherwise} \end{cases}$$

2. Create the following gray-scale image M^*:

$$\forall p, \qquad M^*(p) = \begin{cases} h_{\min} & \text{if } M(p) = 1 \\ h_{\max} & \text{otherwise} \end{cases}$$

where $h_{\max}$ is chosen such that $\forall p, J(p) < h_{\max}$.

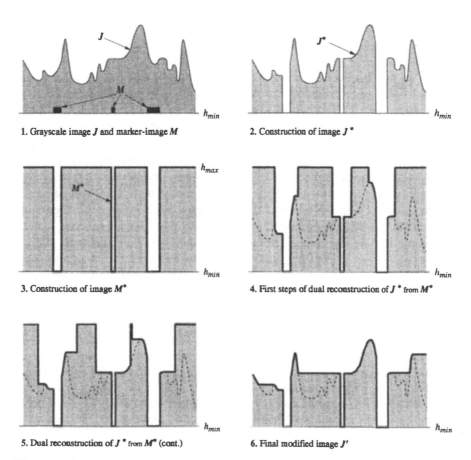

Figure 45 Use of dual gray-scale reconstruction to ''impose'' a set M of minima to a gray-scale image J.

3. Use M^* to remove all the unwanted minima of J^* while preserving its highest crest lines between markers. This is done using the dual gray-scale reconstruction operation ρ^*:

$$\forall p, \qquad J'(p) = \rho^*_{J^*}(M^*) \tag{37}$$

This process is illustrated by Fig. 45. The watersheds of the resulting image J' provide the desired segmentation. The entire procedure presented above is often referred to as *marker-driven watershed segmentation*. It is extremely powerful in a number of complex segmentation cases, where it mostly reduces the segmentation task to (1) the choice of a gradient and (2) the extraction of object markers (the latter task can itself be very complex in some cases).

2. Electrophoresis Example

Let us illustrate this segmentation paradigm on the two-dimensional electro-phoresis image of Fig. 46a (see also [4,52]). The standard morphological gra-dient of this image is shown in Fig. 46b. As mentioned above, if we simply compute the watersheds of Fig. 46b, the result is clearly disappointing (see Fig. 46c). Indeed, the gradient exhibits a large number of minima, due mainly to the presence of noise in the original image. Nevertheless, one can notice that *all* the spots are marked by these minima and hence all the correct contours are present in Fig. 46c. The watershed simply produces *oversegmentation*.

Avoiding oversegmentation requires the prior extraction of correct spot markers. Since the spots constitute the dark part of the image, they should be interpreted as the image minima. Yet direct extraction of the initial image's min-ima is not a satisfactory solution, as illustrated in Fig. 46d: once again, many of these minima are due to acquisition noise in our data. Here, however, filtering the image shown in Fig. 46a (with a morphological filter called an *alternating sequential filter* (ASF), see [46, Chap. 10] or [47]) is sufficient to produce an image whose minima correctly mark the spots (see Fig. 46e and f).

In fact, this marker extraction step is followed by binary watershed segmen-tation in order to cut markers like the one in the upper right corner, which clearly should mark two different spots. The final image of object markers is shown in Fig. 47a. As concerns the background marker, it is extracted as the set of the highest crest lines of the original image that separate the spot markers. This is the best way to assure that it will be located in the lightest areas of the image and separate all the object markers. Its determination is done similar to the method for gradient modification [see Eq. (37) and Fig. 45]. It is shown in Fig. 47b.

Both sets of markers are then combined in a final marker image, shown in Fig. 47c. It is used to modify the gradient of Fig. 46b, this resulting in Fig. 47d. The watersheds of the latter image provide the desired segmentation, as shown in Fig. 47e and f.

The result is in accordance with our expectations: each spot has a unique con-tour that is located on the inflection points of the initial luminance function (i.e., the original image). Given the set of markers extracted, we found the op-timal gradient crest lines (i.e., the best possible segmentation for these markers and, to a lesser extent, this gradient).

3. Difficult Segmentations: Recent Developments

Hierarchical Watershed Segmentations. The method described in Section III.C.2 can be applied to a wide range of problems as long as the preliminary marker extraction step can be performed with sufficient accuracy. However, in some difficult segmentation cases, it is impossible to find markers of the regions or of the objects to extract, since these objects or regions are not themselves well

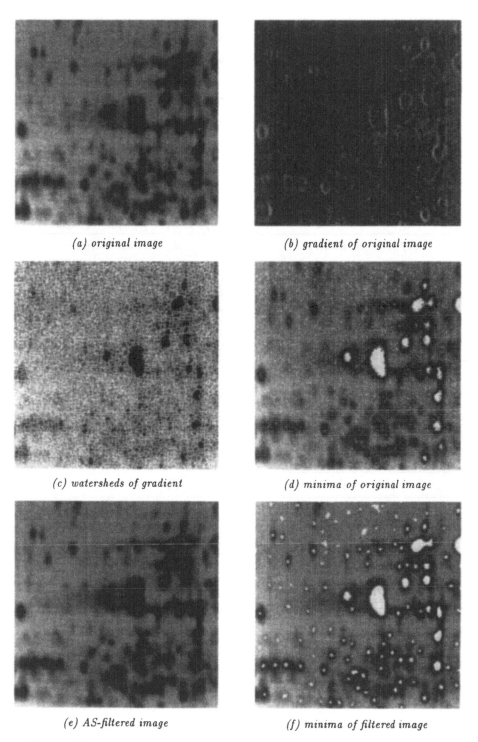

(a) original image

(b) gradient of original image

(c) watersheds of gradient

(d) minima of original image

(e) AS-filtered image

(f) minima of filtered image

Figure 46 Segmentation of electrophoresis gels.

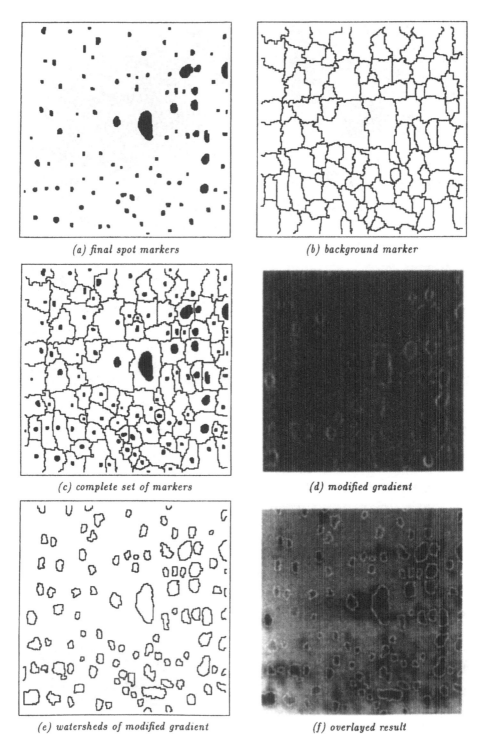

(a) final spot markers

(b) background marker

(c) complete set of markers

(d) modified gradient

(e) watersheds of modified gradient

(f) overlayed result

Figure 47   Segmentation of electrophoresis spots.

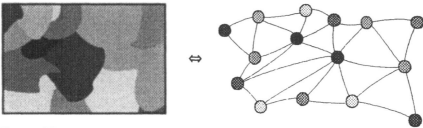

Figure 48 Mosaic image and associated adjacency graph.

defined. This type of situation often occurs, for instance, with remote sensing images, where the large variety of zones (fields, roads, houses, towns, lakes, and so on, under different lighting conditions) makes it almost impossible to design robust marking procedures.

In the latter case, region-growing types of techniques (see the beginning of Section III.C.1) may provide an appropriate solution. In fact, a few morphological region-growing techniques based on the watershed transformation have recently been proposed and seem very promising:

1. In the first, "raw" watershed segmentation is applied to the gradient of the original image, resulting in an oversegmented image of catchment basins. Each catchment basin C is assigned a uniform gray level corresponding, for example, to the mean gray-level of the pixels of the original image over C. In a second step, this "mosaic image" is regarded as a graph (the dual adjacency graph, see Fig. 48) and morphological operations are performed on this graph [50] to merge adjacent regions with comparable gray levels. In fact, graph gradients and graph-watersheds can themselves be iteratively applied to the original adjacency graph. This results in an image pyramid containing a hierarchy of contours at different resolutions [51].
2. Starting again from an oversegmented "mosaic" of catchment basins (resembling Fig. 46c), some methods introduced by S. Beucher process the adjacency graph *of the watershed arcs*. Catchment basins are hierarchically merged via recursive removal of these contour elements in the graph [6].
3. Finally, Salembier and Serra [41] approach the problem of general image segmentation via a combination of filters and watersheds that is started at coarse scales and progressively refined into more and more detailed segmentations. Their approach is therefore a region-splitting one, as opposed to the previous region-growing approaches.

Obviously, the watersheds are of enormous interest for complex segmentation problems, and the solutions above are barely starting to explore the extraordinary possibilities of this tool.

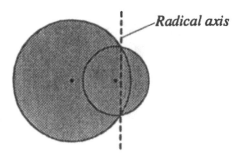

Radical axis

Figure 49   These two disks overlap too much to be both marked by ultimate erosion.

Segmentation of Intricately Overlapping Particles.   In binary segmentation, the marking by ultimate erosion has its limitations [52]: it is efficient only if, on the one hand, the components of the set X under study are "sufficiently blobby," and on the other hand, if they do not overlap too much. For instance, when X is composed of two overlapping disks, they are both marked by ultimate erosion if and only if their centers are located on either sides of the radical axis (see Fig. 49).

However, these exist more sophisticated marking tools. Among them, let us mention a transformation called *conditional bisector* (see [37, p. 55] or [49]), in which the marking is no longer related to the maxima of the quench function but to the extrema of its derivative. For $\theta \in [0, \pi/2]$, $X \in \mathbb{R}^2$, and assuming that the skeleton $S(X)$ is continuously differentiable, the θ-conditional bisector of X can be defined as follows:

Definition 23: Conditional Bisector.   The θ-conditional bisector of X, denoted $CB_\theta(X)$, is the set of points of $S(X)$ where the derivative of the quench function q_x along the skeleton is within $[-\tan(\theta), \tan(\theta)]$.

An efficient algorithm for computing discrete approximations of $CB_\theta(X)$ for any θ has been proposed in [49]. It is extremely useful for marking "sharp" portions of sets.

Similarly, when the set to be segmented is made up of elongated particles, it may be of interest to mark their extremities. This can be achieved either via techniques based on geodesic ultimate erosions or by means of the maxima of the *propagation function* [31,44].

Let us illustrate how the conditional bisector can be used to solve a complex binary segmentation application. Figure 50a is a binary image of a cross section of vitreous fibers. These fibers are overlapping and need to be separated. A very acceptable marking of these fibers was proposed in [48] (Fig. 50b), resulting in the first watershed segmentation shown in Fig. 50c. However, since some intricately overlapping fibers were not marked, they are not separated in Fig. 50c (these particles are pointed at by arrows).

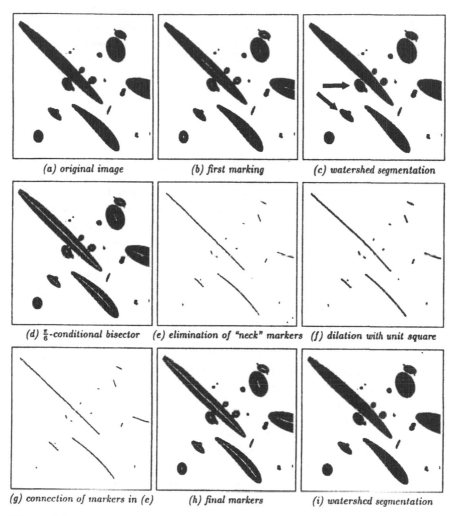

(a) original image   *(b) first marking*   *(c) watershed segmentation*

(d) $\frac{\pi}{6}$-conditional bisector   *(e) elimination of "neck" markers*   *(f) dilation with unit square*

(g) connection of markers in (e)   *(h) final markers*   *(i) watershed segmentation*

Figure 50 Use of conditional bisector and watersheds for the segmentation of binary images of glass fibers.

To mark them, a $\pi/6$-conditional bisector was used (see Fig. 50d). Not only does it mark these fibers, it also marks some "necks" between fibers. The latter are eliminated as crossing the separating lines of Fig. 50c (see Fig. 50e). Now, since some of the resulting markers are still slightly disconnected, a connection technique described in [25,49] was used: Fig. 50e is dilated by a unit-size disk, resulting in Fig. 50f. The latter image is then skeletonized with the constraint that the pixels of Fig. 50e belong to the resulting skeleton (see [53] for more details on constrained skeletons). This results in Fig. 50g. Adding the markers of Fig. 50b to this image yields the final marker image shown in Fig. 50h. At this stage, a new watershed segmentation provides the (almost) perfect result shown in Fig. 50i.

IV. CONCLUSIONS

In this chapter we dealt first with the segmentation of images of textures. We showed that granulometries and their moments constitute a general approach to the segmentation and classification of such images. To keep the chapter simple, only the binary case was presented, but the methods extend to gray-scale images, as illustrated by the example of Fig. 10. This section showed that adequate combinations of simple morphological operations (i.e., openings and closings in the present case) can lead to powerful segmentation algorithms.

This statement becomes even clearer in the case of object segmentation problems: we started from elementary operations and progressively constructed a set of more and more elaborate tools that finally led to watersheds and gray-scale reconstruction. As illustrated by numerous examples, they are extremely powerful operations for binary and gray-scale object segmentation problems. In addition, throughout the second part of this chapter, we attempted to derive a general philosophy of object segmentation using mathematical morphology. The outcome of this can be summarized in two key words: *markers* and *watersheds*.

The examples presented have each illustrated a different aspect of this morphological approach to segmentation. However, in no way are these examples full-size applications. There is practically no trivial segmentation problem, and with each new problem comes a load of new difficulties to overcome: this leads to new ways to combine existing operations or even to completely new transformations. The segmentation philosophy described above should only be considered as a guideline for the image analyst and should not put any constraints on his or her creativity. In this way, each new application will lead to new advances and the field of morphology will continue to progress and be enriched with new operations.

REFERENCES

1. B. Bettoli and E. R. Dougherty, Linear granulometric moments of noisy binary images, *J. Mathematical Imaging and Vision*, 2(4), pp. 299–319 (1993).
2. B. Bettoli and E. R. Dougherty, Linear granulometric moments of noise-degraded images, *Proc. EURASIP Workshop on Mathematical Morphology and Its Applications to Signal Processing*, Barcelona, Spain, May 1993.

3. S. Beucher and Ch. Lantuéjoul, Use of watersheds in contour detection, *Proc. International Workshop on Image Processing, Real-Time Edge and Motion Detection/ Estimation*, Rennes, France, 1979.

4. S. Beucher, Analyse automatique des gels d'électrophorèse bi-dimensionnels et morphologie mathématique, *Technical Report CMM*, École des Mines, Paris, 1982.

5. S. Beucher, Watersheds of functions and picture segmentation, *Proc. IEEE International Conference on Acoustics, Speech and Signal Processing*, Paris, May 1982, pp. 1928–1931.

6. S. Beucher, Segmentation d'images et morphologie mathématique, Ph.D. thesis, École des Mines, Paris, June 1990.

7. S. Beucher and F. Meyer, The morphological approach to segmentation: the watershed transformation, in *Mathematical Morphology in Image Processing*, E. R. Dougherty ed., Marcel Dekker, New York, Sept. 1992, pp. 433–481.

8. C. Bhagvati, D. A. Grivas, and M. M. Skolnick, Morphological analysis of surface condition, in *Mathematical Morphology in Image Processing*, E. R. Dougherty ed., Marcel Dekker, New York, Sept. 1992, pp. 121–150.

9. H. Blum, An associative machine for dealing with the visual field and some of its biological implications, in *Biological Prototypes and Synthetic Systems*, Vol. 1, pp. 244–260, *Proc. 2nd Annual Bionics Symposium*, Cornell University, Ithaca, N.Y., 1961, E. E. Bernard and M. R. Kare, eds., Plenum Press, New York, 1962.

10. J. W. Brandt and V. R. Algazi, Lossy encoding of document images using the continuous skeleton, *Proc. SPIE*, Vol. 1818, *Visual Communications and Image Processing*, Boston, Nov. 1992.

11. G. Borgefors, Distance transformations in digital images, *Comput. Vision Graphics Image Process.*, 34:334–371, 1986.

12. L. Calabi and W. E. Harnett, Shape recognition, prairie fires, convex deficiencies and skeletons, *Scientific Report 1*, Parke Mathematical Laboratories, Inc., Carlisle, Mass., 1966.

13. Y. Chen and E. R. Dougherty, Texture classification by gray-scale morphological granulometries, *Proc. SPIE*, Vol. 1818, *Visual Communications and Image Processing*, Boston, Nov. 1992.

14. Y. Chen and E. R. Dougherty, Gray-scale morphological granulometric texture classification, *Report MIL-16-92*, Morphological Imaging Laboratory, Rochester Institute of Technology, Rochester, N.Y., Nov. 1992.

15. Y. Chen, E. R. Dougherty, S. Totterman, and J. Hornak, Classification of trabecular structure in magnetic resonance images based on morphological granulometries, *J. Magn. Reson. Med.* 29(3), pp. 358–370 (1993).

16. P. E. Danielsson, Euclidean distance mapping, *Comput. Graphics Image Process.* 14:227–248 (1980).

17. E. R. Dougherty and C. R. Giardina, *Image Processing: Continuous to Discrete*, Prentice Hall, Englewood Cliffs, N.J., Jan. 1987.

18. E. R. Dougherty and J. Pelz, Morphological granulometric analysis of electrophotographic images: size distribution statistics for process control, *Opt. Engrg.*, 30(4) pp. 438–445 (1991).

19. E. R. Dougherty, J. Pelz, F. Sand, and A. Lent, Morphological image segmentation by local granulometric size distributions, *J. Electron. Imag.*, 1(1) pp. 46–60 (1992).
20. E. R. Dougherty, *An Introduction to Morphological Image Processing*, SPIE Press, Bellingham, Wash., Feb. 1992.
21. E. R. Dougherty, Euclidean gray-scale granulometries: representation and umbra inducement, *J. Math. Imag. Vision*, 1(1) pp. 7–21 (1992).
22. E. R. Dougherty, J. Newell, and J. Pelz, Morphological texture-based maximum-likelihood pixel classification based on local granulometric moments, *Pattern Recognition*, 25(11) pp. 1181–1198 (1992).
23. C. R. Giardina and E. R. Dougherty, *Morphological Methods in Image and Signal Processing*, Prentice Hall, Englewood Cliffs, N.J., 1988.
24. M. Golay, Hexagonal pattern transforms, *Proc. IEEE Trans. Comput.*, C-18(8) (1969).
25. G. G. Gordon and L. Vincent, Application of morphology to feature extraction for face recognition, *Proc. SPIE/SPSE*, Vol. 1658, *Nonlinear Image Processing III*, San Jose, Calif., Feb. 1992, pp. 151–164.
26. M. Grimaud, A new measure of contrast: dynamics, *Proc. SPIE*, Vol. 1769, *Image Algebra and Morphological Processing III*, San Diego, Calif., July 1992, pp. 292–305.
27. R. Haralick and L. Shapiro, Survey: image segmentation techniques, *Comput. Vision Graphics Image Process.*, 29:100–132 (1985).
28. Ch. Lantuéjoul, Skeletonization in quantitative metallography, in *Issues of Digital Image Processing*, R. M. Haralick and J.-C. Simon, eds., Sijthoff en Noordhoff, Groningen, The Netherlands, 1980.
29. Ch. Lantuéjoul and S. Beucher, On the use of geodesic metric in image analysis, *J. Microsc.*, 121:39–49 (1981).
30. Ch. Lantuéjoul and F. Maisonneuve, Geodesic methods in image analysis, *Pattern Recognition*, 17:117–187 (1984).
31. F. Maisonneuve and M. Schmitt, An efficient algorithm to compute the hexagonal and dodecagonal propagation function, *Acta Stereol.*, 8(2), *Proc. 5th European Congress for Stereology*, Freiburg-im-Breisgau, Germany, 1989, pp. 515–520.
32. P. Maragos, Pattern spectrum and multi-scale shape representation, *IEEE Trans. Pattern Anal. Mach. Intelligence*, PAMI-11(7): 701–716 (1989).
33. P. Maragos and R. D. Ziff, Threshold superposition in morphological image analysis, *IEEE Trans. Pattern Anal. Mach. Intelligence*, PAMI-12(5) (1990).
34. G. Matheron, *Random Sets and Integral Geometry*, Wiley, New York, 1975.
35. G. Matheron, Examples of topological properties of skeletons, in *Image Analysis and Mathematical Morphology*, Part II: *Theoretical Advances*, J. Serra, ed., Academic Press, London, 1988.
36. F. Meyer, Contrast feature extraction, in *Quantitative Analysis of Microstructures in Material Sciences, Biology and Medicine*, J.-L. Chermant, ed., Special issue of *Practical Metallography*, Riederer Verlag, Stuttgart, Germany, 1978.
37. F. Meyer, Cytologie quantitative et morphologie mathématique, Ph.D. thesis, École des Mines, Paris, 1979.
38. I. Pitas and A. N. Venetsanopoulos, Morphological shape representation, *IEEE Trans. Pattern Anal. Mach. Intelligence*, PAMI-11(7) (1989).

39. J.-F. Rivest, P. Soille, and S. Beucher, Morphological gradients, *Proc. SPIE*, Vol. 1658, *Nonlinear Image Processing III*, San Jose, Calif., Feb. 1992, pp. 139–150.

40. A. Rosenfeld and J. L. Pfaltz, Distance functions on digital pictures, *Pattern Recognition*, 1:33–61 (1968).

41. Ph. Salembier and J. Serra, Morphological multiscale image segmentation, in *Proc. SPIE*, Vol. 1818, *Visual Communications and Image Processing*, Boston, Nov. 1992.

42. F. Sand and E. R. Dougherty, Asymptotic normality of the morphological pattern-spectrum moments and orthogonal granulometric generators, *J. Visual Comm. Image Representation*, 3(2) pp. 203–214 (1992).

43. F. Sand and E. R. Dougherty, Statistics of the morphological pattern-spectrum moments for a random-grain model, *J. Math. Imag. Vision*, 1(2) (1992).

44. M. Schmitt, Variations on a theme in binary mathematical morphology, *J. Visual Comm. Image Representation*, 2(3): 244–258 (1991).

45. J. Serra, *Image Analysis and Mathematical Morphology*, Academic Press, London, 1982.

46. J. Serra, ed., *Image Analysis and Mathematical Morphology*, Part II: *Theoretical Advances*, Academic Press, London, 1988.

47. J. Serra and L. Vincent, An overview of morphological filtering, *Circuits Systems Signal Process.*, 11(1): 47–108 (1992).

48. H. Talbot, Binary image segmentation using weighted skeletons, *Proc. SPIE*, Vol. 1769, *Image Algebra and Morphological Image Processing III*, San Diego, Calif., July 1992, pp. 393–403.

49. H. Talbot and L. Vincent, Euclidean skeletons and conditional bisectors, *Proc. SPIE*, Vol. 1818, *Visual Communications and Image Processing*, Boston, Nov. 1992.

50. L. Vincent, Graphs and mathematical morphology, *Signal Process.*, 16:365–388 (1989).

51. L. Vincent, Mathematical morphology for graphs applied to image description and segmentation, *Proc. Electronic Imaging West '89*, Pasadena, Calif., Vol. 1, 1989, pp. 313–318.

52. L. Vincent and S. Beucher, The morphological approach to segmentation: an introduction, *Technical Report CMM*, School of Mines, Paris, 1989.

53. L. Vincent, Efficient computation of various types of skeletons, *Proc. SPIE Medical Imaging V*, San Jose, Calif., 1991.

54. L. Vincent and P. Soille, Watersheds in digital spaces: an efficient algorithm based on immersion simulations, *IEEE Trans. Pattern Anal. Mach. Intelligence*, PAMI-13(6): 583–598 (1991).

55. L. Vincent, Exact Euclidean distance function by chain propagations, *Technical Report 91-4*, Harvard Robotics Laboratory, Cambridge, Mass.; *Proc. IEEE Computer Vision and Pattern Recognition '91*, Maui, Hawaii, 1991, pp. 520–525.

56. L. Vincent and B. Masters, Morphological image processing and network analysis of corneal endothelial cell images, *Proc. SPIE*, Vol. 1769, *Image Algebra and Morphological Image Processing III*, San Diego, Calif., July 1992, pp. 212–226.

57. L. Vincent, Morphological algorithms, in *Mathematical Morphology in Image Processing*, E. R. Dougherty, ed., Marcel Dekker, New York, Sept. 1992, pp. 255–288.
58. L. Vincent, Morphological area openings and closings for grayscale images, *Proc. NATO Shape in Picture Workshop*, Driebergen, The Netherlands, Sept. 1992.
59. L. Vincent, Morphological grayscale reconstruction in image analysis: applications and efficient algorithms, *IEEE Trans. Image Process.*, 2(2) pp. 176–201 (1993).
60. P. D. Wendt, E. J. Coyle, and N. C. Gallagher, Stack filters, *IEEE Trans. Acoustics Speech Signal Process.*, ASSP-34(4): 898–911 (1986).

3

Multispectral Image Segmentation in Magnetic Resonance Imaging

Joseph P. Hornak and Lynn M. Fletcher

Rochester Institute of Technology
Rochester, New York

I. MAGNETIC RESONANCE IMAGING

In this chapter we describe the principles upon which magnetic resonance imaging is based. A microscopic molecular description is adopted so as to give the reader both (1) an appreciation of the mechanism by which contrast is achieved and manipulated in magnetic resonance images, and (2) a good foundation for the understanding of the processing of the raw data into an image. Applications of digital image processing in magnetic resonance imaging are discussed and a specific example of image processing in magnetic resonance imaging employing multispectral image segmentation is presented in detail. A comprehensive treatise on magnetic resonance imaging cannot be offered in the space of a book chapter; therefore, the reader is encouraged to seek more information from the literature referenced throughout this chapter.

Magnetic resonance imaging is a tomographic imaging modality based on the principles of nuclear magnetic resonance, a spectroscopic technique used by chemists and physicists to gain physical and structural information on chemical compounds. The name of the imaging technique might have more appropriately been called nuclear magnetic resonance imaging. Owing to the negative connotations associated with the word *nuclear* in the 1980s, when the technique was developed, the word *nuclear* was dropped and the technique became known as magnetic resonance imaging or by the acronym MRI. This story is to this day a source of amusement to those who realize that MRI has nothing to do with nuclear energy. Since its development, MRI has been used primarily as a diagnostic medical imaging technique competing with and often winning out over x-ray-based computed tomography (CT) because of its excellent image contrast between soft tissues. This excellent contrast is not based on conventional optical properties. Rather, this excellent contrast is due to the fact that the modality creates images of the nuclear magnetic resonance signal in the imaged slice. This signal is based not only on density, but also on intrinsic dynamic properties of signal-bearing nuclei in the tissue. These properties are explained in detail in Section I.A. Like CT, MRI is capable of producing transaxial tomographic images of the human body. Unlike CT, MRI can also produce coronal, sagittal, and oblique plane images without the need first to collect a volume image or to physically change the orientation of the patient relative to the imager. Unlike CT, MRI does not use ionizing x-ray radiation, but rather nonionizing radiofrequency radiation.

MRI is still in its infancy with new applications continually being developed. A few of the more notable applications are magnetic resonance (MR) angiography, microscopy, volume imaging, fluoroscopy, and materials imaging. MR angiography produces velocity-encoded images of flowing blood, compared to blood vessel volume as produced by x-ray techniques, without the introduction of radiopaque dyes. MR microscopy is producing 10μm in plane resolution images of microscopic samples. Volume MRI is being used to produce three-dimensional images of the body with the same excellent tissue contrast as in conventional tomographic MRI. MR fluoroscopy holds the promise of producing routine real-time MR video images once faster two-dimensional Fourier transform procedures are available. A newly developing field of materials imaging is opening up new applications for MRI in oil production and polymer chemistry. As a consequence of these developments and the advantages of MRI over CT mentioned previously, MRI is rapidly replacing x-ray-based imaging as the modality of choice for several radiological investigations, with its primary restricting aspect being the cost of a scan. The cost factor, however, is expected to diminish as more specialized magnetic resonance imagers complement the present hardware suitable for all anatomies.

A. NMR Physics

Magnetic resonance imaging is based on the principles of nuclear magnetic resonance (NMR); therefore, any description of MRI should be prefaced by a description of the principles of nuclear magnetic resonance [1]. This section contains a brief synopsis of the underlying physics of nuclear magnetic resonance. The reader unfamiliar with NMR is encouraged to skim through the section once before reading for comprehension.

NMR is a spectroscopy based on a property of the nucleus of an atom called *spin*. Spin is a fundamental property of nature, which is observable in nuclei

Table 1 Biologically Relevant Nuclei

Nucleus	Gyromagnetic ratio (MNz/T)
^{1}H	42.58
^{23}Na	11.27
^{31}P	17.25

with unpaired protons or neutrons. Spin can be thought of as a simple magnetic moment possessed by the nucleus. When placed in an external magnetic field the magnetic moment can take on one of two possible orientations, one aligned with the field and the other opposing the field. A photon with an energy equal to the energy difference, ΔE, between the two orientations or states will cause a transition between the states. The greater the magnetic field, the greater the energy difference and hence the frequency of the absorbed photon. The relationship between the applied magnetic field B_0 and the frequency of the absorbed photon v is linear.

$$v = \gamma B_0 \tag{1}$$

The proportionality constant γ is called the gyromagnetic ratio. The gyromagnetic ratio is a function of the magnitude of the nuclear magnetic moment. Each isotope with a net nuclear spin possesses a unique γ. Three of the most abundant spin-bearing nuclei found in the human body are hydrogen, ^{1}H; sodium, ^{23}Na; and phosphorus, ^{31}P (Table 1). The magnetic field experienced by a nucleus is altered by the electron cloud around it. As a consequence, the resonant frequency of a nucleus is dependent on the chemical compound with which the nucleus is associated. The variation in v from chemical compound to compound is on the order of parts per million of v and is called chemical shift. A ^{31}P NMR spectrum of muscle tissue will possess five absorptions due to the five types of phosphorus: inorganic phosphate, phosphor creatine, and α, β, and γ adenosine triphosphate.

The signal in a magnetic resonance experiment is proportional to the difference between the number of spins in the lower spin energy state, N_-, and the number in the upper state, N_+. This population difference, $(N_- - N_+)$, may be calculated using Boltzmann statistics. Boltzmann statistics state that for a group of spins, the population difference between the states increases as the energy difference between the states increases:

$$\frac{N_+}{N_-} = e^{-\Delta E / kT_a} \tag{2}$$

where k is Boltzmann's constant and T_a is the absolute temperature. Since ΔE is proportional to B_0, the higher the field, the better the signal-to-noise ratio

(S/N). A spin system that has been perturbed by the absorption of photons such that $N_+ = N_-$ is said to be saturated. Unless the system relaxes toward equilibrium, no further absorption of energy or NMR signal will be observed.

An NMR spectrum is a plot of the absorbed energy versus magnetic field or frequency. The location of the absorption lines in the spectrum is indicative of the chemical composition of the sample. Based on the description of NMR thus far, an NMR spectrum could be recorded by holding B_0 constant and sweeping v or by holding v constant and sweeping B_0. Both techniques of recording an NMR spectrum were used up until the development of pulsed Fourier transform NMR spectrometers. In pulsed Fourier transform NMR a pulse of radio frequency (RF) containing a band of frequencies is introduced into the sample, and the entire spectrum is recorded simultaneously in the time domain and Fourier transformed into the frequency domain.

A macroscopic picture of the spin system is needed to understand the operation of present-day pulsed Fourier transform NMR spectrometers. Groups of nuclei experiencing the same B_0 are called spin packets. When placed in a B_0 field, spin packets precess about the direction of B_0 just as a spinning top precesses about the direction of the gravitational field. The precessional frequency, also called the Larmor frequency, ω is equal to $2\pi v$. The direction of the precession is clockwise about B_0 and the symbol ω_0 is reserved for spin packets experiencing exactly B_0. A rotating frame of reference is usually adopted to simplify the description of the precessional motion. This frame rotates about the z axis at ω_0 such that a spin packet precessing at a frequency $\omega < \omega_0$ in the laboratory frame of reference will appear to be traveling counterclockwise at $\omega - \omega_0$ in the rotating frame. Similarly, a precessional motion at a frequency $\omega > \omega_0$ in the laboratory frame will appear as a clockwise rotation at $\omega - \omega_0$ in the rotating frame. The rotating frame is usually denoted by primes on the x and y axes, (x' and y').

An NMR sample contains millions of spin packets, each with a slightly different Larmor frequency. The magnetization vectors from all these spin packets form a cone of magnetization about the z axis. At equilibrium, the net magnetization vector M from all the spins in a sample lies in the center of the cone along the z axis. Therefore, the longitudinal magnetization M_z equals M and the transverse magnetization M_{xy} equals zero at equilibrium. Net magnetization perturbed from its equilibrium position will want to return to its equilibrium position. This process is called spin relaxation.

The return of the z component of magnetization to its equilibrium value is called spin-lattice relaxation. The time constant that describes the rate at which M_z returns to its equilibrium value M_{z0} is called the spin-lattice relaxation time, T_1. The equation governing this behavior as a function of the time t after its displacement is

$$\frac{dM_z}{dt} = \frac{-(M_z - M_{z0})}{T_1} \tag{3}$$

Integration of Eq. (3) reveals that the spin-lattice relaxation time is the time to reduce by a factor of e the difference between the longitudinal magnetization and its equilibrium value.

Spin-lattice relaxation is caused by time-varying magnetic fields at the Larmor frequency. These variations in the magnetic field at the Larmor frequency cause transitions between the spin states and hence change M_z. Time-varying fields are caused by the random rotational and translational motions of the magnetic moment possessing molecules in the sample. The frequency distribution of random motions in a solution varies with the temperature and viscosity of the solution. Therefore, T_1 will not only vary as a function of temperature and the solvent but as a function of B_0. In general, relaxation times tend get longer as B_0 and v increase because there are fewer relaxation-causing frequency components present in the random motions of the molecules as v increases.

At equilibrium, the transverse magnetization M_{xy} equals zero. A net magnetization vector rotated off the z axis creates transverse magnetization. This traverse magnetization decays exponentially with a time constant called the spin-spin relaxation time T_2.

$$\frac{dM_{xy}}{dt} = \frac{-M_{xy}}{T_2} \tag{4}$$

The definition of T_2 is therefore the time to reduce the transverse magnetization by a factor of e. Spin-spin relaxation is caused by fluctuating magnetic fields that perturb the energy levels of the spin states and dephase the transverse magnetization. T_2 is inversely proportional to the number of molecular motions less than and equal to the Larmor frequency. Scientists break down T_2 further into a pure T_2 due to molecular interactions and one due to inhomogeneities in the B_0 field.

$$\frac{1}{T_2^*} = \frac{1}{T_2} + \frac{1}{T_{2_{\text{inhomogeneous}}}} \tag{5}$$

The overall T_2^* is referred to as "T_2 star."

In pulsed NMR spectroscopy RF energy is put into a spin system by sending RF into a resonant LC circuit, the inductor of which is placed around the sample. The inductor must be oriented with respect to the B_0 magnetic field so that the oscillating RF field created by the RF flowing through the inductor is perpendicular to B_0. The RF magnetic field is called the B_1 magnetic field. When the RF inductor, or coil as it more often called, is placed around the x axis in the laboratory frame, the B_1 field will oscillate back and forth along the $\pm x$ axis. In the rotating frame of reference rotating about z at ω_0 an RF magnetic field from the coil will be stationary along the x' axis.

In pulsed NMR spectroscopy it is the B_1 field that is pulsed. In the laboratory frame of reference turning on a B_1 field for a period of time τ will cause the net magnetization vector to begin to precess about B_0 in ever-widening angles. Eventually, the vector will reach the xy plane. If B_1 is left on longer, the net magnetization vector will reach the negative z axis.

In a frame of reference rotating about z at the the Larmor frequency, a B_1 field along x' will cause the net magnetization vector to rotate clockwise about the x' axis. The rotation angle θ, which is measured clockwise away from the z axis in radians, is proportional to γ, B_1, and τ. τ is the time the B_1 field is left on, or in other words, the length of the RF pulse.

$$\theta = 2\pi\gamma B_1\tau \tag{6}$$

Any transverse magnetization will rotate about the direction of B_0. An NMR signal is generated from transverse magnetization rotating about the z axis. This magnetization will induce a current in a coil of wire placed around the x or y axis. As long as there is transverse magnetization that is changing with respect to time there will be an induced current in the coil. For a group of nuclei with one identical chemical shift, the signal will be an exponentially decaying sine wave. The sine wave decays with time constant T_2^*. It is predominantly the in-homogeneities in B_0 that cause the spin packets to dephase. Net magnetization which has been rotated away from its equilibrium position along the z axis by exactly 180° will not create transverse magnetization and hence not give a signal. The time-domain signal from a net magnetization vector in the xy plane is called a free induction decay (FID). This time-domain signal must be converted to a frequency-domain spectrum to be interpreted for chemical information. The conversion is performed using a Fourier transform. The hardware in most NMR spectrometers is capable of detecting both M_x' and M_y' simultaneously. This detection scheme is called quadrature detection. These two signals are equivalent to the real and imaginary signals; therefore, the input to the Fourier transform will be complex. Sampling theory tells us that one need only digitize the FID at frequency f complex points per second to obtain a spectrum of frequency width f.

In pulsed Fourier transform NMR spectroscopy, short bursts of RF energy are applied to a spin system to induce a particular signal from the spins within a sample. A pulse sequence is a description of the types of RF pulses used and the response of the magnetization to the pulses. In other words, a pulse sequence is a description of the type of NMR experiment being performed.

A timing diagram for a pulse sequence is a plot of RF energy and signal as a function of time. A timing diagram shows the time when RF pulses are applied, their shape and size, and the period when a signal is recorded. When a pulse sequence is repeated, for signal-averaging purposes, the period between repetitions is called the repetition time and is often given the symbol TR. To maximize the signal in a set of repeated pulse sequences the net magnetization should be allowed to recover to its equilibrium value along the z axis before the next sequence is applied. There are literally hundreds of different type of pulse sequences in pulsed NMR spectroscopy [2]. Each one is designed to measure either a specific chemical property or type of spectra. The two most useful sequences for MRI are the 90-FID and the spin-echo sequence.

The simplest and most widely used pulse sequence for routine NMR spectroscopy is the 90-FID pulse sequence. As the name implies, the pulse sequence is a 90° pulse followed by the acquisition of the time-domain signal called an FID. The net magnetization vector, which at equilibrium is along the positive z axis, is rotated by 90° down into the $x'y'$ plane. The rotation is accomplished by choosing an RF pulse width and amplitude so that the rotation equation [Eq. (6)] for a 90° pulse is satisfied. If the RF pulse is chosen so as to set up a B_1 field along x', the net magnetization vector along z will be rotated clockwise about x' down to the y' axis. The net magnetization vector now behaves as all transverse magnetizations; it begins to precess about the direction of the applied magnetic field B_0. Assuming that $T_2 \ll T_1$, the net magnetization vector begins to dephase as the vectors from the individual spin packets in the sample precess at their own Larmor frequencies. Eventually, M_{xy} will equal zero and the net magnetization will return to its equilibrium value along z. A pulse sequence that causes the net magnetization to equal zero and then follows its recovery to equilibrium is called a *saturation recovery pulse sequence*.

The signal that is detected by the spectrometer is the decay of the transverse-magnetization as a function of time. This signal is once again called a free induction decay. The FID decays exponentially with a time constant T_2^* as can be seen by integrating Eq. (4) to yield the equation

$$M_{xy} = e^{-t/T_2^*} \tag{7}$$

The FID or time-domain signal is Fourier transformed to yield the frequency-domain NMR spectrum.

The spin-echo sequence is a two-RF pulse sequence containing a 90° and a 180° pulse. The 90° pulse sets up a B_1 field along the x' axis which rotates the net magnetization vector from along the z axis to along the y' axis. The net magnetization vector begins to dephase. Those spin packets with Larmor frequency greater than ω_0 precess clockwise and those with Larmor frequency less than ω_0 counterclockwise. After a period of time equal to TE/2, a 180° pulse is applied, which sets up a B_1 field along the x' axis and rotates all the y' and z components of magnetization by 180°. X' magnetization is unchanged by a 180° pulse applied along x'. After the 180° pulse the slower vectors are still traveling counterclockwise and those faster clockwise. As a result, the vectors from all the spin packets refocus at a point in time TE after the 90° pulse, where TE is called the echo time.

The 90° pulse creates an FID, but this is not the primary signal in the sequence. The 180° pulse refocuses the magnetization and thus creates another signal called an echo. The echo looks like a set of back-to-back free induction decays, one growing and the other decreasing with respect to time. Fourier transforming the echo gives a frequency-domain NMR signal. The spin-echo sequence is used because often one cannot record an FID immediately after the 90° pulse. The spin-echo sequence moves the signal away from the RF pulses. Both the FID and the echo shapes decay exponentially as T_2^*. However, the maximum echo height decreases with time constant T_2 as TE increases.

$$M_{xy} = e^{-TE/T_2} \tag{8}$$

A spin-echo sequence could therefore be used to measure T_2. It is possible to produce multiple echoes by applying additional 180° pulses, each separated by TE. Such a pulse sequence is called a multiecho sequence.

B. Imaging

Magnetic resonance imaging is a tomographic imaging modality that produces pictures of the NMR signal from selected slices through the human body [3]. The selected slice in the body is said to be made up of several volume elements or voxels. The NMR signal from a given voxel gives rise to an intensity in the corresponding picture element or pixel in the magnetic resonance image. In routine clinical imaging the thickness of a slice is approximately 3 mm and the in-plane resolution approximately of 0.8 mm for a 20-cm-field-of-view (FOV) 256×256 pixel image. The volume of a voxel is therefore approximately 2 mm^3. Taking into account the size of organs and composition of tissues in the body, it is clear to see that a voxel is often comprised of more than one substance. As a consequence, the NMR signal from a voxel is a summation of the NMR signals from the substances found in the voxel. Any variation in an imagable quantity (signal, T_1, T_2, etc.) due to the relative amounts of the components found in a voxel is referred to as a *partial volume effect*.

The most abundant spin-bearing nucleus in the human body is hydrogen. The two most abundant forms of hydrogen are adipose (fat) tissue and water hydrogens. The adipose tissue hydrogens are bonded primarily to carbon atoms in long lipid molecules, and the water hydrogens are bonded to oxygen in a small molecule. Other hydrogens associated with proteins and the other building blocks of tissues have very short T_2 relaxation times and do not contribute directly to the NMR signal. The fat and water hydrogens in the body have slightly different chemical shifts, a fact that can lead to an artifact present in certain magnetic resonance images [3].

The most commonly used type of imaging pulse sequence is the spin-echo sequence. The spin-echo NMR signal from a voxel in the body will therefore be equal to a summation over all the different types of spins, i, in the voxel.

$$S = \sum_i \rho_i \left(1 - 2e^{-(TR/T_{1i} - TE/2T_{2i})} + e^{-TR/T_{1i}}\right) e^{-TE/T_{2i}} \tag{9}$$

The quantity ρ_i is the spin density of i in the voxel. The quantity $(1 - 2e^{-(TR/T_{1i} - TE/2T_{2i})} + e^{-TR/T_{1i}})$ enters into the equation because the spin-echo sequence is repeated, approximately 128 to 256 times, to obtain an image and the spin system experiences the cumulative effects of all the 90° and 180° pulses. By examining this equation it should be apparent that by varying TR and TE the clincan has a tremendous amount of flexibility in selecting the contrast between tissue in an image. This point is developed in detail later.

The equation behind which all MRI is based is Eq. (1), which states that the resonance frequency of a nucleus is proportional to the magnetic field it is experiencing. The proportionality constant is called the gyromagnetic ratio. If one could set up a spatially varying magnetic field across a sample the nuclei within the sample would resonate at a frequency related to their position. For example, assume that a one-dimensional linear magnetic field gradient G_z is set up in the B_0 field along the z axis. The resonant frequency v will be equal to

$$v = \gamma (B_0 + zG_z) \tag{10}$$

(see Fig. 1). The origin of the xyz coordinate system is taken to be the point in the magnet where the field is exactly equal to B_0 and spins resonate at v_0. This point is referred to as the isocenter of the magnet. Equation (10) explains how a simple one-dimensional imaging experiment can be performed. The sample to

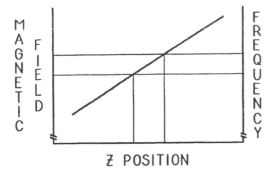

Z POSITION

Figure 1 Graphical representation of the relationship between the magnetic field, position of a sample, and resonant frequency. The diagonal line represents the applied magnetic field gradient.

be imaged is placed in a magnetic field B_0. A 90° pulse of RF energy is applied to rotate magnetization into the *xy* plane. A one-dimensional linear magnetic field gradient G_z is turned on after the RF pulse, and the FID is recorded immediately. The Fourier transform of the FID yields a frequency spectrum $S(\nu)$ which can be converted to $S(z)$, the spectrum as a function of z, by

$$z = \frac{\nu - \nu_0}{\gamma G_z} \tag{11}$$

This simple concept of a one-dimensional image can be expanded to a two-dimensional image employing the concept of backprojection imaging, similar to that used in CT imaging. If a series of one-dimensional images, or projections of the signal in a sample, are recorded for linear one-dimensional magnetic field gradients applied along several different trajectories in a plane, the spectra can be tranformed into a two-dimensional image using an inverse radon transform or backprojection algorithm. The difficulty with this procedure is that it is cumbersome to apply a linear combination of two perpendicular gradients to achieve the different projections. Second, the image quality tends to be less than that available from Fourier-based imaging procedures, described next.

A Fourier-based imaging technique collects data from the *k* space of the imaged object [4]. There *k*-space data are then Fourier transformed into a image using a two-dimensional Fourier transform. Figure 2 depicts a timing diagram for a Fourier imaging sequence using simple 90-FID sequence. The timing diagram describes the application of RF and three magnetic field gradients called the slice selection (G_s), phase-encoding (G_p), and frequency-encoding (G_f) gradients. The first step in the Fourier imaging procedure is slice selection of those spins in the object for which the image is to be generated. Slice selection is accomplished by the application of a magnetic field gradient at the same time as the RF pulse is applied. An RF pulse with frequency width $\Delta\nu$ centered at ν, will, when applied in conjunction with a field gradient G_z, excite spins centered at $z = (\nu - \nu_0)/\gamma G_z$, with a spread of spins at z values $\Delta z = \Delta\nu/\gamma G_z$. Spins experiencing a magnetic field strength not satisfying the resonance condition will not be rotated by the RF pulses, and hence slice selection is accomplished. The image slice thickness (Thk) is given by Δz. For a clean slice (i.e., all spins along the slice thickness are rotated by the prescribed rotations) the frequency content of the pulse must be equal to a rectangular-shaped (rect) function. Therefore, the RF pulse must be shaped as an apodized sinc function in the time domain.

The next step in the Fourier imaging procedure is to encode some property of the spins as to the location in the plane selected. One could easily encode spins as to their *x* position by applying a gradient G_x after the RF pulse and during the acquisition of the FID. The difficulty is in encoding the spins with information as to their *y* location. This is accomplished by encoding the phase of the pre-

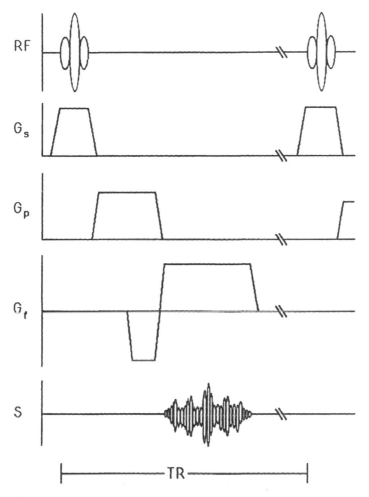

Figure 2 Timing diagram for a 90-FID Fourier transform imaging sequence depicting the RF, slice selection gradient (G_s), phase-encoding gradient (G_p), frequency-encoding gradient (G_f), and signal.

cessing spin packets with the y position (see Fig. 3). Phase encoding is accomplished by turning on a gradient in the y direction immediately after the slice selection gradient is turned off and before the frequency-encoding gradient is turned on. The spins in the excited plane now precess at a frequency dependent on their y position. After a period of time τ the gradient is turned off and the spins have acquired a phase ϕ equal to

$$\phi = 2\pi\gamma\tau y G_y \tag{12}$$

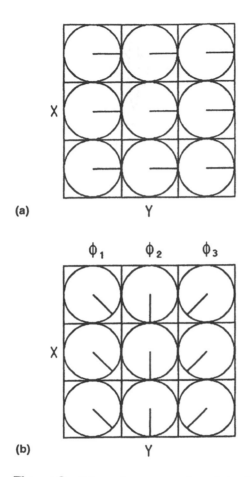

Figure 3   Schematic representation of nine hypothetical spin vectors in an *xy* plane experiencing a 90-FID Fourier imaging sequence. The planes represent the positions of the magnetization vectors (a) after the 90 pulse, (b) after the phase-encoding gradient, and (c) during the frequency-encoding gradient.

Figure 3 describes for nine magnetization vectors the effect of the application of a phase-encoding gradient G_y and a frequency-encoding gradient G_x. The phase-encoding gradient assigns each *y* position a unique phase. The frequency-encoding gradient assigns each *x* position a unique frequency. If one had the capability of independently assessing the phase and frequency of a spin packet, one could assign its position in the *xy* plane. Unfortunately, this cannot be accomplished with a single pulse and signal. The phase-encoding gradient must be

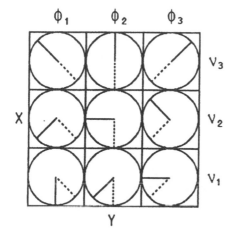

(c)

varied in amplitude so as to achieve a 2π phase variation between the isocenter and the first resolvable point in the y direction, and therefore a 256π variation from center to edge of the imaged space for 256-pixel resolution in the phase-encoding direction. The result is to traverse the k space of the image line by line. The negative lobe on the frequency-encoding gradient (see Fig. 2), which was not described previously, shifts the center of k space to the center of the signal acquisition window.

The k-space data are recorded in quadrature. The FOV is dependent on the sampling rate, R_s, during G_f and the magnitude of G_f:

$$\text{FOV} = \frac{R_s}{\gamma G_f} \tag{13}$$

The k-space data are typically multiplied by an exponentially decaying cone to achieve an effect equivalent to convoluting the image with a Lorentzian lineshape function to reduce noise. The resolution of the image is not affected significantly in this operation provided that the time constant of the exponentially decaying cone is greater than T_2^*. The two-dimensional k-space data set is then Fourier transformed and the magnitude image generated from the real and imaginary outputs of the Fourier transform. Data points are then interpolated to expand the 256×256 image to a 512×512 pixels.

A problematic artifact associated with MRI arises when the imaged subject moves during acquisition of the k-space data. Such a motion may result in a step in the frequency-encoded direction data or in the phase-encoding direction data of k space. When Fourier transformed, such a step causes a blurred band across the image corresponding to the object that moved. Such an artifact in an image is referred to as a motion artifact.

In the previous example of Fig. 3, the slice selection gradient was applied along the z axis and the phase- and frequency-encoding gradients along the y and x axes, respectively. In practice the gradients can be applied along any three orthogonal axes, the only restrictions being that the slice selection gradient be perpendicular to the imaged plane.

The most routinely used imaging sequence is the spin-echo imaging sequence. Its popularity is attributable to its ability to produce images that display both T_1 and T_2 contrast, as was seen in Eq. (8). This sequence consists of the required 90 and 180° RF pulses, as seen in Fig. 4. These pulses are applied in conjunction with the slice selection gradient. The phase-encoding gradient is applied between the 90 and 180° pulses. The frequency-encoding gradient is turned on during the acquisition of the echo.

The goal of the MR imaging scientist is not to maximize the S/N ratio but to maximize the contrast-to-noise ratio between the tissues in an image. The contrast between any two tissues may be maximized by prudent choice of TR and TE. This choice requires a knowledge of the T_1, T_2, and ρ values of the tissues in an image. MR clinicians have adopted a nomenclature for describing the type of spin-echo image being produced. A T_1 weighted image is one in which image contrast displays differences in T_1 of the tissues. A T_1 weighted image is one where TR $\leq T_1$ and TE $\ll T_2$. A T_2 weighted image is one in which contrast between the tissues is due primarily to differences in T_2 of the tissues. A T_2 weighted image is one where TR $\gg T_1$ and TE $\geq T_2$. Spin density weighting is, as expected, an image whose contrast displays differences in spin density of the tissues. A spin density weighted image is obtained with a spin-echo sequence by setting TR $\gg T_1$ and TE $\ll T_2$. A T_1, T_2, or ρ weighted image will always have some dependance on the other two quantities.

The clinician has another means of inducing contrast in an image, the contrast agent. A contrast agent is usually a paramagnetic substance that is induced into the body and which has an affinity to certain tissue types. The contrast agent changes the T_1 of the tissue with which it is in contact and hence its signal and contrast properties relative to other tissues with a lesser affinity to the contrast agent.

The first decade of magnetic resonance imaging has seen a great deal of scientific effort directed toward improving the imaging hardware and developing new pulse sequences. These efforts have yielded tremendous advances in image quality, to the point where most magnetic resonance images are not noise limited. With the tremendous advances on the hardware and pulse sequence front, very little effort has been directed, relatively speaking, to developing novel image-processing techniques for MRI. There are several applications of image processing in magnetic resonance imaging.

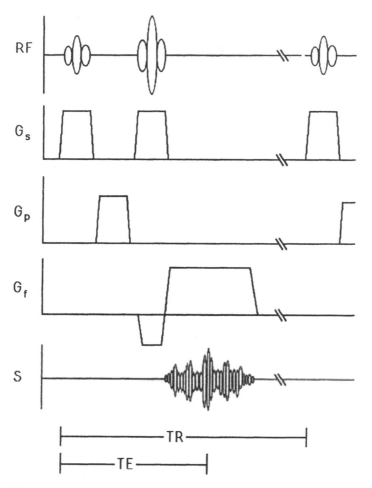

Figure 4 Timing diagram for a spin-echo Fourier transform imaging sequence depicting the RF, slice selection gradient (G_s), phase-encoding gradient (G_p), frequency-encoding gradient (G_f), and signal.

II. MULTISPECTRAL IMAGE SEGMENTATION

The tissues in an imaging space may be classified and segmented from one another in an image with the aid of an image-processing technique called multispectral image segmentation (MIS). Multispectral image segmentation is the classification of objects within an image space based on the properties of the image space in several spectral regions [5]. For example, in remote sensing where the technique has been used successfully for decades, satellite images

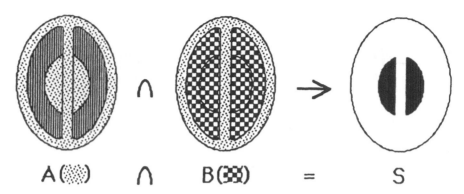

Figure 5 Schematic representation of the multispectral image segmentation process. The segmented image S is achieved by taking the intersection of the speckled region of A with the checkerboard region of B.

from the visible, near infrared, and far infrared are used to discriminate between several types of vegetation which might otherwise be indistinguishable if only a visible spectrum image was used. Successful classification of imaged objects with MIS depends on the prudent choice of independent or orthogonal spectral regions which provide unique information on the features to be segmented. Using a remote-sensing example again, it is easier to distinguish between a blue house and green grass with a blue-wavelength image than with a black-and-white image containing information from all visible wavelengths.

A simple illustration of MIS in MRI can be seen in Fig. 5. Image A contains two tissues with a speckled pattern. If the goal is to segment the two half-oval speckled regions from surrounding tissues, a segmentation could not be performed. Similarly, in image B the smaller half-oval regions have the same checkerboard pattern as the larger half-oval regions. Taking the intersection of the speckled region of A and the checkerboard region of B yields the desired region of the image. Of course, this may be applied to any number of independent spectral regions. The likelihood of two tissues possessing similar spectral properties in m spectral regions decreases as m increases.

A. Spectral Regions

In MRI, some of the possible spectral regions satisfying the independence criterion include spin-lattice relaxation time [6], spin-spin relaxation time [6], spin density [6], magnetization transfer [7], translational diffusion [8], flow [9], and texture [10] for each nucleus and chemical shift. Clearly, magnetic resonance images of ^1H, ^{23}Na, and ^{31}P would constitute independent spectral regions. Certain physiological conditions are known to be associated with changes in the concentration of the nuclei. For each nucleus one might obtain a chemical shift image [11]. For example, in hydrogen MRI a water image would provide distinctly different information from a fat image. For each chemical shift of each type of nucleus, one can create a T_1, T_2, and ρ image. The ρ image will be unique for each nucleus and chemical shift. T_1 and T_2, which reflect molecular motions, will provide unique information as the Larmor frequency is different for each nucleus, and therefore the molecular motions which they probe will reflect different information. The texture in an image is indicative of certain tissue types and pathologies. For example, muscle displays a fibrous texture, trabecular bone a spongy texture, and adipose tissue a globular texture. Texture images can be produced of the spectral regions previously mentioned, with a smaller slice thickness. Diffusion images are a direct measure of the mobility of a nucleus. Whereas T_1 and T_2 probe primarily rotational motions at specific frequencies, diffusion images probe translational motions. Flow images are similar to diffusion images except that they are tuned to measure faster translational motions, such as vascular flow. Magnetization transfer is a quantum-mechanical phenomenon based on a property called spin exchange. A magnetization transfer image is therefore a measure of the ability of one nucleus type to transfer its energy to another. The transfer rate is dependent on such properties as the physical distance between the exchanging nuclei and the strength of the nuclear moment. Such an image therefore gives information on the proximity of one relative to the other nucleus and the local magnetic environment.

B. Computing the Spectral Regions

Creating a spectral region for a region other than that for hydrogen, for which most imagers are tuned, is a matter of changing the operating frequency of the imager. The RF electronics on most imagers can accommodate changes such as this. The imaging coil, in contrast to the RF electronics, must be changed or retuned to the new resonant frequency. Images of different nuclei at different chemical shifts may be achieved by a chemical shift imaging sequence [11]. Such a procedure is applied routinely to image fat or water in a proton MRI. Texture images require the acquisition of thin-slice high-resolution images followed by morphological image processing. The slice thickness of the initial MRI is typically ≤ 1 mm and the in-plane resolution is ≤ 0.3 mm. Diffusion and flow images are produced with a special pulse sequence which applies an additional set of gradient pulses, called bipolar gradient pulses, to the sample. Spins that are moving during the bipolar gradient pulse do not completely refocus and hence do contribute less to the echo signal. By subtraction of the appropriate images recorded with opposite phase bipolar gradient pulses it is possible to obtain an image of just diffusing or flowing spins. Magnetization transfer images may be produced with a pulse sequence that applies an additional RF pulse to the spin system. The frequency content of this pulse is tailored to saturate the spins at a chemical shift corresponding to a substance in the tissue: for example, protein. The resultant image intensity for water shows a difference over a regular image because of the transfer of magnetization from the saturated spins to the water nucleus hydrogens.

Although a plethora of spectral regions are available in magnetic resonance, the ease with which the spectral regions may be obtained varies enormously. The following more detailed discussion is of the generation of three of the more readily available spectral regions T_1, T_2, and ρ. Several pulse sequences and protocols have been designed to calculate T_1, T_2, and ρ images. This description is restricted to a procedure utilizing the more readily available spin-echo pulse sequence.

The most commonly used spectral regions for MIS in MRI are T_1, T_2, and ρ. True T_1, T_2, and ρ images are difficult to obtain. For this reason T_1, T_2, and ρ weighted images are often used [12–14] even though they do not satisfy the independence criterion. Some of the problems that make the generation of pure T_1, T_2, and ρ images difficult are as follows. Eddy currents in the metal bore of the magnet caused by unshielded gradient coils have caused T_2 to be a measure of the imaging instrument rather than the imaged tissues. Inhomogeneities in the radio-frequency field, B_1, across the image plane will cause errors in T_1 and T_2 [15]. The effects of inhomogeneities in B_1 are most pronounced when multiecho spin-echo images are used to calculate T_2 [16,17]. Translational diffusion affects a T_2 calculation more with longer-echo-time (TE) values than with shorter ones [1]. Multislice sequences tend to lengthen a calculated T_1 and shorten a calculated T_2 because spins are excited by adjacent slice radio-frequency pulses. Both partial volume effects and magnetization transfer processes cause the signal to be multiexponential [7,18,19], causing an average relaxation time to be measured when single exponential fits are used. The measurement of T_1, T_2, and ρ therefore requires a carefully planned and executed set of experiments. This study utilized T_1, T_2, and ρ images, which were calculated from a rather lengthy imaging protocol. This protocol minimizes the dependence of one spectral region on the remaining two, but as a consequence introduced a flow dependence on the three spectral regions.

The spin-echo signal equation [Eq. (9)] may be simplified to

$$S(\text{TR, TE}) = \rho \left(1 - e^{-\text{TR}/T_1}\right) e^{-\text{TE}/T_2} \tag{14}$$

assuming that TR $\gg$ TE and that the image voxel contains one component. Having made these assumptions, T_1, T_2, and ρ may readily be calculated from a spin-echo sequence. T_2 images may be obtained rapidly from variable-echo-time (TE) MR images using a linear least-squares algorithm [20]. Holding TR constant in Eq. (14) gives

$$S(\text{TE}) = ce^{-\text{TE}/T_2} \tag{15}$$

and taking the natural logarithm yields

$$\ln(S) = \frac{-TE}{T_2} + \ln(c) \tag{16}$$

where c is a constant. Plotting $\ln(S)$ versus TE should yield a linear relationship with a slope equal to $-1/T_2$.

Unfortunately, both the acquisition and postprocessing of data to generate T_1 images is not as rapid. T_1 data acquisition may take as long as 1 h using a saturation recovery sequence, although shorter sequences are available [21]. Calculation of T_1 images from magnetic resonance images is a computationally intensive procedure, requiring for a 256 × 256 pixel image that 65,536 T_1 values be determined using a nonlinear least-squares algorithm. The postprocessing of MR images to generate a T_1 image could take up to several hours, depending on the specific T_1 algorithm utilized, the number of images used to calculate T_1, and the computer hardware. Rapid non-least-squares techniques have been described in the literature [22,23] which generally sacrifice accuracy for speed. A fast nonlinear least-squares technique for calculating a T_1 image in less than 1 min from saturation recovery images has recently been published [24]. The details of this technique are described next.

Under ideal conditions (i.e., the absence of spin exchange, primarily one component filling an image voxel, and perfect radio-frequency B_1 pulses) the magnetic resonance signal from a saturation recovery imaging experiment may be calculated using

$$S(TR_i) = k \, (1 - e^{-TR_i/T_1}) \tag{17}$$

where TR is the repetition of the experiment and k is a proportionality constant dependent on such factors as the spin density and amplifier gains. In the case of the commonly used spin-echo sequence, k is also a function of the spin-spin relaxation time and TE. A T_1 value is typically calculated by fitting Eq. (17) to imaging data obtained at several values of TR. The imaging data are the actual MR signal S_i at TR_i. A least-squares technique seeks to minimize $\phi(T_1, k)$ for the n data points, where

$$\phi(T_1, k) = \sum_i^n [S(TR_i) - S_i]^2$$

Therefore,

$$\frac{\partial\phi}{\partial k} = \sum_{i=1}^n [k(1 - e^{-TR_i/T_1}) - S_i] \, (1 - e^{-TR_i/T_1}) = 0 \tag{19}$$

and

$$\frac{\partial \phi}{\partial T_1} = \sum_{i=1}^{n} [k (1 - e^{-TR_i/T_1}) - S_i] TR_i e^{-TR_i/T_1} = 0 \tag{20}$$

Expanding k explicitly yields

$$\frac{\sum_{i=1}^{n} S_i(1 - e^{-TR_i/T_1})}{\sum_{i=1}^{n} (1 - e^{-TR_i/T_1})^2} = \frac{\sum_{i=1}^{n} S_i TR_i e^{-TR_i/T_1}}{\sum_{i=1}^{n} TR_i e^{-TR_i/T_1}(1 - e^{-TR_i/T_1})} \tag{21}$$

It should be noted that Eq. (21) contains one unknown, T_1. To find the root solution T_1 of Eq. (21), it should be rearranged by cross-multiplying the numerators and denominators. After reordering the summations, one obtains

$$\sum_{i=1}^{n} \sum_{j=1}^{n} S_i (1 - e^{-TR_i/T_1})TR_j e^{-TR_j/T_1}(1 - e^{-TR_j/T_1})$$

$$= \sum_{i=1}^{n} \sum_{j=1}^{n} (1 - e^{-TR_i/T_1})^2 S_j TR_j e^{-TR_j/T_1} \tag{22}$$

Let the difference of the two sides be a function $Z(T_1)$:

$$Z(T_1) = \sum_{i=1}^{n} \sum_{j=1}^{n} TR_j e^{-TR_j/T_1} (1 - e^{-TR_i/T_1})$$

$$[S_i (1 - e^{-TR_j/T_1}) - S_j (1 - e^{-TR_i/T_1})] \tag{23}$$

The function $Z(T_1)$, shown in Fig. 6, goes to zero in the limit as $T_1 = 0$ and ∞. The single root of Eq. (23) is the T_1 value sought.

The fast T_1 algorithm determines the root T_1 by the controlled iterative process depicted in Fig. 6. Two T_1 values, T_{1a} and T_{1b}, are used to seed the algorithm. T_{1a} is chosen to be less than the minimum T_1, and T_{2b} greater than the maximum T_1 in the image. A straight line is drawn between $Z(T_{1a})$ and $Z(T_{1b})$. If the $Z(T_1)$ value at the zero crossing, T_{1c}, of this line, $Z(T_{1c})$, is less than an arbitrarily chosen minimum ϵ, then T_{1c} is the sought-after root; otherwise, T_{1c} becomes T_{1b} (or T_{1a}) and the process is repeated. This process is repeated for each pixel in the image space to yield a T_1 image.

The fast T_1 algorithm is inherently quicker than the general Levenberg–Marquardt [6] nonlinear least-squares method at finding the T_1 data based on Eq. (17) because the fast T_1 algorithm lacks the computational overhead necessary to make the Marquardt method general. The ability of the algorithm to find T_1 from data with noise is governed by the signal-to-noise ratio (S/N)and the TR values [26,27]. The longest TR should be approximately three times the longest T_1, while the shortest TR may be approximately equal to the shortest T_1. Increasing the value of n improves the accuracy of the measured T_1 but increases the computation time.

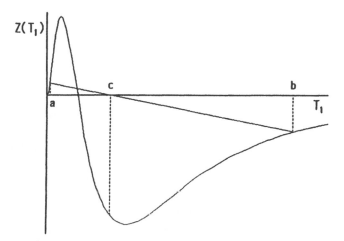

Figure 6 Plot of the function $Z(T_1)$ as defined by Eq. (23) for an arbitrary set of data. T_{1a} and T_{1b} are starting points for the fast T_1 algorithm. T_{1c} is the zero crossing of the line joining T_{1a} and T_{1b}.

Background noise located in those regions of the T_1 and T_2 image devoid of signal may be eliminated by calculating T_1 only for those pixels where the average pixel intensity in the n images is greater than a noise level. Both algorithms treated as noise, and subsequently set to zero, any j,k pixel in the image plane that had an intensity S_{jk} less than a noise level N_{jk}. N_{jk} was defined for the constant TE series of images as the maximum ($\vee$) signal over a background image space,

$$N_{jk} = n \bigvee_{i=1}^{n} \bigvee_{j=10}^{20} \bigvee_{k=10}^{20} S(TR_i, j, k) \qquad (24)$$

and

$$S_{jk} = \sum_{i=1}^{n} S(TR_i, j, k) \qquad (25)$$

where $S(TR_{i,j,k})$ is the signal intensity of pixel j,k at TR_i and n is the number of images with constant TR. N_{jk} for the constant TR series used to calculate T_2 was similarly defined except that the summation was over TE_i rather than TR_i.

The spin density may be determined once T_1 and T_2 are known. Starting with Eq. (19), one obtains

$$k = \frac{\sum_{i=1}^{n} S_i (1 - e^{-TR_i/T_1})}{\sum_{j=1}^{n} (1 - e^{-TR_j/T_1})(1 - e^{-TR_j/T_1})} \qquad (26)$$

and ρ is therefore

$$\rho = \frac{k}{e^{-TE/T_2}} \tag{27}$$

C. Segmentation Aids and Procedures

Multispectral image segmentation is usually accomplished with the aid of a multidimensional histogram. Similar tissues tend to form clusters in multidimensional histograms. A two-dimensional histogram of the images of Fig. 5 will display three clusters. The desired cluster will be located at the cross between speckled A and checkerboard B. The dimensionality of the histogram is equal to the number of spectral regions. Visualizing a cluster in a two dimensions is straightforward as an image of the space with pixel intensity corresponding to number of pixels at a given spectral intensity. Visualizing a cluster in a three-dimensional histogram on a two-dimensional surface is more difficult. Rotatable stereoscopic images are possible on video terminals. Hardcopy display of three-dimensional histograms is usually accomplished either by presenting perspective views of the histogram, stacked two-dimensional sections of the three-dimensional histogram, or projections of the three-dimensional histogram information onto the faces of the cube of the histogram. Higher dimensionalities are difficult to display, and therefore the determination of cluster locations relies primarily on computer algorithms. Several clustering algorithms are available to the imaging scientist and are discussed in detail elsewhere [28,29].

III. MULTISPECTRAL IMAGE SEGMENTATION OF BRAIN TISSUE

As an example of the utility of multispectral image segmentation in magnetic resonance imaging, an application of the technique to the segmentation of normal tissues found in an axial slice through the head is presented. The slice passes through the lateral ventricles and contains six major tissues. The six tissues are gray matter (GM), white matter (WM), cerebrospinal fluid (CSF), meninges and skin (MG), adipose (AT), and muscle (MT). Two additional image planes were recorded for comparison purposes, one passing through the eyes and another through the legs. For a more rigorous demonstration of the capabilities of the technique as applied to a large set of individuals, the reader is directed to the literature [30].

A. Experimental Procedure

Multispectral image segmentation was applied to magnetic resonance imaging data from three normal volunteers: (1) a 25-year-old man, (2) a 25-year-old woman, and (3) a 37-year-old man. Magnetic resonance images were acquired with a 1.5-T General Electric (Milwaukee, WI) Signa imager with shielded gradient coils. Images through the lateral ventricles and the eyes were obtained with a standard quadrature birdcage-style RF head coil. A lateral ventricle slice through the brain was recorded on all volunteers. The volunteers were imaged in the supine position for both head image planes with a single-echo, spin-echo sequence. Imaging parameters for the 14 image sequence were TR/TE = 4000, 3000, 2000, 1500, 1000, 750, 500, 250/15 and 1000/25, 50, 75, 100, 150, 200 ms, 256 × 192 matrix, 5-mm Thk, 1 excitation (*NEX*), and 22 cm FOV. The 5-mm slice thickness was chosen as a compromise to maximize the signal-to-noise ratio and minimize partial volume effects. An imaging phantom was placed in the FOV of each image. The phantom consisted of two poly(vinyl chloride)–rubber copolymer tubes, one filled with 12 and the other 35 m*M* aqueous $NiCl_2$, and served as a ρ standard. The eyes were immobilized with pressure pads to prevent motion during the long imaging sequence.

The foregoing sequence of 14 images were also recorded for an axial plane passing through the legs of a 25-year-old woman approximately 25 cm from the hips. The volunteer was imaged in the prone position using the standard body coil on the imager. Images were recorded using the same imaging protocol with the exception of the FOV, which equaled 48 cm.

All magnetic resonance images were transferred on tape as 256 × 256 16-bit images to a DEC (Maynard, MA) VAX-4300 computer for generation of T_1, T_2, and ρ images and histograms. Two computer algorithms were used for generation of T_1, T_2, and ρ images, as described previously [20,24]. T_2 images were generated from the seven constant TR images using a linear least-squares procedure described in Section II. T_1 and ρ images were generated from the eight constant TE images with a nonlinear least-squares procedure, also described in Section II. The signal intensity of the 12 m*M* $NiCl_2$ phantom in the TR = 4 s image was used to normalize all ρ values in the resultant ρ image. This standard had a mean ρ value of 111 on a scale of 0 to 256. The 35 m*M* standard was not used because its signal was too low to pass the noise discrimination test described in Section II.

B. Results and Discussion

Three independent T_1, T_2, and ρ images were created as described above, and used for the multispectral analysis. A considerable amount of knowledge can be gained about the segmentation process by examining the images of the three spectral ranges shown in Fig. 7. In the T_1 image one can easily differentiate between the gray matter and the white matter of the brain, due to the differences in intensities. However, it would be more difficult to tell the difference between the gray matter and the CSF based on the intensities of those tissues since they

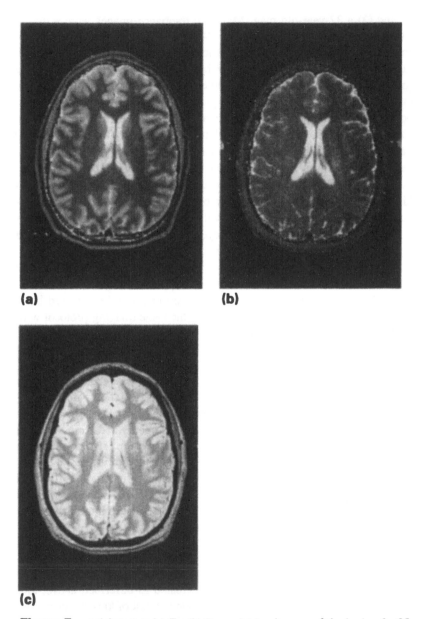

(a)   **(b)**

(c)

Figure 7 Axial proton (a) T_1, (b) T_2, and (c) ρ images of the brain of a 25-year-old normal male volunteer at the level of the lateral ventricles. (Used with permission from Ref. 30.)

appear quite similar. Given this observation, it becomes apparent that this one spectral region, T_1, would be inadequate in segmenting all of the necessary tissue types. Unlike a T_2 weighted image [3], the measured T_2 image has very little contrast between the white and gray matter. This may be attributed to the dependence of a T_2 weighted image on spin density information. Again, if spectral regions are dependent on one another, it becomes difficult to justify their use, due to the presence of repetitive information. Also, the repetitive information may be harmful to the data set since unnecessary noise will be included. The T_2 image shows further that a great deal of contrast exists between the CSF and the gray and white matters. This makes up for the inability of the T_1 image to differentiate between the gray matter and the CSF. The spin density image adds additional information that allows for a more accurate segmentation of the tissues.

The way in which the segmentation process can be visualized is by the use of histograms. Examples of two-dimensional T_1–T_2, T_1–ρ, and T_2–ρ histograms are shown in Fig. 8. It is obvious that it would be difficult to see where the cluster boundaries could be drawn to achieve a reasonable segmentation of tissues. In addition, there is a very long stream of pixels extending toward the upper T_1, T_2, and ρ corners which seemed to belong to the same cluster. The streaming was found to be caused by partial voluming, a digital imaging problem described previously, and the flow of the cerebrospinal fluid in the brain. Therefore, other ways were explored to try to improve the separation of clusters and compress the stream of pixels. The way in which this was achieved was to use the reciprocal data of the relaxation times T_1 and T_2, creating relaxation rate constants T_1^{-1} and T_2^{-1}. Other displays, such as the logarithm of the relaxation times, may be equally successful.

An example of a two-dimensional histogram of pixels with specific T_1^{-1} and T_2^{-1} values is shown in Fig. 9. On the sides of the two-dimensional histogram are projections of both the T_1 and T_2 information onto one dimension. This was done to show clearly the ability of more than one spectral region to separate the clusters of pixels over just one spectral region. In other words, the segmentation can be improved as various spectral regions become available. Five clusters of pixels are evident in the two-dimensional histogram, whereas only two peaks in T_1^{-1} or one peak in T_2^{-1} can be seen in the one-dimensional histograms. The five clusters that are visible in the two-dimensional histogram are adipose, muscle, gray matter, white matter, and CSF. Their exact location will be identified shortly. Clearly, peaks for these five clusters cannot be seen in the one-dimensional projections.

There are other observations to be made from Fig. 9. For example, it was mentioned briefly that the inverse ranges of T_1 and T_2 were chosen for the histogram space, but not the inverse range of spin density. This choice displayed clearer tissue clusters, which was made difficult before by the problem of partial voluming, flow, and the large relaxation times for CSF relative to the other tissues.

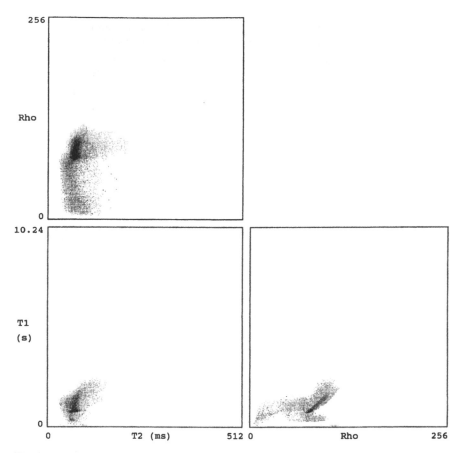

Figure 8 T_1-T_2, T_1-ρ, and T_2-ρ histograms of the images of the 25-year-old volunteer shown in Fig. 7. Each histogram is the equivalent of a projection of the data within the three-dimensional histogram into the corresponding two-dimensional space. Histograms do not display clear clusters.

With the increased use of digital imaging, one must realize that there are consequences of analyzing images made up of volume elements with finite sizes. Although it is true that in an image space, like pixels will tend to be next to each other, it is also true that some pixels will mark the barrier between tissues, thereby making it possible to have a pixel contain information on more than one type of tissue. These barrier pixels are affected by partial voluming, which makes it difficult to define the tissue type of these pixels due to mixed spectral values in those pixels, or in this case voxels.

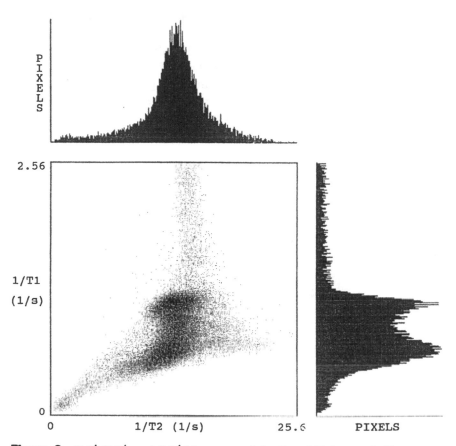

Figure 9 $T_1^{-1}-T_2^{-1}$, and T_2^{-1} histograms of the T_1 and T_2 images in Fig. 7. The $T_1^{-1}-T_2^{-1}$ histogram displays distinguishable clusters of pixels for five of the six tissue types, meninges excluded, in the image space, whereas all the clusters overlap in the one-dimensional histograms. (Used with permission from Ref. 30.)

The T_1 of CSF was found to be exceptionally long and vary considerably over the image [30]. The long measured in vivo T_1 is due to the flow of the CSF in the brain. The flow velocity of CSF has been reported to be as fast as 30 mm/s [31]. The magnetization of CSF flowing into a slice will cause the magnetization of the CSF in the slice to appear less saturated. The observable quality will be a signal that appears to get longer and longer as TR increases. This effect will yield longer measured T_1. In fact, the longer the largest TR value in the T_1 sequence, the longer the measured T_1 for CSF. This effect contributed significantly to the streaming problem in the histogram space.

The actual segmentation was done by viewing the histograms and determining the most appropriate cuttoff points of T_1, T_2, and ρ for each tissue type. A simple way of transforming the display from three dimensions to two dimensions is to calculate three projections of the volume of histogram data and display them as the faces of a cube. Three histograms were generated by this technique: a T_1^{-1}–T_2^{-1}, T_1^{-1}–ρ, and T_2^{-1}–ρ (see Fig.10). One can readily see clusters in the T_1^{-1}–T_2^{-1} histogram, but it is difficult to visualize more than two dense re-

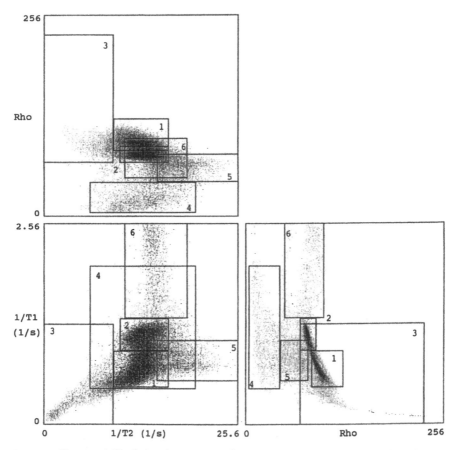

Figure 10 T_1^{-1}–T_2^{-1}, T_1^{-1}–ρ, and T_2^{-1}–ρ histograms of the images of a 25-year-old volunteer in Fig. 7. Each histogram is the equivalent of a projection of the data within the three-dimensional histogram into the corresponding two-dimensional space. Boxes within each two-dimensional histogram represent the range of T_1^{-1}, T_2^{-1} and ρ used to segment the tissues (see Table 1). The six boxes are for (1) gray matter, (2) white matter, (3) CSF, (4) meninges and skin, (5) muscle, and (6) adipose tissues. (Used with permission from Ref 30.)

gions, the white and the gray matter, in the T_1^{-1}–ρ and T_2^{-1}–ρ histograms. Therefore, other display options became more appropriate.

An informative way of looking at the three-dimensional version of the histograms is through stacked slices through the three-dimensional histogram space shown in Fig. 11. The boxes in the histograms show the boundaries used for the segmentation of each of the six major tissue types. The sixth tissue type, meninges and skin, could not be seen in the two-dimensional T_1^{-1}–T_2^{-1} histogram and is just barely visible in the T_1^{-1}–ρ and ρ–T_2^{-1} histograms. The slices through the ρ axis of the volume histogram were necessary to locate this group of pixels conclusively.

The clustering of pixels in the histogram has proven to be reproducible [30] and found not to be the result of an instrumental artifact. A convenient way of seeing the latter (i.e., that the data are truly representative of the tissues present in the imaging slice) is to change the imaged slice through the body. This could be done in MRI by moving the slice to a different location in the body. Figure 12 depicts T_1, T_2, and ρ images of a slice location 5 cm lower than the original slice through the lateral ventricles. This new slice location includes the original six tissues, admittedly in different proportions, and one new tissue type, vitreous humor, from the eyes. Vitreous humor is a viscous jellylike liquid and is expected to have a long relaxation time. Figure 13 shows a comparison of T_1^{-1}–T_2^{-1} histograms from the two slice locations through the head. The histogram in Fig. 13a was computed from the data in the original slice location through the lateral ventricles and the one in Fig. 13b from the new slice location. When comparing the histograms, one can see that they both have the same general shape and cluster positions; however, the relative sizes of the clusters have changed. The most obvious differences are with the white and gray matter clusters. Of the two dense cluster regions, the one on the top represents the white matter in the brain. This cluster has diminished in size from the first histogram. When referring back to the spectral region images of the location through the eyes (Fig. 12) it is obvious that there is much less white matter tissue in these images. This fact is reflected in the histogram. Also, there is much more gray matter in the new slice location, reaffirmed by the smaller size of the gray matter cluster in the corresponding histogram. Looking closely, there in an additional cluster of pixels with very small T_1^{-1} and T_2^{-1} values. These pixels represent the vitreous humor in the eyes. Since the changes in the histograms have been predicted by actual changes in the image slices, it is probable that instrumental artifacts are not causing reproducible histograms of the same slice in the head. This, however, may be a less than convincing demonstration. Therefore, a completely different slice location was chosen.

T_1, T_2, and ρ images of an axial slice location through the midthigh region are shown in Fig. 14. These images contain primarily two tissue types, adipose and muscle. From the T_1, T_2, and ρ images a T_1^{-1}–T_2^{-1} histogram was created

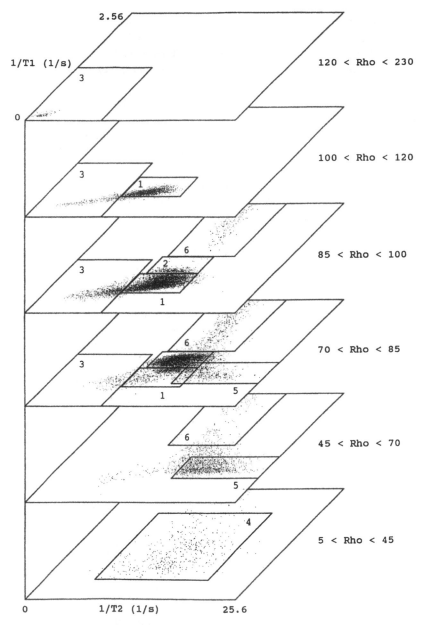

Figure 11 $T_1{}^{-1}$–$T_2{}^{-1}$–ρ stacked histogram of the T_1, T_2, and ρ images found in Fig. 7. Rectangular boxes in each plane represent the six segmentation ranges found in Table 2. The six boxes are for (1) gray matter, (2) white matter, (3) CSF, (4) meninges and skin, (5) muscle, and (6) adipose tissues. (Used with permission from Ref. 30.)

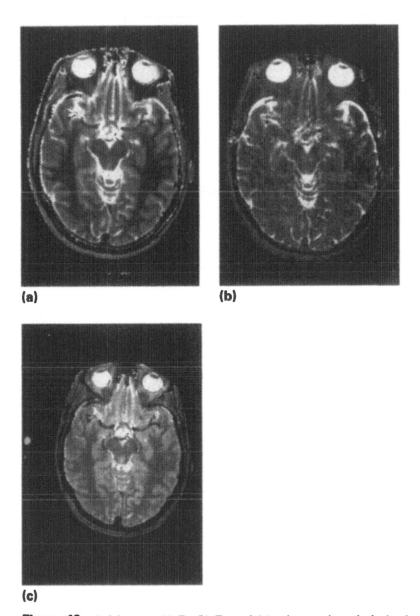

Figure 12 Axial proton (a) T_1, (b) T_2, and (c) ρ images through the head of a 37-year-old normal male volunteer passing through the eyes. This space contains the tissue types found in the previous head images in addition to vitreous humor found in the eyes.

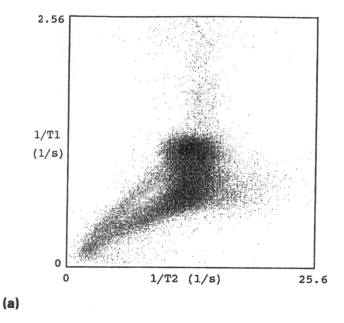

(a)

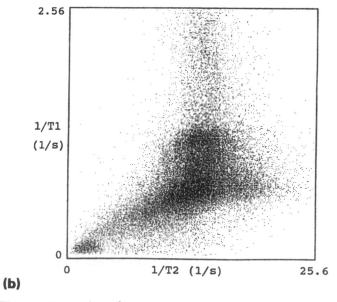

(b)

Figure 13 $T_1{}^{-1}$–$T_2{}^{-1}$ histograms of the 37-year-old normal male volunteer for (a) an axial slice through the head at the level of the lateral ventricles, and (b) an axial slice passing through the head at the level of the eyes as shown in Fig. 12.

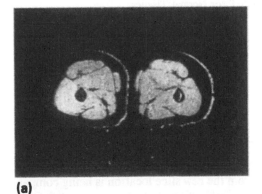

(a)

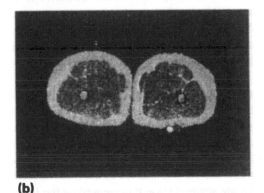

(b)

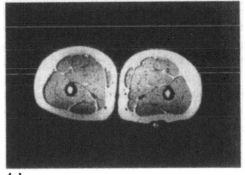

(c)

Figure 14 Axial proton (a) T_1, (b) T_2, and (c) ρ images of the legs of a 25-year-old normal female volunteer passing through the legs approximately 25 cm from the hip. This space contains primarily two tissue types: adipose and muscle tissue. (Used with permission from Ref. 30.)

Table 2 Segmentation Ranges at 1.5 T for Normal Tissues

Tissue	T_1 (s)	T_2 (ms)	ρ
CSF	0.80–20	110–2000	70–230
White matter	0.76–1.08	61–100	70–90
Gray matter	1.09–2.15	61–109	85–125
Meninges	0.50–2.20	50–165	5–44
Muscle	0.95–1.82	20–67	45–80
Adipose	0.20–0.75	53–94	50–100

(see Fig. 15). Again the histogram from the new slice location is being compared to the histogram of the original slice location through the lateral ventricles of the brain. As expected, the leg histogram displays only two clusters, corresponding to the adipose and muscle tissues. Note that the cluster locations for the adipose and the muscle in the legs matched the cluster locations of the adipose and the muscle in the head. This is proof that the clusters of pixels do represent the tissue types in the image space.

Having shown that the histograms are representative of the image slice and the tissues within the slice, the actual segmentation of the tissues may now be performed confidently. As mentioned previously, boundaries were set for the three spectral regions to separate the six tissue types: adipose, muscle, gray matter, white matter, CSF, and the meninges and skin (see Table 2). The segmentation from just this simple use of perpendicular boundaries in the three spectral regions can be seen in Fig. 16. The tissue segmentation represented by the tissue map was achieved by turning on pixels from within each box of the histogram in the image slice. Box locations were determined by cluster shape, size, and location, and knowledge of the anatomy of the segmented tissue. Clearly, each set of boxed pixels is a cluster and the three properties T_1, T_2, and ρ will identify a specific tissue type. In this demonstration, successful segmentation has been achieved by a simple approach that is by no means intended to represent optimal segmentation. A variation on the method that would improve the segmentation is to segment regions using geometrical-shaped objects more closely matching the shapes of the clusters, such as ellipsoids.

IV. SUMMARY

Magnetic resonance imaging is a relatively new diagnostic medical imaging modality which is becoming very popular. The large number of spectral regions available to the magnetic resonance scientist makes magnetic resonance imaging an excellent application for the image-processing technique called multispectral image segmentation. With the appropriate choice of orthogonal spectral

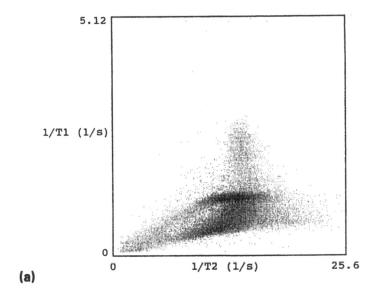

(a)

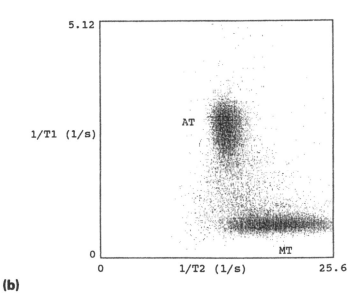

(b)

Figure 15 T_1^{-1}–T_2^{-1} histograms of a 25-year-old normal female volunteer for (a) an axial slice through the head at the level of the lateral ventricles, and (b) an axial slice passing through both legs as shown in Fig. 14. Only two major clusters exist in the histogram of the legs: muscle (MT) and adipose (AT) tissue. (Used with permission from Ref. 30.)

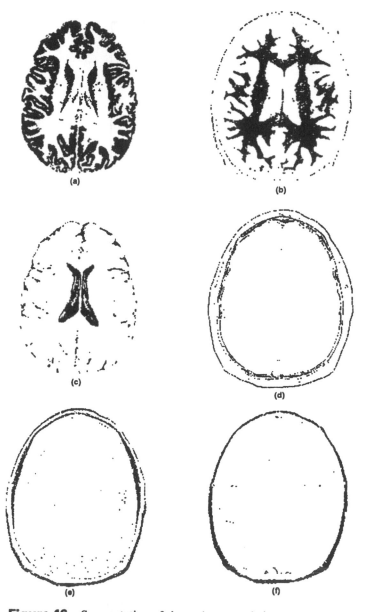

Figure 16 Segmentation of the major normal tissue types present in an axial slice through the head, shown in Fig. 7, at the level of the lateral ventricles using T_1, T_2, and ρ. Tissue types are predominantly (a) gray matter, (b) white matter, (c) CSF, (d) meninges and skin, (e) muscle, and (f) adipose tissue. Tissue types were segmented with the ranges of T_1, T_2, and ρ listed in Table 2. (Used with permission from Ref. 30.)

regions, such as spin-lattice relaxation time, spin-spin relaxation time, and spin density, multispectral image segmentation can classify normal tissues of the human body and may be capable of classifying pathology. A long-standing dream of the magnetic resonance scientist, which stimulated the development of MRI, has been to identify and diagnose pathology with nuclear spin relaxation times [32,33]. This dream may soon be realized with the aid of multispectral image segmentation.

REFERENCES

1. C. P. Schlichter, *Principles of Magnetic Resonance*, Springer-Verlag, Berlin, 1980.
2. T. C. Farrar, *An Introduction to Pulse NMR Spectroscopy*, Farragut Press, Chicago, 1987.
3. D. D. Stark and W. G. Bradley, *Magnetic Resonance Imaging*, C. V. Mosby, St. Louis, Mo., 1988.
4. S. L. Smith, Nuclear magnetic resonance imaging, *Anal. Chem.*, 57:595A–608A (1985).
5. T. M. Lillesand and R. W. Kiefer, *Remote Sensing and Image Interpretation*, Wiley, New York, 1979.
6. M. Just and M. Thelen, Tissue characterization with T_1, and T_2, proton density values: results in 160 patients with brain tumors, *Radiology*, 169:779–785 (1988).
7. G. Grad and R. G. Bryant, Nuclear magnetic cross-relaxation spectroscopy, *J. Magn. Reson.*, 90:1 (1990); D. A. Mendelson, J. F. Heinsberger, S. D. Kennedy, L. S. Szczepaniak, C. C. Lester, and R. G. Bryant, Comparison of agarose and crosslinked protein gels as MRI phantoms, *Magn. Reson. Imag.*, 9:975–978 (1991).
8. M. E. Moseley, Y. Cohen, J. Kucharczyk, J. Mintorovitch, H. S. Asgari, M. F. Wendland, J. Tsuruda, and D. Norman, Diffusion weighted MR imaging of anisotropic water diffusion in cat central nervous system, *Radiology*, 176:439–445 (1990).
9. C. L. Dumoulin, S. P. Souza, and H. R. Hart, Rapid scan magnetic resonance angiography, *Magn. Reson. Med.*, 5:238–245 (1987).
10. Y. Chen, E. R. Dougherty, S. M. Totterman, and J. P. Hornak, Classification of trabecular structure in magnetic resonance images based on morphological granulometries, *Magn. Reson. Med.*, 29: 358–370 (1993).
11. J. P. Hornak, A. C. Smith, and J. Szumowski, Spin-lattice relaxation time measurements using hybrid CSI-phantom study, *Magn. Reson. Med.*, 13:398–406 (1990).
12. M. W. Vannier, C. M. Speidel, and D. L. Rickman, Magnetic resonance imaging multispectral tissue classification, *News in Phys. Sci.*, 3:148–154 (1988).
13. J. K. Gohagan, E. L. Spitznagel, W. A. Murphy, M. W. Vannier, W. T. Dixon, D. J. Gersel, S. L. Rossnick, W. G. Totty, J. M. Destouet, D. L. Rickman, T. A. Sraggins, and R. L. Butterfield, Multispectral analysis of MR images of the breast, *Radiology*, 163:703–707 (1987).
14. H. P. Higerand and G. Bielke, eds., *Tissue Characterization in MR Imaging*, Springer-Verlag, New York, 1990.

15. R. K. Breger, F. W. Wehrli, H. C. Charles, J. R. MacFall, and V. M. Haughton, Reproducibility of relaxation and spin-density parameters in phantoms and the human brain measured by MR imaging in 1.5 T, *Magn. Reson. Med.*, 3:649–662 (1986).

16. S. Majumdar, S. C. Orphanoudakis, A. Gmitro, M. Odonnell, and J. C. Gore, Errors in the measurements of T_2 using multiple echo MRI techniques, I: Effects of radiofrequency pulse imperfections, *Magn. Reson. Med.*, 3:397–417 (1986); II: Effects of static field homogeneity, *Magn. Reson. Med.*, 3:562–574 (1986).

17. R. H. Darwin, B. P. Drayer, S. J. Riederer, H. Z. Wang, and J. R. MacFall, T_2 estimates in healthy and diseased brain tissue: a comparison using various pulse sequences, *Radiology*, 160:375–381 (1986).

18. R. M. Kroekeer and R. M. Henkelman, Analysis of biological NMR relaxation data with continuous distributions of relaxation times, *J. Magn. Reson.*, 69:218–235 (1986).

19. R. M. Kroekeer, C. A. Stewart, M. J. Bronskill, and R. M. Henkelman, Continuous distributions of NMR relaxation times applied to tumors before and after therapy with x-rays and cyclophosphamide, *Magn. Reson. Med.*, 6:24–36 (1988).

20. J. P. Hornak, A. Blaakman, D. Rubens, and S. Totterman, Multispectral image segmentation of breast pathology, *SPIE Image Process.*, 1445:523–533 (1991).

21. A. P. Crawley and R. M. Henkelman, A comparison of one-shot and recovery methods in T_1 imaging, *Magn. Reson. Med.*, 7:23–34 (1988).

22. G. O. Sperber, A. Ericsson, and A. Hemmingson, A fast method for T_1 fitting, *Magn. Reson. Med.*, 9:113–117 (1989).

23. A. Jesmanowicz, E. C. Wong, J. C. Wu, and J. S. Hyde, Fast algorithm for computation of T_1, T_2, and diffusion coefficient images, Abstract 740, *10th Annual Meeting of the Society of Magnetic Resonance in Medicine*, San Francisco, Aug. 1991.

24. J. Gong and J. P. Hornak, A fast T_1 algorithm, *Magn. Reson. Imag.* 10: 623–626 (1992).

25. W. H. Press, B. P. Flannery, S. A. Teukolskv, and W. T. Vetterling, *Numerical Recipes: The Art of Scientific Computing*, Cambridge University Press, Cambridge, 1988.

26. E. A. Guggenhem, The determination of the velocity constant for a unimolecular reaction, *Philos. Mag.* [7], 2:538–543 (1926).

27. J. R. Wolberg, *Prediction Analysis*, D. Van Nostrand, Princeton, N.J., 1967.

28. M. W. Vannier, R. L. Butterrfield, D. L. Rickman, D. M. Jordan, W. A. Murphy, and P. R. Biondetti, Multispectral magnetic resonance image analysis, *CRC Crit. Rev. Biomed. Engrg.*, 15:117–144 (1987).

29. R. L. Cannon, J. V. Dave, and J. C. Bezdek, Efficient implementation of the fuzzy *c*-means clustering algorithms, *IEEE Trans. Pattern Anal. Mach. Intelligence*, PAMI-8:248–255 (1986).

30. L. M. Fletcher, J. B. Barsotti, and J. P. Hornak, A multispectral analysis of brain tissue, *Magn. Reson. Med.* 29: 623–630 (1993).

31. F. Stahlberg, J. Mogelvang, C. Thomsen, B. Nordell, M. Staubgaard, A. Ericsson, G. Sperber, D. Greitz, H. Larsson, O. Henriksen, and B. Persson, A method for

magnetic resonance quantification of flow velocities in blood and CSF using interleaved gradient-echo pulse sequences, *Magn. Reson. Imag.* 7:655–667 (1989).

32. R. Damadian, Tumor detection by nuclear magnetic resonance, *Science*, 171:1151–1153 (1971).

33. P. A. Bottomley, T. A. Foster, R. E. Argersinger, and L. M. Pfeifer, A review of normal tissue hydrogen NMR-relaxation times and relaxation mechanisms from 1-100 MHz: dependence on tissue type, NMR frequency, temperature, species, excision, and age, *Med. Phys.*, 11:425–448 (1984).

4

Thinning and Skeletonizing

Jennifer L. Davidson

Iowa State University
Ames, Iowa

I. INTRODUCTION

A general approach in analyzing images is to transform the given image to another where the information represented in the transformed image is more easily understood. For example, investigations of the shape of objects in a binary image often use thinning algorithms. Reduction of an object to a minimal set of pixels representing an invariant of the object's geometrical shape is called *thinning*, and we are interested in the minimal set of pixels called the *skeleton* of the object. These concepts are defined formally later. The image resulting from the thinning process has many fewer black pixels representing the object and is therefore easier to manipulate. Uses of thinning include image coding and data compression, although these assume that the original image or a close facsimile can be generated from the skeleton image, which is not always the case. Another use is for object recognition, where the skeleton is used as a representation of the object against which to match. If the main goal of thinning is for data reduction and exact reconstruction of the original image is not essential, many techniques are available that yield an acceptable skeleton representation. However, if close or exact reconstruction is desired, care must be taken in choosing an appropriate algorithm. Other applications include optical character recognition [1,10,11,13,30,36]; chromosome analysis [15], fingerprint classification [24], biological shape description [3,5,6], pattern recognition [31], image coding [12,22,27], quantitative metallography [18,19], and automated industrial inspection [21]. The main purpose of this chapter is to provide an introduction to thinning/skeletonizing concepts and to familiarize the reader with the various techniques available and their limitations and capabilities.

The remainder of the chapter is divided as follows. First, a history of the concepts of thinning and skeletons in binary images is given. In Section III we give the mathematical definitions that lay the foundation for precisely defining a skeleton of an image, including Air Force image algebra as a mode of expressing the image-processing algorithms. In Section IV, three thinning algorithms are given and their advantages and disadvantages compared. In Section V we briefly review other skeletonizing algorithms from the literature. In Section VI we present a general skeletonizing technique that is applicable to gray-scale images as well as binary ones. Finally, in Section VII, a concluding overview is given, along with some open problems in this area.

II. HISTORY AND BACKGROUND

The concept of the skeleton of a continuous binary image was introduced by Blum [2]. Montanari [25] also discusses skeletonizing techniques. Blum's interest was to find possible physiological mechanisms for explaining how an animal's vision system extracted global geometrical shape information. Several years later he introduced the concept of a "medial axis function" [3]. The medial axis function or transform (MAT) operates on the medial axis or skeleton of the object. If A is an object in a binary image, a skeleton of A is a subset I of A such that if $x \in I$, there are at least two boundary points of A that are equidistant to x. The set I is often described by using a "brushfire" analogy [4]: if a "fire" is set on the boundary locations of A and burns inward at an equal rate, the points interior to A, where the fronts of the brush fire meet is the skeleton of A. The MAT takes a point in I to a positive real number that is its distance to the boundary of A. Given the MAT of a continuous object, the object can be recovered completely. Define a disk of radius r centered at location y to be the set $D_r(y) = \{z \in \mathbb{R}^2 : |z - y| \leq r\}$. If MAT $(x) = r_x$, the original object A, with skeleton I, occupies the spatial locations in $\mathbb{R}^2$ given by the set $\cup_{x \in I} D_{r_x}(x)$. The boundary of A is the envelope of $\cup_{x \in I} D_{r_x}(x)$.

Thus the MAT is a representation for A that is unique and from which A may be reconstructed exactly. There is a much different situation using a boundary representation of an image, of course. The boundary does not necessarily describe an object uniquely, especially if there are "holes" in the object. If the object is well behaved, its skeleton I is like a stick figure, and graph representations of skeletons have been used in various applications [15,31]. There is more local information about the shape represented in the MAT than in a boundary representation.

Figure 1   (a) Medial axis of a 3 × 3 square, each leg of length $3\sqrt{2}$; (b) same as (a) but rotated 45°.

In the discrete case, however, we are not guaranteed that the medial axis (MA) of a discrete set approaches that of the corresponding continuous MA. Apart from the obvious results of the discretization process, we have geometrical distortions such as differences in lengths. An easy example of this is to see that the MA of a solid square having sides of length 3 is the "cross" inside it; see Fig. 1a, where each leg of the MA has length $3\sqrt{2}$. Rotated by 45°, the result is still the same; see Fig. 1b. However, in the discrete case, the discrete version of Fig. 1a, which is shown in Fig. 2a, has a diagonal of length $3\sqrt{2}$, while rotated by 45° gives diagonals of length 9 (see Fig. 2b). Other effects of discretization of a continuous object include disconnection of an originally connected object and the inability to recover the original object exactly.

Thinning algorithms usually "peel off" layers from the outer boundary of an object until a set of pixels remain that approximate a stick figure representation of the original object. As there is often no distinction in the literature between a thinning and a skeletonizing algorithm, and in essence they are the same, we use the terms interchangeably.

Figure 2   (a) Medial axis of discrete version of Fig. 1a; diagonals have length $3\sqrt{2}$; (b) same as (a) but rotated 45°; now diagonals have length 9.

Thinning the background of a set of objects can separate the objects. If the background and object pixel values are exchanged, the thinning algorithm can be performed on the background and usually results in a separation of the objects. Curvature information about the original is also available from its skeleton.

III. DEFINITIONS

In this section we give several definitions that will be used in describing the thinning algorithms. A mathematical environment called (Air Force) *image algebra* is used to express the algorithms. Image algebra is a high-level image-processing language that provides a common mathematical environment for algorithm development, optimization, comparison, coding, and performance evaluation. The main reason for using such a construct is to provide the reader with an environment in which any image-processing algorithm can be expressed, and quite often in an easily understandable way. The algorithms expressed in image algebra most often closely follow their implementation in computer code. For examples of preprocessors and other compiler codes, see [16,23,39]. In particular, the image algebra can facilitate mapping of many algorithms to parallel machines, including mesh-connected architectures and pipeline architectures [32], helping to decrease the computation time of many computationally intensive image-processing algorithms. Here we given only an abbreviated description of the image algebra. For a more complete description, see [8,33].

Since we will be using mostly morphological-type operations, our prime interest will be in describing the part of image algebra that generalizes morphology. For our purposes, an image is viewed as a function on a subset X of Euclidean space that is, an image is a function with its domain in $\mathbb{R}^n$, a rectangular grid, for example. A template is viewed as a collection of images on the same domain as our input image. When we combine an image **a** and a template **t** using the *additive maximum* operation, which generalizes the mathematical morphology operation of dilation, the result is an image that is defined as follows:

$$\mathbf{a} \oplus \mathbf{t} \equiv \mathbf{b} \in \mathbb{R}^Y, \qquad \mathbf{b} = \{(y, b(y)) : b(y) = \bigvee_{x \in X} \mathbf{a}(x) + \mathbf{t}_y(x), \, y \in Y\}$$

Here the operation $\oplus$ is *not* the usual notation in image algebra for additive maximum (as described in [33], for example,) but denotes the image algebra operation $\boxed{\vee}$. The reason for this notational change is to allow the reader to incorporate more easily the ideas presented in this chapter with the ideas in the other chapters, which use the notation $\oplus$ for mathematical morphology dilation. This set X is the subset of Euclidean space over which the input image **a** is defined and the object **t** is a template (or a structuring element). The set Y is the domain of the output image **b**. For a more complete and precise description of these operands and operations, see [33] or [8]. For the remainder of this chapter, we use the symbol $\oplus$ to denote the image algebra generalization of the morphology operation of dilation, the symbol ⊖ to denote the image algebra generalization of the morphology operation of erosion, and the symbol ⑦ to denote the image algebra operation of (generalized) linear convolution.

To facilitate the discussion with a minimum of notational overhead, we focus primarily on translation or shift-invariant templates. This is because a structuring element corresponds to a shift-invariant template, that is, one that "looks" the same everywhere (see Fig. 3).

The additive maximum has a dual operation, the additive minimum, defined by

$$\mathbf{a} \ominus \mathbf{t} = -(-\mathbf{a} \oplus -\mathbf{t}')$$

Again, the notation ⊖ is not the usual image algebra notation; the symbol ⋀ is. The prime notation on the template denotes the reflection of the template with respect to the origin. This generalizes the mathematical morphology operation of erosion. The values that template takes are in the lattice $\mathbb{R}_{\pm\infty} = \mathbb{R} \cup \{-\infty,\infty\}$. This allows the lattice algebra (a lattice-ordered group) $\mathbb{R}_{\pm\infty}$ to be consistent. For a more detailed description, see [7].

The concept of a template unifies and generalizes the notion of a window, mask, and structuring element. When computing the linear operation $\Sigma_{x\in X} \mathbf{a}(x)\mathbf{t}_y(x)$, many of the template values are zero and do not contribute to the sum. Similarly, if $\mathbf{t}_y(x) = -\infty$, the value of $\mathbf{a}(x) + \mathbf{t}_y(x) = \mathbf{a}(x) + (-\infty) = -\infty$ does not contribute to the calculation of $\vee_{x\in X} \mathbf{a}(x) + \mathbf{t}_y(x)$. Thus the symbol $-\infty$ in the lattice-ordered group $\mathbb{R}_{\pm\infty}$ acts like the zero in the field of real numbers.

It is important to note that the additive maximum and minimum operations can describe operations more general than morphology dilation and erosion, respectively. This is because the image algebra templates allow transformations between images with different domains, whereas morphology allows only transformations from one domain to itself. In addition, an image algebra template need not be invariant, whereas morphology structuring elements are always invariant. Image algebra templates can vary in support and values from pixel to pixel, resulting in spatially variant templates.

An important morphological operation we will need later is the hit-or-miss transform, which is performed on binary images. Skeletonizing can be described in terms of this operation. Let **a** be a binary image and **s** and **t** be two binary

$$t_y = \begin{array}{|c|c|c|} \hline 1 & 1 & 1 \\ \hline 1 & (10) & 1 \\ \hline 1 & 1 & 1 \\ \hline \end{array}$$

Figure 3 Invariant template.

templates that do not overlap. Then the hit-or-miss transform, abbreviated by HMT, is described in image algebra by

$$\mathbf{a} \circledast (\mathbf{s}, \mathbf{t}) = (\mathbf{a} \ominus -\mathbf{s}') * (\bar{\mathbf{a}} \ominus -\mathbf{t}')$$

where

$$\bar{\mathbf{a}} = \left\{ \mathbf{x}, (\mathbf{b}(\mathbf{x})) : \mathbf{b}(\mathbf{x}) = \begin{cases} 1 & \text{if} \quad \mathbf{a}(\mathbf{x}) = 0 \\ 0 & \quad \mathbf{a}(\mathbf{x}) = 1 \end{cases} \right\}$$

is the binary complement of **a**.

Another concept important to skeletonizing is connectivity. The connectivity describes the relationship in a binary image between the object pixels (value 1) and the background pixels (value 0). Given a pixel location $\mathbf{x} = (i, j)$ in **X**, the 4-*neighborhood* of (i, j) comprises the five pixel locations $N_4(i, j) = \{(i, j), (i - 1, j), (i + 1, j), (i, j + 1), (i, j - 1)\}$ This is also called the *von Neumann* neighborhood of (i, j). The 8-*neighborhood* of (i, j), also called the *Moore* neighborhood of (i, j), comprises the eight pixel locations $N_8(i, j) = N_4(i, j) \cup \{(i - 1, j - 1), (i - 1, j + 1), (i + 1, j - 1), (i + 1, j + 1)\}$. We abuse notation and often say that the pixel *value* $a(i, j)$, or simply the pixel $a(i, j)$ has 4-neighborhood $N_4(i, j)$ instead of the pixel location (i, j). However, this is not unreasonable since the pixel value has the spatial location included in its notation. We say that two pixels $\mathbf{a}(\mathbf{x}_1)$ and $\mathbf{a}(\mathbf{x}_n)$ are *4 (8)-connected* if there exists a sequence of pixels $\{\mathbf{a}(\mathbf{x}_1), \mathbf{a}(\mathbf{x}_2), \ldots, \mathbf{a}(\mathbf{x}_n)\}$ such that $\mathbf{a}(\mathbf{x}_i) = 1$ for each $i = 1, \ldots, n$ and $\mathbf{a}(\mathbf{x}_i)$ is in the 4 (8)-neighborhood of $\mathbf{a}(\mathbf{x}_{i+1})$ for $i = 1, \ldots, n - 1$. In this case we call the sequence of pixels a *4 (8)-path* between $\mathbf{a}(\mathbf{x}_1)$ and $\mathbf{a}(\mathbf{x}_n)$. An *object A* in a binary image **a** is a collection of pixels $\mathbf{a}(\mathbf{x})$ in **a** such that $\mathbf{a}(\mathbf{x}) \neq 0$. An object A is said to be *4 (8)-connected* if for all $\mathbf{a}(\mathbf{x})$, $\mathbf{a}(\mathbf{y})$ in A, there exists a 4 (8)-path from $\mathbf{a}(\mathbf{x})$ to $\mathbf{a}(\mathbf{y})$.

We say that a pixel $\mathbf{a}(\mathbf{x})$ in an image **a** is a boundary point of the object A if

1. $\mathbf{a}(\mathbf{x}) \in A [\mathbf{a}(\mathbf{x}) = 1]$.
2. The 4-neighborhood of $\mathbf{a}(\mathbf{x})$ contains a pixel $\mathbf{a}(\mathbf{y}) = 0$ that is not in A.

To be precise, this definition should be described as a 4-boundary. However, due to the topological nature of the rectangular array (**X** in our case), we want to have objects that are 4-connected and the remaining pixels, typically the background, have 8-connectivity. This allows the topological properties of connectivity to be relatively consistent given that the 4-neighborhood does not define a mathematical topology on the set **X**. For a more complete discussion of this topic, see [34].

IV. GENERAL THINNING ALGORITHM

Due to its uniqueness in representing the skeleton of a continuous object, the MAT is often used as a standard against which to compare the results of thinning algorithms. The MAT can easily be expressed in image algebra. Let $D_n(y)$ represent a digital disk of radius n centered at pixel y. We can view a binary digital disk as an invariant template having value 0 on its support and $-\infty$ outside. Typically, D_0 is a one-point template, D_1 is the von Neumann neighborhood, and D_2 is the von Neumann convolved with itself, but D_n for $n \geq 3$ has several interpretations. For example, we can define $D_n = D_{n-1} \oplus D_1$ in a recursive manner. Or we can define it by

$$D_n(y) = \left\{ (x, d_n(x)) : d_n(x) = \begin{cases} 0 & \text{if } |y - x| \leq n \\ -\infty & \text{otherwise} \end{cases} \right\} \tag{1}$$

See Fig. 4 for D_n, $n = 0, \dots, 3$. In general, we describe the MAT of a binary image by the image algebraa pseudocode shown in Fig. 5.

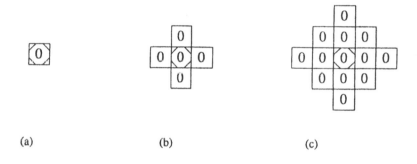

(a)   (b)   (c)

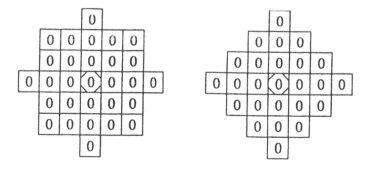

(d)   (e)

Figure 4   Digital disks: (a) zero-disk D_0; (b) one-disk D_1; (c) two-disk D_2; (d) three-disk D_3 described by Eq. (1); (e) different three-disk D_3.

```
i = 0

a₀ = a

REPEAT
```

$$a_{i+1} = a_i \ominus -t'$$

$$b_{i+1} = a_{i+1} * \chi_0[(a_{i+1} \ominus -t') \oplus t]$$

```
        i = i + 1

UNTIL  (bᵢ₊₁ = 0)
```

$$\text{MAT}(a) = \sum_{k=1}^{i-1} k b_k$$

Figure 5   Image algebra pseudocode for the MAT algorithm.

The characteristic function applied to an image a, $\chi_r(a)$, results in a binary 0–1 image having value 1 where $a(x) = r$, and 0 where it does not. The medial axis transform of a is given by the last line, $\text{MAT}(a)$, and in general, is a gray-value image. Here the template t is a digital disk, but in principal, any invariant template can be used [22]. The skeleton thus obtained is then with respect to the template t. For a digital disk, reconstruction $\hat{a}$ of a is obtained by calculating

$$\hat{a} = \bigvee_{k=1}^{i-1} [b_k + (D_1)^k] \quad \text{or} \quad \hat{a} = \bigvee_{k=1}^{i-1} [b_k \oplus D_k]$$

depending on whether the von Neumann neighborhood disk or Eq. (1) was used, respectively.

Next, we describe a general thinning algorithm for binary images that is typical of many algorithms currently in use. This method of thinning [17] involves a two-cycle procedure. In the first cycle, 3×3 templates remove boundary points from the northwest (NW), northeast (NE), southeast (SE), and southwest (SW) directions. This corresponds to removing pixels that satisfy a certain HMT. The second cycle removes boundary pixels in the north (N), south (S), east (E), and west (W) directions as well as unnecessary pixels at junctions, such as at the intersection of two lines. To avoid ambiguities, we need to define precisely the skeleton $\mathbf{I}$ of an object A, as well as the concept of homotopy and critical points.

Suppose that A and B are two objects in a binary image. We say that A and B are *homotopic* if A can be deformed continuously to B (and vice versa), preserving connectivity and holes within the object (see Fig. 6 and 7 for some examples). A *critical pixel* is a pixel having value 1 that is part of an object, and

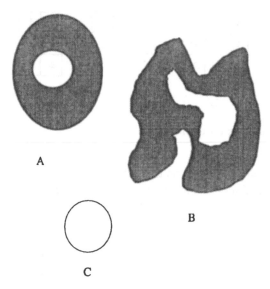

Figure 6   A is homotopic to B, which is homotopic to C.

if its value is changed to 0 (background), the resulting object becomes disconnected or has a newly created hole in it. See Fig. 8 for the configuration surrounding all possible critical pixels.

Let $a \in \{0, 1\}^X$ be a binary image. Here we use the accepted mathematical
notation that

$$A^B = \{\text{functions } f : f \text{ is a function from } B \text{ to } A\}$$

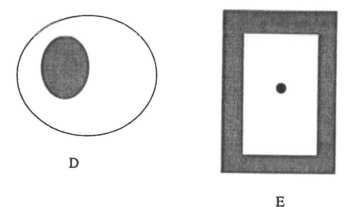

Figure 7   D is homotopic to E.

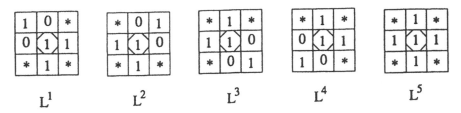

Figure 8 Five possible configurations for a critical pixel.

Thus the set $\{0, 1\}^{\mathbf{X}}$ denotes all binary images on the domain **X**. Without loss of generality we assume for the remainder of this section that the image **a** has just one object, *A*, in it. The results discussed here will extend to an image having more than one object in it as long as the objects are 8-disconnected in **a**. We also assume that the object is 8-connected and the background (including holes that may be contained "inside" *A*) contains 4-connected components. Under these constraints, the image **a** represents the object *A*. Then the skeleton of **a** is a binary image **I** that must satisfy:

1. **a** and **I** are homotopic.
2. **I** is 8-connected and is one pixel wide except at critical points.
3. **I** is "close" to the MA of **a**.

The thinning algorithm proceeds as follows. Let **a** satisfy the constraints noted above, and call **I** the skeleton corresponding to **a**. First, for all configuration given in Fig. 9 found in the image, the pixel marked by hash marks (at the center of the 3 × 3 windows) is set to 0; this removes NW, NE, SE, and SW pixels on the boundary of the object.

The pixel locations having value "*" are "don't care" values and are allowed to take either value 0 or 1. Finding and changing each value corresponding to a specific configuration can be represented by the HMT using the appropriate template pair (**u,w**) and expressed as follows:

$$\mathbf{a} - \mathbf{a} \circledast (\mathbf{u}, \mathbf{w}) \tag{2}$$

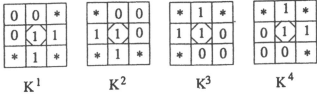

Figure 9 Configurations used to remove NW, NE, SE, and SW boundary pixels.

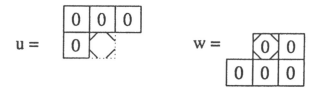

Figure 10 One realization of configuration K^1.

For example, see Fig. 10 for the template pair $(\mathbf{u},\mathbf{w})$ as one of the realizations of the configuration K^1 given in Fig. 9. Each of the eight realizations of configuration K^1 is obtained by replacing the "*" values with all possible variations of 0's and 1's. After Eq. (2) is calculated with one realization, its output is used as input for the next calculation of Eq. (2) with another realization of K^1. This is done sequentially for all K^i, $i = 1, \ldots, 4$.

To represent this in image algebra, we use the concept of a real-valued *census* template, a template having values each a unique power of a prime number, and the operation ⑦(linear convolution). For our purposes, we used the census template in Fig. 11.

Each possible configuration of 0's and 1's in the 3×3 window centered at its target pixel is represented by a unique number between 0 and 255. Performing the image algebra expression

$$\mathbf{b} = \chi_n(\mathbf{a} \, ⑦\, \mathbf{t})$$

will produce an image **b** having values of 1 wherever the number n occurs in the image $\mathbf{a} \, ⑦\, \mathbf{t}$ and 0's everywhere else. If we calculate

$$\mathbf{a} - \chi_n(\mathbf{a} \, ⑦\, \mathbf{t})$$

then the pixels having values of 1 wherever the number n occurs in the image $\mathbf{a} \, ⑦\, \mathbf{t}$ are set to 0, thus removing each pixel having configuration corresponding to the appropriate census template number n. For example, to perform the HMT

$t =$
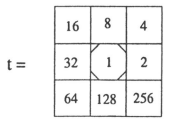

Figure 11 Census template used for HMT.

using the template given in Fig. 10, we first find the number n corresponding to the appropriate configuration (see Fig. 12).

Doing this for all possible configurations of K^1, we find that the set of eight numbers

$$S_{K^1} = \{131, 387, 195, 135, 451, 391, 199, 455\}$$

represents numerically the eight possible configurations. Locating these eight configurations is represented simply in image algebra by the equation

$$\chi_{S_{K^1}} (\mathbf{a} \, \textcircled{v} \, \mathbf{t})$$

To remove those pixels, we calculate

$$\mathbf{a} - \chi_{S_{K^1}} (\mathbf{a} \, \textcircled{v} \, \mathbf{t})$$

This is done for each configuration K^i, $i = 2, 3, 4$, given in Fig. 9, in a sequential manner. The other three sets are

$$S_{K^2} = \{161, 177, 225, 417, 241, 481, 433, 497\}$$

$$S_{K^3} = \{41, 57, 45, 105, 61, 121, 109, 125\}$$

$$S_{K^4} = \{11, 27, 15, 267, 31, 271, 283, 287\}$$

The pseudocode given in Fig. 13 describes the first cycle of the algorithm.

For the moment we assume that this process does stabilize, that is, that it takes a finite number of steps to complete the first cycle. The second cycle performs a similar process, starting with the output of the first cycle, say f, and using the templates corresponding to the configurations given in Fig. 14. This second cycle is used to remove N, S, E, and W boundary pixels.

The corresponding sets are

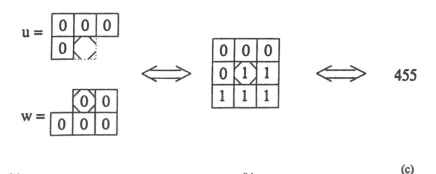

(a)   (b)   (c)

Figure 12   How to derive the census template number for a particular realization: (a) template pair (u, w); (b) image algebra template; (c) number corresponding to (b).

```
input binary image a

b = a
repeat
    c = b
    do i=1 to 4
```
$$b = b - \chi s_{\kappa_i}(b \oplus t).$$
```
    end do
until (b = c)
output f = b
```

Figure 13 Pseudocode for first cycle of general thinning algorithm.

$$S_{J_1} = \{163, 179, 167, 419, 227, 183, 423, 483, 243, 439, 487, 499, 503, 435, 231, 247\}$$

$$S_{J_2} = \{169, 185, 173, 425, 233, 189, 429, 489, 249, 441, 237, 445, 493, 505, 253, 509\}$$

$$S_{J_3} = \{43, 59, 47, 299, 107, 303, 363, 123, 315, 111, 319, 367, 379, 127, 383, 63\}$$

$$S_{J_4} = \{139, 155, 143, 395, 203, 159, 399, 459, 219, 411, 207, 415, 463, 475, 223, 479\}$$

Implementation of this can be performed by using a look-up table. Assuming that this process stabilizes, at the end of the second cycle, the final image is **I**, the skeleton of **a**.

Applying the templates in a different order from the one given above results in a slightly different skeleton. The order of application of the templates must be chosen so that homotopy is preserved on each pass. We state the conditions for this later.

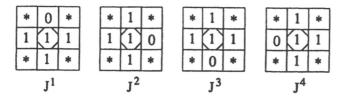

Figure 14 Configurations used to remove N, S, E, and W boundary pixels.

Also, it has been shown [17] that indeed the two-cycle algorithm does converge in a finite number of steps, and to a set *I* satisfying the three properties of a skeleton given earlier. We outline the approach to solving this, as it gives insight as to how to choose an ordering of template application that is more computationally efficient and produces different and perhaps "better" skeletons.

Lemma 1. *Let* $a \in \{0, 1\}^X, a \neq 0$, (s, t) *be a pair of nondegenerate invariant templates to be used in the hit-or-miss transform. Then*

$$a - a \circledast (s, t) \neq 0$$

Corollary 2. *Let* $a \in \{0,1\}^X, a \neq 0$, *and let* (s^i, t^i) *for* $i = 1, \ldots, n$ *be a sequence of template pairs where each pair satisfies the conditions for the template pair given in Lemma 1. Define* $a^0 = a$ *and*

$$a^{i+1} = a^i - a^i \circledast (s^i, t^i)$$

Then for any K,

$$a^K \neq 0$$

Theorem 3. *Given* a, (s^i, t^i) *as in Corollary 2, there always exists a positive integer M such that* $a^{M-1} \neq a^M$ *but* $a^k = a^M$ *for* $k \geq M + 1$, *except for possibly the trivial case where* $a^i = a^0 \; \forall i$.

Theorem 3 states that the thinning algorithm converges to a fixed image, I, the skeleton of a. In particular, for the set of configurations given in Fig. 9, if these templates are applied sequentially to a and produce f at the end, we have $\chi_{S_{Ki}} (f \circledast t) = 0$ for $i = 1,2,3,4$. That is, all NW, NE, SW, and SE nonessential boundary points have been removed. Similarly, $\chi_{S_{Ji}} (I \circledast t) = 0$ for $i = 1,2,3,4$, that is, all N, S, E, and W nonessential boundary pixels have been removed.

To show that I is one pixel thick, an analysis is done on I. If I contains any patterns listed in Fig. 15, either I is not one pixel wide or I has a critical point.

Performing such an analysis and using the results from Theorem 3 with the specified templates corresponding to the configurations given in Fig. 9 and 15, we get an output I at the end of this two-cycle process. It can be shown [17] that any pixel in I having configuration as in Fig. 15 is a critical point (i.e., I is not

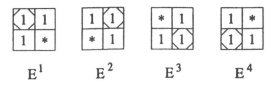

Figure 15

4-connected but is 8-connected, except for critical points). All critical points take the value 1 in the image c,

$$c = \chi_{\geq 1}[\chi_{S_{L^1}} (\mathbf{I} \circledast t) + \chi_{S_{L^2}} (\mathbf{I} \circledast t) + \chi_{S_{L^3}} (\mathbf{I} \circledast t) + \chi_{S_{L^4}} (\mathbf{I} \circledast t)$$
$$+ \chi_{S_{L^5}} (\mathbf{I} \circledast t)] \tag{3}$$

Here the five terms inside the brackets of Eq. (3) represent finding the HMT for the five critical point configurations given in Fig. 8. The sets L^i are

$S_{L^1} = \{147, 151, 211, 403, 215, 467, 407, 471\}$

$S_{L^2} = \{165, 181, 229, 421, 245, 485, 437, 501\}$

$S_{L^3} = \{297, 301, 313, 361, 317, 365, 377, 381\}$

$S_{L^4} = \{75, 79, 91, 331, 95, 347, 335, 351\}$

$S_{L^5} = \{171, 175, 187, 235, 427, 191, 239, 431, 251, 443, 491, 255,$
$\qquad 507, 495, 447, 511\}$

The connectedness of $\mathbf{I}$ can be shown in several steps for the thinning algorithm given above. These results are generalized to obtain a necessary and sufficient condition for the connectivity of thinning algorithms in general [17]. This condition can be used to develop thinning algorithms that produce skeletons that are connected and are one pixel thick. A second thinning algorithm is derived from these necessary and sufficient conditions, and is compared with the first algorithm.

Lemma 4. *If an object A in* **a** *is 8-connected, then*

$$\mathbf{a} - \chi_{S_{K^1}} (\mathbf{a} \circledast t) \tag{4}$$

is 8-connected.

Equation (4) removes NW boundary points. Lemma 4 states that if NW boundary points are removed, the resulting image is still 8-connected. Using the fact that the remaining sets of templates corresponding to the configurations K^i, $i = 2,3,4$, are rotated versions of K^1 by 45°, we have:

Lemma 5. *If* **a** *is 8-connected, then* **f** *is 8-connected, where* **f** *is the image output at the end of the first cycle.*

Theorem 6. *If* **a** *is 8-connected, then* **I** *is also 8-connected.*

Theorem 7. *If* **ä** *is 4-connected then* **Ì** *is 4-connected.*

We must have these last two results because we do not want the thinning process to generate holes.

Define a *skeletal leg* to be a one-pixel-thick branch with one end not connected to any other part of the object.

The generalization of Theorems 6 and 7 is:

Theorem 8. *If **a** is 8-connected, then the thinning process described in Theorem 3 preserves homotopy and skeletal legs if and only if*

$$\mathbf{a} - \mathbf{e} \le \chi_{S_{\kappa^i}} (\mathbf{a} \, \textcircled{Y} \, \mathbf{t}) + \chi_{S_{\kappa^{(i+1)(mod\ 4)}}} (\mathbf{a} \, \textcircled{Y} \, \mathbf{t}) + \chi_{S_{ji}} (\mathbf{a} \, \textcircled{Y} \, \mathbf{t})$$

*for each i = 1,2,3,4, and where **e** is the stabilized output to Theorem 3.*

From this condition, the thinning algorithm shown in Fig. 16 was devised [17], where the output **I** is the skeleton.

Note that Eq. (5) (Fig. 16) can be computed in parallel, first by computing **b** $\textcircled{Y}$ **t** in parallel and then by computing each individual term in parallel. The image algebra captures the spirit of these parallel operations in its notation very succinctly.

Results.   The foregoing two algorithms were run on several images and compared with the medial axis and MAT. The results are produced in Fig. 17. For comparison, the MA is also given. A measure of "closeness" of the resulting skeletons to the MA that could be used is

$$\epsilon = \frac{\Sigma \, \mathbf{I}}{\Sigma \, \chi_{\geq 1} \, [MA(\mathbf{a})]}$$

Here $\Sigma \, \mathbf{b} \equiv k = \Sigma_{\mathbf{x} \in \mathbf{X}} \, \mathbf{b}(\mathbf{x})$. For the image data displayed in Fig. 17, image (a), the panda image, has measure $\epsilon_a(1) = 1.53$, calculated using the output from the first thinning algorithm, and $\epsilon_a(2) = 1.38$, calculated using the output from the second thinning algorithm. As for image (b), the tree image has measure

```
input binary image a


 b = a

 repeat

    c = b

    do i=1 to 4
```

$$\mathbf{b} = [\mathbf{b} - \chi_{S_{\kappa^i}} (\mathbf{b} \, \textcircled{Y} \, \mathbf{t})] * \left[\mathbf{b} - \chi_{S_{\kappa^{(i+1)(mod\ 4)}}} (\mathbf{b} \, \textcircled{Y} \, \mathbf{t})\right] * [\mathbf{b} - \chi_{S_{ji}} (\mathbf{b} \, \textcircled{Y} \, \mathbf{t})]$$

(5)

```
    end do

 until (b = c)

 I = b; output I
```

Figure 16   Pseudocode for second thinning algorithm.

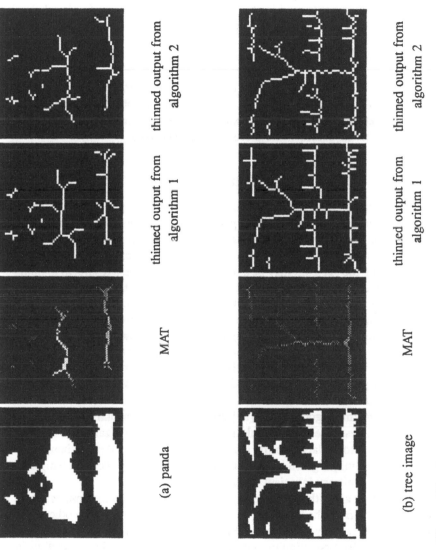

thinned output from algorithm 2

thinned output from algorithm 1

MAT

(a) panda

thinned output from algorithm 2

thinned output from algorithm 1

MAT

(b) tree image

Figure 17 Output of the MAT and two skeletonizing algorithms: (a) panda image; (b) tree image.

$\epsilon_b(1) = 1.32$, calculated using the output from the first thinning algorithm, and $\epsilon_b(2) = 1.30$, calculated using the output from the second thinning algorithm. For both images, the output of the second algorithm gives a closer "fit" to the MAT than does the first algorithm.

A consideration not treated in this analysis is the trimming of noisy skeletal legs. Skeletal legs can be erased completely by iteratively performing

$$\mathbf{I} - \chi_{2^i+1}(\mathbf{I} \oslash \mathbf{t})$$

for $i = 1, \ldots, 8$ until the process converges. However, in cases where the skeletal leg is an integral part of the skeleton, this is not a useful procedure. Other information, such as the approximate size of the image, can be used to limit the number of passes while preserving skeletal legs.

V. OTHER THINNING ALGORITHMS

The common theme to most thinning algorithms is the notion of connectivity of the skeleton and its thin width (1 to 2 pixels wide), and the local neighborhood over which the computation is done. The general approach to developing a thinning algorithm given in Section IV is done for 3×3 templates. The three conditions defining the skeleton may be sufficient to extend these results to thinning with a template of arbitrary size $k \times k$ (k odd). The advantage of using a larger window is, of course, increased speed in implementation. A general thinning algorithm using a $k \times k$ widow is described in [29]. By removing more pixels in fewer iterations, computation time may be decreased. However, it may also be a cost of increased coarseness in the final skeleton. This particular algorithm has not been described in the context of the mathematical formalism for thinning algorithms as given in Section IV, and an investigation would probably yield interesting results.

The skeletonizing algorithms in [22] are basically the medial axis used with arbitrary templates instead of digital disks. This is a generalization of the MA that gives a skeleton relative to the template shape used for skeletonizing. The skeletonizing technique given in Section VI for gray-value images and templates also allows for arbitrary invariant templates.

Many thinning algorithms can be implemented in a parallel fashion, as often, removal of pixels from the boundary of the object does not depend on other pixel values spatially far away. The second thinning algorithm in Section IV is a parallel version of the first thinning algorithm. The $k \times k$ thinning algorithm in [29] also has a parallel version. In [35] Rosenfeld establishes necessary and sufficient conditions for a very simple class of parallel thinning algorithms. Improvements on speed and performance of some parallel thinning algorithms are discussed in [14]. A good basic review of 14 thinning algorithms is provided in

[28], as well as a comparison of the 14 with one particular thinning algorithm put forth by the authors.

Finally, thinning algorithms on arrays other than rectangular are discussed in [11]. Thinning for hexagonal and triangular arrays presents difficulties not encountered on rectangle arrays. It turns out that the hexagonal array is easiest to deal with and that the triangular array is the most complex of the three. This is related to the connectivity schemes possible in each array type. The rectangular array has two connectivity schemes, four and eight connectivity. The hexagonal array has only one, whereas the triangle array has two possibilities, each having 12 neighbors.

VI. GRAY-SCALE SKELETONIZING

There are various ways to extend skeletonizing to gray-valued images and templates. One method defines a *gray-weighted distance* between two pixel locations as the smallest sum of gray levels along any 8-path joining the points. Then a binary image is produced, where a black pixel at location x corresponds to a pixel from the original image, $a(x)$, whose gray-weighted distance to the boundary of the object is a local maximum. This algorithm, by Levi [20], must generally use a segmented image for input and may not produce connected skeletons. Also, reconstruction of the original image may not be as good as from other algorithms.

Another approach segments the image into blocks of different sizes where each block satisfies a homogeneity criterion and discards blocks if they are contained in other such blocks. This results in a set of maximal blocks with respect to the homogeneity measure. The skeleton consists of the centers of these blocks. If other information, such as the centers, radii, and mean gray levels of these blocks are known, the original image can be approximated by superimposing blocks with these gray values. However, this method is computationally expensive, and reconstruction of the original image may not be accomplished easily.

A third method [37] uses gradient magnitudes at pairs of point. Let x be a pixel location, and assign a score to x based on the gradient magnitudes of pairs of points that have x as its midpoint. The skeleton consists of pixels whose scores satisfy a certain requirement. The coordinates and gradient magnitudes of all the pairs of pixels y contributing to the score of x are also stored at x. Using this information, several ways to reconstruct the image can be performed [38]. This method is particularly sensitive to noise.

A fourth method is based on mathematical morphology operations of dilation and erosion [38]. Let t be the 3×3 invariant template with nine zeros. The skeleton of a gray-value image is a vector-valued image on the image domain X. Let m be the radius of the largest digital disk that will fit inside any object in the image. Using the fact that for $1 \leq k \leq m$, we have

$\mathbf{a} \ominus (-\mathbf{t}')^{k-1} \geqslant (\mathbf{a} \ominus (-\mathbf{t}')^k) \oplus \mathbf{t}$

and can define the nonnegative image $\mathbf{d}_k$ as follows:

$\mathbf{d}_k = \mathbf{a} \ominus (-\mathbf{t}')^{k-1} - (\mathbf{a} \ominus (-\mathbf{t}')^k) \oplus \mathbf{t}$

Doing this for all $1 \leqslant k \leqslant m$, we construct a vector-valued image Δ by

$\Delta(\mathbf{x}) = (\mathbf{d}_1(\mathbf{x}), \mathbf{d}_2(\mathbf{x}), \ldots, \mathbf{d}_m(\mathbf{x}))$

This is the skeleton of the original image $\mathbf{a}$. Exact reconstruction of $\mathbf{a}$ is given by the pseudocode in Fig. 18.

This procedure can be computationally expensive. However, in [38], several shortcuts are given to produce good approximate reconstructions of $\mathbf{a}$ using only one to three of the vector values in $\Delta(\mathbf{x})$.

Next we present a general skeleton algorithm for gray-value images that is a direct generalization of the binary medial axis transform. The template is allowed to be any invariant one and thus gives a skeleton with respect to a certain template. The image algebra code for the gray-value case is exactly the same as for the binary case. One interesting result of this algorithm is that it produces "remainder" images; that is, after the template reduces the boundary of the three-dimensional object, there are some residual pixels. In the binary case, the last remainder image is the null image $-\infty$ (no remainder). Also, given the immediate images, the original gray-scale image one can be reconstructed exactly.

Let $\mathbf{t}$ be a nonempty invariant template, and $\mathbf{a} \in \mathbb{R}^{\mathbf{x}}$. Then for the images $\mathbf{a}^1$ and $\mathbf{r}$ defined by $\mathbf{a}^1 = \mathbf{a} \ominus -\mathbf{t}'$ and

$\mathbf{r} = \mathbf{a} + \chi_{>0}^{-\infty} [\mathbf{a} - (\mathbf{a}^1 \oplus \mathbf{t})]$

we can always write

$$\mathbf{a} = (\mathbf{a}^1 \oplus \mathbf{t}) \vee \mathbf{r} \tag{6}$$

Here the characteristic function $\chi_{>r}^{-\infty}$ when applied to an image $\mathbf{b}$ results in an image with values 0 or $-\infty$ depending on whether $\mathbf{b}(\mathbf{x})$ is greater than $\mathbf{r}$ or not, respectively.

```
input gray scale images dₖ, k = 1,...,m, and
a⊖(-t')ᵐ
b = a⊖(-t')ᵐ
do i = m to 1 (step -1)
     b = (b ⊕ t) + dᵢ
end do
output b, which is original input a
```

Figure 18   Pseudocode for Wang's algorithm.

The proof of this can be found in [9].

If we apply the foregoing procedure to $\mathbf{a}^1$, we get

$$\mathbf{a}^1 = (\mathbf{a}^2 \oplus t) \vee \mathbf{r}^1 \tag{7}$$

where $\mathbf{a}^2 = \mathbf{a}^1 \ominus -t'$, and

$$\mathbf{r}^1 = \mathbf{a}^1 + \chi_{>0}^{-\infty} [\mathbf{a}^1 - (\mathbf{a}^2 \oplus t)]$$

Plugging (7) back into (6), we get

$$\mathbf{a} = (\mathbf{a}^2 \oplus t^2) \vee (\mathbf{r}^1 \oplus t) \vee \mathbf{r}^0$$

where $\mathbf{r}^0 = \mathbf{r}$ from Eq. (6). Continuing in this manner, we obtain

$$\mathbf{a} = \left\{ \bigvee_{i=0}^{k} [\mathbf{r}^i \oplus t^i] \right\} \vee [\mathbf{a}^{k+1} \oplus t^{k+1}] \tag{8}$$

where t raised to the zeroth power, t^0, is the identify template. The image $\mathbf{a}^{k+1} \oplus t^{k+1}$ is called the *residual* image, while the r's are called *remainder* images.

In the boolean case, there exists an integer m such that $\mathbf{a}^{m-1} = \mathbf{a}^m = 0$, so that the expression for **a** becomes

$$\mathbf{a} = \bigvee_{i=0}^{m-1} \mathbf{r}^i \oplus t^i$$

In Fig. 19, an example of computing Eq. (8) is presented. The remainder and residual images are shown. The parameters used for these data are $k = 3$ and the template t as shown in Fig. 20. In general, it is typical that the residual image $\mathbf{a}^{k+1} \oplus t^{k+1}$ represents a significant part of the data in the original image, so, unfortunately, this term cannot simply be dropped. Otherwise, this method would present a useful compression technique.

VII. CONCLUSIONS

Thinning algorithms remain a useful tool in image processing and computer vision. The skeleton can give useful characterizations of an object, providing features for pattern recognition as well as reducing the amount of memory necessary for storing and transmitting data. Fast implementations on parallel machines make thinning algorithms an even more attractive and useful transform.

Currently, there is no unified theory of thinning. The work by Jang and Chin [17] comes close, in that specific properties that a skeleton need satisfy are formalized mathematically. One key point in that work is that the thinning algorithm is "close" to the MA in some sense. The MAT is unique for continuous images and in a sense, finding a discrete version that is closest is a natural

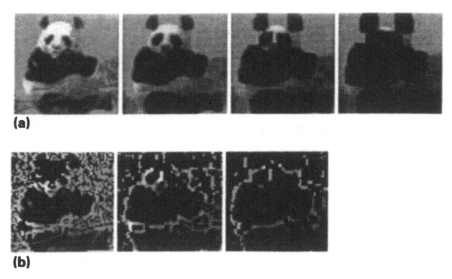

(a)

(b)

Figure 19 (a) Original image; images a^1 (b), a^2 (c), and residual image a^3 (d) for input image (a); remainder images r^0 (e), r^1 (f) and r^2 (g) for input images (a).

$$t_y = \begin{array}{|c|c|c|} \hline 0 & 0 & 0 \\ \hline 0 & 0 & 0 \\ \hline 0 & 0 & 0 \\ \hline \end{array}$$

Figure 20 Template t used in Eq. (8) for image in Fig. 19a.

approach. However, since the MA is not unique for digital images [26], this may not be the most appropriate approach in the digital case. This remains an open problem.

Also, as mentioned in Section V, a generalization of the approach to thinning given in Section IV to a $k \times k$ window has not been investigated. Results in this area could possibly give improvements of the overall theory for Jang's 3×3 approach. In addition, investigation of the three-dimensional MAT given in Section VII as a particularly useful applications would be of interest.

REFERENCES

1. M. Beun, A flexible method for automatic reading of hand-written numerals, *Philips Tech. Rev.*, 31(4):89–101, 130–137 (1973).

2. H. Blum, An associative machine for dealing with the visual field and some of its biological implications, in *Biological Prototypes and Synthetic Systems*, Vol. 1, E. E. Bernard and M R. Kare, eds., Plenum Press, NY, 1962.

3. H. Blum, A transformation for extracting new descriptors of shape, in *Models for the Perception of Speech and Visual Forms*, W. Wathen-Dunn, ed., MIT Press, Cambridge, Mass., 1967.

4. H. Blum, Medial axis transformations, in *Picture Processing and Psychopictorics*, B. S. Lipkin and A. Rosenfield, eds., Academic Press, New York, 1970.

5. H. Blum, Biological shapes and visual sciences (part I), *J. Theor. Biol.*, 38:205–287 (1973).

6. H. Blum and R. N. Nagel, Shape description using weighted symmetric axis features, *Pattern Recognition*, 10:167–180 (1978).

7. R. Cuninghame-Green, *Minimax Algebra: Lecture Notes in Economics and Mathematical Systems* 166, Springer-Verlag, New York, 1979.

8. J. L. Davidson, Lattice structures in the image algebra and applications to image processing, Ph.D. thesis, Department of Mathematics, University of Florida, Aug. 1989.

9. J. L. Davidson, Foundation and applications of lattice transforms in image processing, *Adv. Electron. Electron Phys. Vol. 84*, pp. 61–130, Academic Press, London, (1992).

10. E. R. Davies and A. P. N. Plummer, Thinning algorithm, a critique and new methodology, *Pattern Recognition*, 14:53–63 (1981).

11. E. S. Deutsch, Thinning algorithms on rectangular, hexagonal and triangular arrays, *Comm. ACM*, 15(9):827–837 (1972).

12. A. J. Frank, J. D. Daniels, and D. R. Unangst, Progressive image transmission using a growth-geometry coding, *Proc. IEEE*, 68:897–909 (July 1980).

13. A. Gudsen. A quantitative analysis of preprocessing techniques for the recognition of handprinted characters, *Pattern Recognition*, 8:219–227 (1976).

14. Z. Guo and R. W. Hall, Parallel thinning with two-subiteration algorithms, *Comm. ACM*, 32(3) (1989).

15. J. Hilditch, Linear skeletons from square cupboards in *Machine Intelligence*, Vol. 4, B. Meltzer and D. Michie, eds., American Elsevier, New York, 1969.

16. IVS, Image algebra FORTRAN version 2.0 language description and implementation notes, *Technical report*, IVS, Inc., Gainesville, Fla., May 1988.

17. B. K. Jang and R. T. Chin, Analysis of thinning algorithms using mathematical morphology, *IEEE Trans. Pattern Anal. Mach. Intelligence*, PAMI-12(6):541–551 (1990).

18. C. Lantuéjoul, La Squelettisation et son application aux mesures topologiques des mosaïques polycristallines, Ph.D. thesis, School of Mines, Paris, 1978.

19. C. Lantuéjoul, Skeletonization in quantitative metallography, in *Issues of Digital Image Processing*, R. M. Haralick and J. C. Simon, eds., Sijthoff and Noordhoff, Groningen, The Netherlands, 1980.

20. G. Levi and U. Montanari, A grey-weighted skeleton, *Inform. Control*, 17:62–91 (1970).

21. J. R. Mandeville, Novel method for analysis of printed circuit images, *IBM J. Res. Develop.*, 29(1):73–86 (1985).

22. P. A. Maragos and R. W. Schafer, Morphological skeleton representation and cod-
     ing of binary images, in *Proc. IEEE International Conference on Acoustics,
     Speech, and Signal Processing*, San Diego, Calif., Apr. 1984, pp. 29.2.1–29.2.4.
23. T. E. Meyer, Image algebra preprocessor for the MasPar parallel computer, *Proc.
     SPIE*, Vol. 1568, San Diego, Calif., July 1991, pp. 125–136.
24. B. Moayer and K. S. Fu, A tree system approach for fingerprint pattern recogni-
     tion, *IEEE Trans. Comput.*, C-25:262–275 (1976).
25. U. Montanari, Continuous skeletons from digitized images, *Comm. ACM*,
     16(4):534–549 (1969).
26. J. C. Mott-Smith, Medial axis transformations, in *Picture Processing and Psy-
     chopictorics*, B. S. Lipkin and A. Rosenfeld, editors, Academic Press, New York,
     1970.
27. J. C. Mott-Smith and T. Baer, Area and volume coding of pictures, in *Picture Band-
     width Compression*, T. S. Huang and O. J. Tretiak, eds., Gordon and Breach, New
     York, 1972.
28. N. J. Naccache and R. Shinghal, Spta: a proposed algorithm for thinning binary
     patterns, *IEEE Trans. Systems Man Cybernet.*, SMC-14(3):409–418 (1984).
29. L. O'Gorman, $k \times k$ thinning, *Comput. Vision Graphics Image Process.* 51:195–
     215 (1990).
30. T. Pavlidis, *Algorithms for Graphics and Image Processing*, Computer Science
     Press, Rockville, Md., 1982.
31. O. Philbrick, Shape recognition with the medial axis transformation, in *Pictorial
     Pattern Recognition (Proc. Symposium on Automated Photointerpretation)*, G. C.
     Cheng et al., eds., Washington, D.C., 1968.
32. G. X. Ritter and P. D. Gader, Image algebra techniques for parallel image process-
     ing, *J. Parallel Distribut. Comput.*, 4(5):7–44 (1987).
33. G. X. Ritter, J. N. Wilson, and J. L. Davidson, Image algebra: an overview, *Com-
     put. Vision Graphics Image Process.*, 49:297–331 (Mar. 1990).
34. A. Rosenfeld, Connectivity in digital pictures, *J. Assoc. Comput. Mach.*,
     17(1):146–160 (1970).
35. A Rosenfeld, A characterization of parallel thinning algorithms, *Inform. Control*,
     29:286–291 (1975).
36. R. Stefanelli and A. Rosenfeld, Some parallel thinning algorithms for digital pic-
     tures, *J. Assoc. Comput. Mach.*, 18(2):255–264 (1971).
37. S. Wang, A. Rosenfeld, and A. Y. Wu, A medial axis transformation for gray-scale
     pictures, *Technical Report*, Computer Vision Laboratory Computer Science Center,
     University of Maryland, College Park, Md., Dec. 1979.
38. S. Wang, A. Y. Wu, and A. Rosenfield, Image approximation from gray scale me-
     dial axis, *IEEE Trans. Pattern Anal. Mach. Intelligence*, PAMI-3(6):687–696
     (1981).
39. J. N. Wilson, Introduction to image algebra ada, *Proc. SPIE, vol. 1568*, San Diego,
     Calif., July 1991, pp. 101–112.

5

Syntactic Image Pattern Recognition

Edward K. Wong

Polytechnic University
Brooklyn, New York

I. INTRODUCTION

In machine recognition of image patterns and shapes, features are extracted and subject to statistical analysis; or primitives are selected and subject to syntax analysis. The former is called the *statistical* or *decision-theoretic* approach [1], and the latter is called the *syntactic* or *structural* approach [2]. The statistical or decision-theoretic approach is the traditional approach to pattern recognition that has been studied since the 1960s [1,3–6]. A block diagram using this approach is shown in Fig. 1. The system consists of two parts: analysis part and recognition part. In the analysis part, a set of image features that are judged to be nonoverlapping (or as widely apart as possible) in the feature space are chosen. A statistical classifier is designed and the chosen set of features are trained to obtain the appropriate classifier parameters. In the recognition part, an unknown image is filtered or enhanced in the preprocessing stage, followed by feature detection and classification.

The statistical or decision-theoretic approach to pattern recognition, however, does not describe or represent the structural information in a pattern, which is often desirable or necessary for certain applications. For example, in Chinese character and fingerprint recognition, the number of classes is so large that it is almost impossible to use a traditional statistical approach directly. In other cases, the given pattern is so complex that it is not easy to reliably extract meaningful features for classification. In both circumstances, the number of features

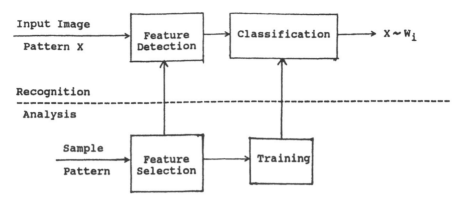

Figure 1   Block diagram of a statistical pattern recognition system.

required is probably very large, which makes the statistical approach impractical. It is therefore more attractive to decompose the complex patterns into subpatterns, and recursively into sub-subpatterns, and so on, until we can reliably
extract meaningful *primitive patterns* (analogous to features in the statistical approach) from them. We also seek to determine the relation between subpatterns
at the same level as well as the relation between a subpattern and its child subpatterns (see Fig. 2) This approach allows us to *describe* and *represent* the input
pattern, in addition to classifying it into a specific class. An example showing
the hierarchical decomposition of a submedium chromosome is shown in Fig. 3.
Another example for the numeral "8" is shown in Fig. 4. Note that in both examples, the primitives are simple curved or linear segments that can be easily
extracted from the input pattern. In the chromosome example, it is a small set of
curved boundary segments, while in the character example, it is the vertical or
horizontal strokes. They can, however, be concatenated (putting the head of one
primitive after the tail of another) to form subpatterns, and in turn, to form the

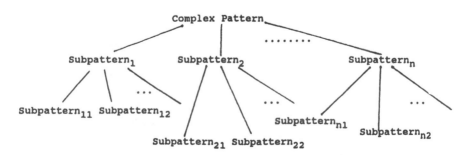

Figure 2   Decomposition of a complex pattern into subpatterns.

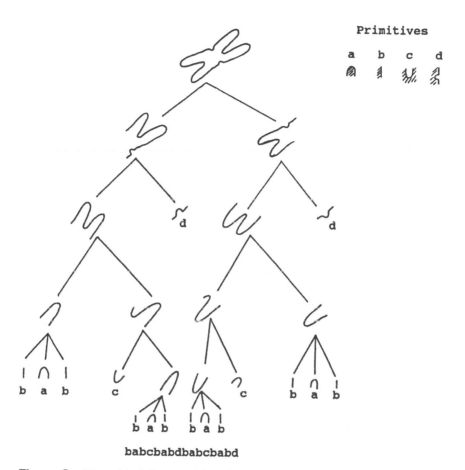

Primitives

a   b   c   d

Figure 3   Hierarchical decomposition of a submedium chromosome.

whole input pattern. This approach, called syntactic or structural pattern recognition, has attracted much attention in later development of pattern recognition research [7–13].

It is observed that the hierarchical structure of a pattern is analogous to the syntax of English or programming languages. Many theories developed in formal languages can equally be applied. For example, an English sentence or phrase can be decomposed into a noun phrase and a verb phrase; a noun phrase into an article and a noun; a verb phrase into a verb and an adverb (see Fig. 5). The article, noun, verb, and adverb are then replaced by different vocabularies to form different English sentences. For example, the leave nodes in the decomposition tree of Fig. 5 form the sentence "A man reads carefully." Note that if we replace ⟨verb⟩ with the vocabulary "looks," we generate another English

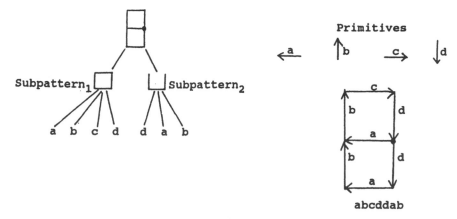

Figure 4 Hierarchical decomposition of the numeral "8."

sentence: "A man looks carefully," and if we replace ⟨noun⟩ with "woman," we get "A woman reads carefully." Many more sentences can be generated using this hierarchical tree structure by replacing ⟨article⟩, ⟨noun⟩, ⟨verb⟩, and ⟨adverb⟩ with different vocabularies. The leave nodes of the tree are called *terminal* symbols and the nonleave nodes (including the root node) are called *nonterminal* symbols. We also refer to the root node as the *starting* symbol.

The structural information contained in the tree structure of Fig. 5 can also be represented by a set of *rewriting* (or *production*) rules:

(1) S→⟨NP⟩⟨VP⟩
(2) ⟨NP⟩→⟨A⟩⟨N⟩
(3) ⟨VP⟩→⟨V⟩⟨ADV⟩
(4) ⟨A⟩→A
(5) ⟨N⟩→man
(6) ⟨V⟩→reads
(7) ⟨ADV⟩→carefully

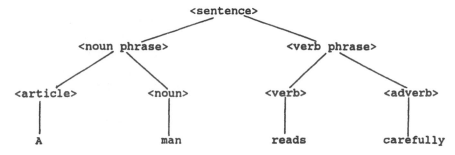

Figure 5 Hierarchical decomposition of an English sentence.

A rewriting rule can be interpreted as a replacement rule in a decomposition process, indicating that the symbols(s) on the left-hand side can be decomposed (or replaced by) symbols on the right-hand side. For example, starting with production rule (1), we get

$$S \overset{(1)}{\to} \langle NP \rangle \langle VP \rangle$$

Note that the number (in parentheses) above the arrow indicates the production rule being applied. Applying rule (2), we get

$$\langle NP \rangle \langle VP \rangle \overset{(2)}{\to} \langle A \rangle \langle N \rangle \langle VP \rangle$$

Then applying rules (4), (5), (3), (6), and (7), we get

$$\langle A \rangle \langle N \rangle \langle VP \rangle \overset{(4)}{\to} A \langle N \rangle \langle VP \rangle \overset{(5)}{\to} A \text{ man} \langle VP \rangle$$
$$\overset{(3)}{\to} A \text{ man } \langle V \rangle \langle ADV \rangle \overset{(6)}{\to} A \text{ man reads } \langle ADV \rangle \overset{(7)}{\to} A \text{ man reads carefully}$$

The above is called a *derivation* of the sentence by applying the production rules.

The block diagram of a syntactic pattern recognition system is shown in Fig. 6, which is analogous to that of the statistical pattern recognition system in Fig. 1. The system can also be divided into two parts: analysis and recognition. In the analysis part, a set of primitives (and their relation) is selected from a given set of sample patterns. A grammar is then inferred to describe and represent the relation between subpatterns and primitives. The construction of a grammar for

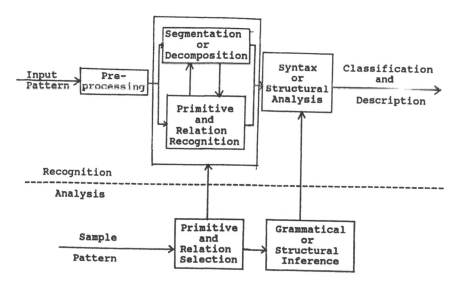

Figure 6 Block diagram of a syntactic pattern recognition system.

a pattern class is often done manually. For automated grammatical inference techniques, the reader is referred to [2]. In the recognition part, image processing (such as filtering and enhancement) is first performed, followed by segmentation or decomposition of the complex image pattern. Primitive patterns, and the relation between them, are often recognized or identified in conjunction with the segmentation or decomposition process. Depending on the image pattern, different segmentation or decomposition processes can be applied. Once the primitives are extracted, they are represented symbolically using a *string* or a *sentence*, or other pattern representation languages (e.g., a *tree* or a *graph*). The symbolic representation of the pattern is then subject to syntax (or structural) analysis. A parser (e.g., a string or tree parser) can be used for parsing the input pattern and a classification, as well as a description of the input pattern, is obtained.

It should be emphasized that image preprocessing is an important component in a syntactic image recognition system. For example, proper image enhancement and filtering techniques will remove or reduce image noise and distortion introduced in the image acquisition process. Enhancement and filtering techniques such as histogram equalization, local enhancement, neighborhood averaging, median filtering, low-pass and high-pass filtering, least-squares filtering, and maximum entropy restoration can be found in textbooks on digital image processing [14,15]. Another class of image filtering techniques based on mathematical morphology can be found in [16–18]. Other preprocessing techniques, such as thinning and pruning [15], are often needed for certain applications. For example, thinning operation is usually applied on the scanned image of a character or a numeral before we can extract the horizontal and vertical strokes (see Fig. 4).

Another important component is the segmentation (or decomposition) process. This is a technique for obtaining the constituent parts of an image, by which we extract primitive patterns (and relation) in the syntactic recognition process. Segmentation techniques can generally be classified as *boundary-based* or *region-based* [19]. In the boundary-based approach, edges of an object are obtained based on the intensity difference of neighboring pixels. The edges are then linked together to form a continuous or closed boundary. For example, each chromosome in Fig. 7 is represented using a closed boundary, and the primitive patterns are boundary segments of different shapes. In the region-based approach, pixels that belong to the same object are identified and grouped together based on a homogeneity criterion. Both approaches are useful for extracting primitive patterns (and their relation) in a syntactic recognition system, and the choice of which to apply is problem dependent.

The remainder of the chapter is organized as follows. In Section II we describe string grammers for pattern description/analysis. In Section III we describe syntactic pattern recognition procedures, including parsing, error-correcting

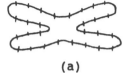

(a)

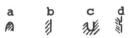

Primitives

a   b   c   d

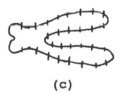

(b)

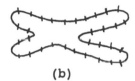

(c)

Figure 7 Three classes of chromosome patterns: (a) median; (b) submedian; (c) arcocentric.

parsing, and sentence-to-sentence clustering techniques. Two-dimensional grammars for image patterns are described in Section IV and some application examples in Section V. Finally, in Section VI we provide some concluding remarks.

II. STRING GRAMMARS FOR PATTERN DESCRIPTION/ANALYSIS

For many image patterns, the relation between the primitives can be described by a one-dimensional string. For example, the submedium chromosome pattern in Fig. 3 can be represented by the one-dimensional (1D) symbolic string *babcbabdbabcbabd*, and the numeral in Fig. 4 can be represented by the string *abcddab*. Similarly, the English sentence in Fig. 5 can be considered to consist of a string of English vocabularies. In the chromosome example, the class of all 1D strings that represent submedium chromosomes is said to belong to the language

$$L(G_s) = \{x \mid x \in V_T^* \text{ such that } S \to x\}$$

where x is a 1D string composed of symbols from the set of *terminal symbols* V_T; V_T^* denotes the set containing all strings over V_T, including the empty string λ; and G_s is a *phrase-structure* grammar with *production rules* that describe the syntax of the submedium chromosome pattern. Then x is the set of 1D strings that can be derived from a *starting symbol* S by using the production rules in G_s.

Formally, a phrase-structure grammar [20,21] can be defined by a four-tuple

$$G = (V_N, V_T, P, S)$$

where V_N and V_T are two disjoint sets representing the nonterminal and terminal symbols of G, respectively; P is a set of production rules of the form $\alpha \to \beta$ with $\alpha \in (V_N \cup V_T)^*V_N(V_N \cup V_T)^*$ and $\beta \in (V_N \cup V_T)^*$; and $S \in V_N$ is the starting symbol. Also, V is often used to denote $V_N \cup V_T$.

There are in general, four types of phrase-structure grammar. *Type 0* (*unrestricted*) grammars have no restrictions on the strings that appear on the left- and right-hand sides of a production. This type of grammar, however, is too general to be of any practical value. *Type 1* (*context-sensitive*) grammars have productions of the form

$$\zeta_1 A \zeta_2 \to \zeta_1 \beta \zeta_2$$

where $A \in V_N$, ζ_1, ζ_2, $\beta \in V^*$, and $\beta \neq \lambda$. This implies that A can be replaced in the context of ζ_1 and ζ_2. *Type 2* (*context-free*) grammars have productions of the form

$$A \to \beta$$

where $A \in V_N$ and $\beta \in V^+$. Here V^+ denotes $V^* - \{\lambda\}$ or that V^+ contains nonempty strings. Hence nonterminal symbol A can be replaced by the string β independent of the context in which A appears. An example of a context-free grammar is

$$G = (V_N, V_T, P, S)$$

where $V_N = \{S, X, Y\}$, $V_T = \{x, y\}$, and production set P:

(1) $S \to yX$, (5) $S \to xY$
(2) $X \to x$, (6) $Y \to y$
(3) $X \to xS$, (7) $Y \to yS$
(4) $X \to yXX$, (8) $Y \to xYY$

The grammar G is context-free since for each production in P, the left part is a single nonterminal and the right part is a nonempty string of terminals and nonterminals. An example of derivation using the production rules is

$$S \xrightarrow{(1)} yX \xrightarrow{(3)} yxS \xrightarrow{(5)} yxxY$$
$$\xrightarrow{(7)} yxxyS \xrightarrow{(1)} yxxyyX \xrightarrow{(2)} yxxyyx$$

where the numbers in parentheses indicate the production rule used. *Type 3 (finite-state or regular)* grammars have productions of the form

$A \rightarrow aB$ or $A \rightarrow b$

where $A, B \in V_N$ and $a, b \in V_T$. It should be noted that finite-state or regular grammars are special cases of context-free grammars in which the right-hand side of a production consists of strings of the form aB or b where a, b and B are all single symbols. An example of a finite-state grammar is

$G = (V_N, V_T, P, S)$

with $V_N = \{S, X\}$, $V_T = \{x, y\}$ and P:

$S \rightarrow xX$
$X \rightarrow xX$
$X \rightarrow b$

This grammar generates the language

$L(G) = \{x^n b \mid n = 1, 2, \ldots\}$

Grammar is an effective tool for syntactic pattern representation because of its capability to represent a large (and possibly infinite) set of sentences using a small set of production rules. This is achieved by using different combinations of production rules in a derivation sequence. Also, a production rule can be made recursive in representing repeating subpatterns or primitive patterns. For example, the recursive production rule

$G \rightarrow bG$

in the submedian chromosome grammar of Section II.A can control the length of one side of a chromosome arm, by repeating the primitive pattern b. For a more detailed treatment of formal languages and theories, the reader is referred to [20–22].

A. Chromosome Example

In this section we present an example of using grammars to describe different classes of chromosome patterns. Although the chromosome patterns are similar in shape, it is possible to construct context-free grammars for the median, submedian, and acrocentric chromosomes as shown in Fig. 7 [23]. Note that we use the same set of primitives for the three grammars.

1. *Median chromosome grammar*:

$G_s = (V_{N_m}, V_{T_m}, P_m, S)$

where $V_{N_m} = \{S, A, B, D\}$, $V_{T_m} = \{a, b, c, d\}$, and the production set P_m:

$S \rightarrow AA$,	$D \rightarrow bDb$,	$B \rightarrow bBb$,
$A \rightarrow cB$,	$D \rightarrow d$,	$B \rightarrow aDa$

2. *Submedian chromosome grammar:*

$G_s = (V_{N_s}, V_{T_s}, P_s, S)$

where $V_{N_s} = \{S, A, B, D, W, G, R, L, M, N\}$, $V_{T_s} = \{a, b, c, d\}$, and P_s:

$S \rightarrow AA,$	$G \rightarrow bG,$	
$A \rightarrow cM,$	$L \rightarrow aNa,$	
$B \rightarrow bBb,$	$L \rightarrow bL,$	$G \rightarrow d$
$B \rightarrow bL,$	$M \rightarrow bBb,$	
$B \rightarrow Rb,$	$N \rightarrow bDb,$	
$D \rightarrow bDb,$	$R \rightarrow aNa,$	$W \rightarrow d$
$D \rightarrow bG,$	$R \rightarrow Rb,$	
$D \rightarrow Wb,$	$W \rightarrow Wb$	

3. *Acrocentric chromosome grammar:*

$G_a = (V_{N_a}, V_{T_a}, P_a, S)$

where $V_{N_a} = \{S, A, B, D, E, F, L, R, W, G\}$, $V_{T_a} = \{a, b, c, d\}$, and P_a:

$S \rightarrow AA,$	$D \rightarrow WE,$	$R \rightarrow RE$
$A \rightarrow cB,$	$G \rightarrow FG,$	$W \rightarrow WE$
$B \rightarrow FL,$	$L \rightarrow aDa,$	$G \rightarrow d$
$B \rightarrow RE,$	$F \rightarrow b,$	
$E \rightarrow b,$	$L \rightarrow FL,$	
$D \rightarrow FG,$	$R \rightarrow aDa,$	$W \rightarrow d$

III. SYNTACTIC RECOGNITION PROCEDURES

If the grammar describing each pattern class is known, it is possible to construct a parser or syntax analyzer for each class. An input pattern is said to belong to a pattern class if it is in the set of language accepted by the corresponding parser. This approach requires a parsing procedure that is typically more complex than the direct sentence-to-sentence matching approach described below. It, however, allows us to obtain the structural information of the input pattern.

It is also possible to construct an automaton for a given grammar instead of constructing a parser. However, an automaton can only accept or reject an input sentence; it cannot generate an explicit structural description of the input pattern. An automaton works by reading the input sentence symbol by symbol. A next-state function maps the current input state into the next state depending on the value of the current input symbol. The input sentence is accepted if we reach one of the final states after reading the entire sentence. A deterministic finite-state automaton can be constructed for finite-state grammar. A nondeterministic automaton is usually required for a context-free grammar. Procedures for constructing automata are described extensively in many references [20,24,25] and are not discussed here.

An alternative approach for recognizing a syntactic pattern is based on string matching. In this approach the string representation of an input pattern is matched directly against some reference string(s) of a pattern class. The input pattern is determined to belong to a certain class if the *matching criterion* or the *similarity measure* is higher than a predetermined threshold *T*. This method is analogous to the *template-matching* approach frequently used in image analysis. This approach does not require that a grammar be constructed for each pattern class. It is useful for classifying an unknown pattern into the class to which it belongs but is not useful for obtaining the structural information of the input sentence.

For noisy images, the 1D string obtained for an input pattern may be distorted. In the direct sentence-to-sentence matching approach, the distortion is reflected in the similarity measure used. We can pick a threshold *T* to decide how much distortion is tolerable in classifying an input pattern. When it is desirable or necessary to use a parsing technique for noisy patterns, a parser with error-correcting capability can be used. An error-correcting parser accepts a distorted sentence by incorporating error production rules into its production set.

In the following, we present the recognition procedure using a parser; then we present definitions for sentence-to-sentence distance metrics, which are to be used in subsequent sections where the error-correcting parser and the direct sentence-to-sentence matching procedures are presented.

A. Syntactic Parsing

A straightforward approach to syntactic pattern recognition is to construct a grammar for each class of patterns. A parser can then be constructed based on the grammar G_i for each pattern class i. The parser will recognize a sentence only if it is in $L(G_i)$. For the chromosome example in Section II.A, a parser can be constructed for the medium, submedium, and acrocentric chromosome grammars. The recognition procedure can then be constructed as in Fig. 8.

Depending on the type of grammars, different parsers or recognizers can be constructed. In addition to the decision of accepting or rejecting a given string pattern, a derivation tree or a parse of the string is usually obtained, from which we obtain a complete description of the pattern, subpatterns, and primitives. For general context-free languages, we describe Early's [26] parsing algorithm:

Algorithm: *Early's Parsing Algorithm*

Input: An input string $w = w_1 w_2 \cdots w_n$ and a context-free grammar $G = (V_N, V_T, P, S)$

Output: The parse list $I_0, I_1, \ldots, I_n$ for w

Procedure: I_0 is constructed as follows:

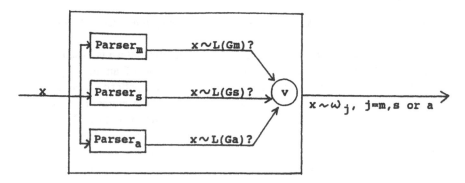

Figure 8   Block diagram of syntactic recognition using parsers.

1a.   If $S \rightarrow \alpha$ is in P, add item $[S \rightarrow \cdot \alpha, 0]$ to I_0.

Repeat (1b) and (1c) until no new items can be added to I_0:

1b.   If $[B \rightarrow \gamma \cdot, 0]$ is in I_0, add $[A \rightarrow \alpha B \cdot \beta, 0]$ for all $[A \rightarrow \alpha \cdot B\beta, 0]$ in I_0.
1c.   For all items $[A \rightarrow \alpha \cdot B\beta, 0]$ in I_0, add the item $[B \rightarrow \cdot \gamma, 0]$ to I_0 for each productions in P of the form $B \rightarrow \gamma$.

Then I_j is constructed from $I_0, I_1, \ldots, I_{j-1}$ as follows:

2a.   For each $[B \rightarrow \alpha \cdot a\beta, i]$ in I_{j-i} such that $\alpha = w_j$, add $[B \rightarrow \alpha a \cdot \beta, i]$ to I_j.

Repeat (2b) and (2c) until no new items can be added:

2b.   Let $[A \rightarrow \alpha \cdot, i]$ be an item in I_j. For each item in I_i of the form $[B \rightarrow \alpha \cdot A\beta, k]$, add $[B \rightarrow \alpha A \cdot \beta, k]$ to I_j.
2c.   For each item $[A \rightarrow \alpha \cdot B\beta, i]$ in I_j, add $[B \rightarrow \cdot \gamma, j]$ to I_j if $B \rightarrow \gamma$ is in P.

It should be noted that w is in $L(G)$ if and only if there exists some item of the form $[S \rightarrow \alpha \cdot, 0]$ in I_n. In general, this algorithm has time complexity proportional to the cube of the input sentence length. If the grammar is unambiguous, the parsing time is proportional to the square of the input sentence length. After constructing the parse lists, the derivation tree or parse for the input string can easily be extracted by a simple procedure [22].

Example.   Given the sentence $w = b \times b$ and the context-free grammar

$$G = (V_N, V_T, P, S)$$

where $V_N = \{S, T, F\}$, $V_T = \{b, +, \times, (,)\}$, and P:

$$S \rightarrow S + T, \quad S \rightarrow T$$
$$T \rightarrow T \times F, \quad T \rightarrow F$$
$$F \rightarrow (S), \quad F \rightarrow b$$

we obtain the following parse lists when Early's algorithm is applied:

I_0	I_1	I_2	I_3
$[S \rightarrow \cdot S + T, 0]$	$[F \rightarrow b \cdot, 0]$	$[T \rightarrow T \times \cdot F, 0]$	$[F \rightarrow b \cdot, 2]$
$[S \rightarrow \cdot T, 0]$	$[T \rightarrow F \cdot, 0]$	$[F \rightarrow \cdot (S), 2]$	$[T \rightarrow T \times F \cdot, 0]$
$[T \rightarrow \cdot T \times F, 0]$	$[S \rightarrow T \cdot, 0]$	$[F \rightarrow \cdot b, 2]$	$[S \rightarrow T \cdot, 0]$
$[T \rightarrow \cdot F, 0]$	$[T \rightarrow T \cdot \times F, 0]$		$[T \rightarrow T \cdot \times F, 0]$
$[F \rightarrow \cdot (S), 0]$	$[S \rightarrow S \cdot + F, 0]$		$[S \rightarrow S \cdot + T, 0]$
$[F \rightarrow \cdot b, 0]$			

Note that $[S \rightarrow T \cdot, 0]$ is in I_3 and therefore $b \times b$ is in $L(G)$.

B. Distance Metrics for Sentences

There are three commonly used distance metrics for syntactic recognition. The first is the Levenshtein distance between two strings [27]:

Definition 1. For two strings $x, y \in \Sigma^*$, we can define a transformation T: $\Sigma^* \rightarrow \Sigma^*$ such that $y \in T(x)$. The following three transformations are defined.

1. *Substitution error transformation*:

$$\omega_1 a \omega_2 \overset{T_S}{\vdash} \omega_1 b \omega_2 \qquad (1)$$

2. *Deletion error transformation*:

$$\omega_1 a \omega_2 \overset{T_D}{\vdash} \omega_1 \omega_2 \qquad (2)$$

3. *Insertion error transformation*:

$$\omega_1 \omega_2 \overset{T_I}{\vdash} \omega_1 a \omega_2 \qquad (3)$$

where $\omega_1, \omega_2 \in \Sigma^*$, $a, b \in \Sigma$, and $a \neq b$. The Levenshtein distance between two strings $x, y \in \Sigma^*$, $d^L(x, y)$, is defined as the smallest number of transformations required to derive y from x.

The second is the weighted Levenshtein distance:

Definition 2. The weighted Levenshtein distance between x and y, denoted as $d^w(x, y)$, is defined as

$$d^w(x, y) = \min_j \{ \sigma k_j + \gamma m_j + \delta n_j \} \qquad (4)$$

where k_j, m_j, and n_j are the number of substitution, deletion, and insertion error transformations, respectively, in J, where J denotes the sequence of transformations used to derive sentence y from sentence x.

The third metric defined below can reflect the difference of the same type of error made on different terminals.

Definition 3. Define the following weights for error transformations on a terminal a where ω_1 and $\omega_2 \in \Sigma^*$,

1. $S(a, b,)$ is the cost of substituting a for b [if $a = b$, then $S(a, b) = 0$]:

$$\omega_1 a \omega_2 \vdash \omega_1 b \omega_2 \quad \text{for } b \in \Sigma, \quad b \nmid a$$

2. $D(a)$ is the cost of deleting a from $\omega_1 a \omega_2$:

$$\omega_1 a \omega_2 \vdash \omega_1 \omega_2$$

3. $I(a, b)$ is the cost of inserting b in front of a:

$$\omega_1 a \omega_2 \vdash \omega_1 b a \omega_2 \quad \text{for } b \in \Sigma$$

4. $I'(b)$ is the cost of inserting a terminal b at the end of a string:

$$x \vdash xb \quad \text{for } b \in \Sigma$$

then the weighted distance between sentences x and y, $d^W(x, y)$, is defined as

$$d^W(x, y) = \{ |J| \} \tag{5}$$

where $|J|$ is the sum of the weights associated with the transformations in J (J is as defined above).

C. Error-Correcting Parsing

Distortion of an input sentence caused by segmentation errors and misrecognition of the primitives (and/or relation) often occur in practical applications. This will lead to the rejection of the input sentence by the parser that describes its pattern class, even when the distorted sentence differs from the correct sentence by one symbol. To resolve this, an error-correcting parser that tolerates distortion in the input sentence has been developed. A pattern grammar is first expanded to include all the possible errors into its productions. The language generated by the expanded grammar therefore covers all possible erroneous sentences in addition to the correct sentences. The distortion of an input sentence can be modeled by the error transformations we defined in Section III.B. Typically, segmentation errors correspond to deletion and insertion errors, and misrecognition of primitives (and/or relations) corresponds to substitution errors. For a given input string y and a given grammar G, a minimum-distance error-correcting parser (MDECP) searches for a sentence z in $L(G)$ such that

$$d(z,y) = \min_i \{d(z_i,y) | z_i \in L(G)\} \tag{6}$$

$d(z, y)$ is also referred to as the distance between $L(G)$ and y and is denoted as $d_1(L(G), y)$. The parser will also generate a parse Π, which consists of the smallest number of error productions. A sentence in $L(G)$ that satisfies the minimum-distance criterion can be generated from Π by eliminating the error productions.

Using the results from the MDECP, a minimum-distance two-class decision rule for classifying syntactic patterns can be obtained:

If $d(L(G_1), y) < d(L(G_2), y)$
$\quad\quad y \in \omega_1$
else
$\quad\quad y \in \omega_2$

This, of course, can be generalized for an m-class problem.

The algorithm for the construction of an expanded grammar is as follows:

Algorithm: Construction of Expanded Grammar
Input: A context-free grammar $G = (V_N, V_T, P, S)$
Output: A context-free grammar $G' = (V'_N, V'_T, P', S')$, where P' is a set of weighted productions
Procedure

1. Expand V'_N such that $V'_N = V_N \cup \{S'\} \cup \{E_a \mid a \in V_T\}$.
2. Let $\alpha_i \in V_N^*$, $b_i \in V_T$ and $E_{a_i} \in \{$new nonterminals in $V'_N\}$. Add $A \rightarrow \alpha_0 E_{b_1} \alpha_1 E_{b_2} \cdots E_{b_m} \alpha_m, 0$ to P' for each production in P of the form $A \rightarrow \alpha_0 b_1 \alpha_1 b_2 \cdots b_m \alpha_m, m \geq 0$, where 0 is the weight of the new production.
3. Add the following productions to P', where the numbers enclosed in parentheses represent the weights assigned to the production rule using the L, w, and W metrics, respectively:
   (a) $S' \rightarrow S$ (0, 0, 0);
   (b) $S' \rightarrow Sa$ (1, δ, $I'(a)$), for all $a \in V'_T$;
   (c) $E_a \rightarrow a$ (0,0,0), for all $a \in V_T$;
   (d) $E_a \rightarrow b$ (1, σ, $S(a, b)$), for all $a \in V_T$, $b \in V'_T$, and $b \neq a$;
   (e) $E_a \rightarrow \lambda$ (1, γ, $D(a)$), for all $a \in V_T$;
   (f) $E_a \rightarrow bE_a$ (1, δ, $I(a, b)$), for all $a \in V_T$, $b \in V'_T$.

In step 3 of the algorithm above, the production rules added in (b), (d), (e), and (f) are called error productions. Each of them represents one type of error transformation on a particular symbol in V_T.

The MDECP can be constructed based on the Earley's parsing algorithm:

Algorithm: MDECP

Input: An input string $w = w_1w_2 \cdots w_m$ in V_T^*; an expanded grammar $G' = (V_N', V_T', P', S')$

Output: (1) The parse list $I_0, I_1, \ldots, I_m$, for w, (2) x: the minimum-distance correction of y, and (3) $d(x, w)$: the distance between sentences x and y

Procedure

1. Add item $[E \rightarrow \cdot S', 0, 0]$ to I_j and let $j = 0$.
2. For all items $[A \rightarrow \alpha \cdot B\beta, i, \xi]$ in I_j, if $B \rightarrow \gamma, \eta$ is a production rule in P', add the item $[B \rightarrow \cdot \gamma, j, 0]$ to I_j.
3. For each item $[A \rightarrow \alpha \cdot i, \xi]$ in I_j, if $[B \rightarrow \beta \cdot A \gamma, k, \xi]$ is in I_i and no item of the form $[B \rightarrow \beta A \cdot \gamma, k, \phi]$ exists in I_j, then add the item $[B \rightarrow \beta A \cdot \gamma, k, \eta + \xi + \zeta]$ to I_j, where ζ is the weight for production rule $A \rightarrow \alpha$. If an item of the form $[B \rightarrow \beta A \cdot \gamma, k, \phi]$ exists in I_j, and if $\phi > \eta + \xi L \zeta$, replace ϕ by $\eta + \xi + \zeta$.
4. Increment j by 1. If $j > m$, go to (6); otherwise, continue.
5. Add $[A \rightarrow \alpha w_j \cdot \beta, i, \xi]$ to I_j for each item in I_{j-1} of the form $[A \rightarrow \alpha \cdot w_j\beta, i, \xi]$. Go to (2).
6. If item $[E \rightarrow S', 0, \xi]$ is in I_m, then $d(x, w) = \xi$, where x is the minimum-distance correction of w.

Note that in the algorithm above, we accumulate the weights associated with productions used in a derivation. The extraction of the parse for w can be done as in Early's algorithm, and x, the minimum-distance correction of w, can then be obtained by deleting all the error productions in w.

D. Direct Sentence-to-Sentence Matching

Using the distance metrics defined in Section III.B, a *nearest-neighbor* classification rule can be defined. Let C_1 and C_2 be two image patterns, represented by sentences $X_1 = \{x_1^1, x_2^1, \ldots, x_{n_1}^1\}$ and $X_2 = \{x_1^2, x_2^2, \ldots, x_{n_2}^2\}$, respectively. Then we can compute the distance between the unknown sentence y and each of the sentences in X_1 and X_2. y is then decided to belong to C_1 if

$$\min_j d(x_j^1, y) < \min_i d(x_i^2, y) \tag{7a}$$

else, y is decided to belong to C_2 if

$$\min_j d(x_j^1, y) > \min_i d(x_i^2, y) \tag{7b}$$

The following algorithm from [28] computes the distance between two strings as defined in Section III.C.

Algorithm: *Distance Between Two Strings*

   Input: Two strings $x = x_1, x_2, \ldots, x_n$, and $y = y_1, y_2, \ldots, y_m$

   Output: $d(x,y)$

   Procedure

1. dist $(0,0) = 0$.
2. For $i = 1$ to n

       dist $(i, 0) = $ dist $(i - 1, 0) + 1$

   For $j = 1$ to m

       dist $(0, j) = $ dist $(0, j - 1) + 1$

3. For $i = 1$ to n

     for $j = 1$ to m

     $e_1 = $ dist $(i - 1, j - 1) + 1$ if $x_i \neq y_j$

     or $e_1 = $ dist $(i - 1, j - 1)$ if $x_i = y_j$

     $e_2 = $ dist $(i - 1, j) + 1$

     $e_3 = $ dist $(i, j - 1) + 1$

     dist $(i,j) = $ min $[e_1, e_2, e_3]$

4. $d(x,y) = $ dist (n,m).

Alternatively, a *K-nearest-neighbor* classification rule can be devised. If you reorder the sentences in X_1 and X_2 such that $\bar{X}_i = \{\bar{x}_1^i, \bar{x}_2^i, \ldots, \bar{x}_n^i\}$ and $d(\bar{x}_j^i, y) \leq d(\bar{x}_l^i, y)$ if $j < l$, for all $1 \leq j, l \leq n_i$, then

$$y \in C_1 \quad \text{if} \quad \sum_{j=1}^{K} \frac{1}{K} d(\bar{x}_j^1, y) < \sum_{j=1}^{K} \frac{1}{K} d(\bar{x}_j^2, y)$$

$$\text{else, } y \in C_2 \tag{8}$$

The nearest-neighbor and the *K*-nearest-neighbor classification rules require that a set of syntactic image samples be grouped into clusters [e.g., clusters $\{x_i^1\}$ and $\{x_i^2\}$ in Eqs. (7) and (8)]. We describe a clustering algorithm below, where t is a design parameter that can be determined experimentally.

Algorithm: *Clustering Algorithm*

   Input: Pattern samples $X = \{x_1, x_2, \ldots, x_n\}$ and threshold t

   Output: Clusters $C_1, C_2, \ldots, C_m$

   Procedure: Let $m = 1$ and assign x_1 to C_1.

   For $j = 2$ to n,

If $\min_i \{ \min_l d(x_l^i, x_j) \} \leq t$

     then assign x_j to C_i

else

     $m = m + 1$ and assign x_j to C_m

Another clustering algorithm based on the definition of a cluster center is given in [7,29]. The cluster center is defined in terms of a β metric for a sentence x_j^i (in cluster C_i):

$$\beta_j^i = \frac{1}{n_i} \sum_{l=1}^{n_i} d(x_j^i, x_l^i) \tag{9}$$

$A_i = x_j^i$ is the cluster center (or representative) of C_i if $\beta_j^i = \min_l \{\beta_l^i \mid 1 < l < n_i\}$. The algorithm is as follows:

Algorithm: *Clustering Based on Cluster Center*
   Input: Pattern samples $X = \{x_1, x_2, \ldots, x_n\}$
   Output: Clusters $C_1, C_2, \ldots, C_m$
   Procedure

1. For each $i = 1$ to m, randomly pick a sample from X, assign it to C_i, and let the sample be the cluster center A_i of C_i.
2. For each $j = 1$ to n,

    if $d(A_i, x_j) = \min_l [d(A_l, x_j]$

    then assign x_j to C_i
3. For each $i = 1$ to m, update the cluster center A_i. If no A_i changes, stop; else, go to step 2.

IV. TWO-DIMENSIONAL GRAMMARS FOR IMAGE PATTERNS

The one-dimensional string representation of an image pattern can only represent one type of relation between primitives, that is, concatenation. This requires that a primitive be either connected at the right or at the left. This is useful for a wide class of problems but for some image recognition problems, it may be more desirable to represent and describe two-dimensional (2D) relationships among primitives. There are several 2D grammars proposed for syntactic analysis: tree grammars [2,30], web grammars [31], and PDL (picture description language) [32,33]. We describe the tree grammar in this section.

A. Tree Grammars

We first present some notation: N^+ is the set of positive integers greater than zero, U the universal tree domain where a node (including root node) in the tree can have an arbitrary number of child nodes, and $\langle \Sigma, r \rangle$ a ranked alphabet where Σ is a finite set of symbols and $r: \Sigma \rightarrow N = N^+ \cup \{0\}$; $r(a)$ is called the rank of a, for some $\alpha \in \Sigma$. Let $\Sigma_N = r^{-1}(n)$. A tree over Σ (i.e., over $\langle \Sigma, r \rangle$) can be defined as a function

$$\alpha : D \rightarrow \Sigma$$

such that D is a tree domain and

$$r\,[\alpha(a)] \;=\; \max\,\{i \mid a \cdot i \in D\}$$

$D\,(\alpha)$ is called the domain of a tree α and T_Σ is the set of all trees over Σ. A subtree of tree α at a can be defined as

$$\alpha/a \;=\; \{(b,x) \mid (a \cdot b, x) \in \alpha\}$$

where a is a member of $D\,(\alpha)$.

A tree grammar can then be defined as follows:

Definition 4. A tree grammar $G_t \;=\; (G, r', P, s)$ is a 4-tuple over $\langle V_T, r \rangle$ such that:

1. $\langle V, r' \rangle$ is a finite ranked alphabet with $V_T \subseteq V$ and $r'/V_T = r$.
2. The set of nonterminals $V_N = V - V_T$.
3. P is the set of productions of the form $\Phi \rightarrow \Psi$, where Φ and Ψ are trees over $\langle V, r' \rangle$.
4. The start symbol S is a subset of T_V, where T_V is the set of trees over alphabet V.

We also have the following definition:

Definition 5. We say that $\alpha \overset{a}{\Rightarrow} \beta$ is in G_t if and only if $\Phi \rightarrow \Psi$ exists in the production set of G_t such that Φ is a subtree of α at a and β is obtained by replacing the occurrence of Φ at a by Ψ. Also, $\alpha \Rightarrow \beta$ in G_t denotes that there exists $a \in D\,(\alpha)$ such that $\alpha \overset{a}{\Rightarrow} \beta$.

Based on the definition above, we say that $\alpha \overset{*}{\Rightarrow} \beta$ is in G_t if and only if there exists $\alpha_0, \alpha_1, \ldots, \alpha_m$, in G_t, with $m > 0$, such that

$$\alpha \;=\; \alpha_0 \Rightarrow \alpha_1 \Rightarrow \cdots \Rightarrow \alpha_m \;=\; \beta$$

The sequence $\alpha_0, \ldots, \alpha_m$ is called a derivative of β from α. The tree language generated by G_t is therefore

$$L(G_t) \;=\; \{\alpha \in T_{V_T} \mid \text{there exists } Y \in s \text{ such that } Y \overset{*}{\Rightarrow} \alpha \text{ in } G_t\}$$

As an example, the following tree grammar generates the rectangular grid patterns shown in Fig. 9:

$$G_t \;=\; (V, r, P, S)$$

where $V = (S, h, v, \$, A, B)$ is the set of symbols $V_T = \{\overset{h}{\rightarrow}, \uparrow v, \cdot \$\}$, is the set of terminal symbols with $r(h) = \{2,1,0\}$, $r(v) = \{2,1,0\}$, and $r(\$) = 2$, and P is the set of productions:

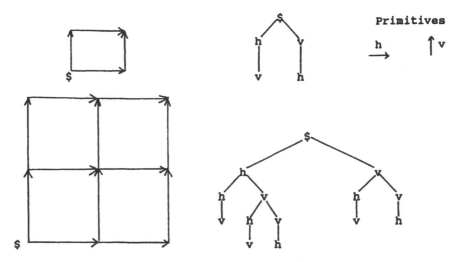

Figure 9   Rectangular grid patterns generated by a tree grammar.

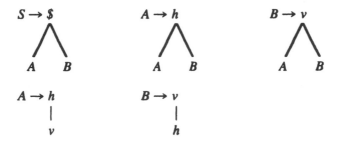

We also say that a tree grammar $G_t = (V,r,P,S)$ is of *expansive* form if (and only if) each production in P is of the special form

$$X_0 \rightarrow \overset{x}{\bigwedge_{X_1 \cdots X_{r(x)}}} \quad \text{or} \quad X_0 \rightarrow x$$

where the X_i's are nonterminal symbols and x is a terminal symbol. It can be shown that we can always construct an equivalent expansive grammar G_t' for every regular tree grammar G_t such that $L(G_t') = L(G_t)$.

For recognizing or accepting regular tree languages, tree automata have been developed. In fact, we can always construct a tree automaton M_t for every regular tree grammar G_t such that $T(M_t) = L(G_t)$ [2], where $T(M_t)$ is the set of trees accepted by M_t and $L(G_t)$ is the set of trees generated by G_t. Formally, a tree automaton can be defined as follows:

Definition 6. A tree automaton $M_t = (Q, f_1, \ldots, f_k, F)$ is a $(k + 2)$-tuple over Σ, where (a) Q is a finite set of states: (b) each f_i is a relation on $Q^{r(\sigma_i)} \times Q$, with $\sigma_i \in \Sigma$, that is, $f_i : Q^{r(\sigma_i)} \to Q$; and (c) F is a set of final states $(F \subseteq Q)$.

With this definition, the set of trees accepted by a tree automaton M_t can be described as

$$T(M_t) = \{\alpha \in T_\Sigma \mid \text{there exists } x \in F \text{ such that } \rho(\alpha) \sim X\}$$

where ρ is the *response relation* of a tree automaton M_t defined as follows:

(a) For $\sigma \in \Sigma_0$, $\rho(\sigma) \sim X$ if and only if $f_\sigma \sim X$; in other words, $\rho(\sigma) = f_\sigma$

(b) For $\sigma \in \Sigma_n$, $n > 0$, $\rho(\sigma, x_0, \ldots, x_{n-1}) \sim X$ if and only if there exists x_i, $0 \le i \le n - 1$ such that $f_\sigma(x_0, \ldots, x_{n-1}) \sim X$ and $\rho(x_i) \sim X_i$, $1 \le i \le n$; in other words, $\rho(\sigma, x_0, \ldots, x_{n-1}) = f_\sigma(\rho(x_{n-1}))$.

To construct a tree automaton M_t from G_t, the following algorithm is presented:

Algorithm: Construction of Tree Automaton M_t

1. Convert $G_t = (V, r, P, S)$ (over alphabet V_T) to expansive form $G_t' = (V', r, P', S)$.
2. Construct $M_t = (V' - V_T, f_1, \ldots, f_k, \{S\})$ where $f_X (X_1, \ldots, X_n) \sim X_0$ if $X_0 \to x X_1, \ldots, X_n$ is in P'.

As an example, the tree automaton for the rectangular-grid patterns in Fig. 9 can be constructed from its tree grammar as

$$M_t = (Q, f_h, f_v, f_s, F)$$

where the set of states $Q = \{q_h, q_v, q, q_F\}$, the set of relations $f =$

$$f_h \sim q_h, \quad f_h(q, q) \sim q, \quad f_h(q_v) \sim q,$$
$$f_v \sim q_v, \quad f_v(q, q) \sim q, \quad f_v(q_h) \sim q, \quad f_s(q,q) \sim q_F$$

and the set of final states $F = \{q_F\}$.

V. APPLICATION EXAMPLES

In this section we present some application examples for syntactic recognition. We first present an example of waveform peak detection in details, then summarize other areas of applications with references.

A. Waveform Peak Recognition

Horowitz [10] presented a syntactic approach to waveform peak recognition. An analog waveform is represented using a sequence of waveform primitives, each a segment from the piecewise linear approximation of the waveform. Based on the mathematical characterization of a waveform peak, we can derive a grammar to represent a waveform containing peaks.

The piecewise-linear approximation of a waveform can be obtained by using a variation of the "split-and-merge" algorithm [34]:

1.   Divide the waveform into n equal-sized intervals.
2.   Calculate the error of linear mean-square approximation for each interval.
3.   Split those intervals so that the greatest decrease in the sum of the squares of the interval error values (E) is obtained. Stop when E is less than some specified error threshold T_E.
4.   Merge neighboring intervals that yield the smallest increase in E. Stop when E is equal to or greater than T_E.
5.   Adjust the interval endpoints to minimize E. Go to step 4 if any adjustments are made; else, each resulting interval is a waveform segment.

1.   Mathematical Characterization

After the piecewise-linear approximation, a waveform can be represented by a discrete set of points (x_1, y_1), (x_2, y_2), . . . , (x_i, y_i), . . . , (x_n, y_n), with each $((x_i, y_i),(x_{i+1}, y_{i+1}))$ pair representing a linear waveform segment. A peak of a waveform can be described as a triple:

$$[\text{left boundary, local peak, right boundary}] = [(x^l,y^l),(x^m,y^m),(x^r,y^r)] \qquad (10)$$

with $x^l < x^m < x^r$. There are a set of constraints that the triple must satisfy simultaneously. We list the constraints for positive peaks below.

1.   Local maximum property: For all i, $x^l \leq x_i \leq x^r$, $y^m \geq y_i$.
2.   Left side of peak has nonnegative slope: For all i, $x^l \leq x_i \leq x^m$, $y_i \leq y_{i+1}$.
3.   Right side of peak has nonpositive slope: For all i, $x^m \leq x_i \leq x^r$, $y_i \geq y_{i+1}$.
4.   Left and right boundaries:
     a.   There exists an i such that $x_i = x^l$, $y_{i-1} \geq y^l$.
     b.   There exists an i such that $x_i = x^r \leq y_{i+1}$.
5.   Flat horizontal peak (x^m equals midpoint of horizontal segment): There exists i, j such that $x_i \neq x_j$, $y_i = y_j = y^m$ $\{(y_{i-1} < y_i)$ and $(y_i > y_{j+1})$, and $x^m = (x_i + x_j)/2\}$.
6.   Flat peak boundary (fix x^l as left endpoint and x^r as right endpoint):
     a.   There exists an i such that $x_i = x^l$, $y^l < y_{i+1}$.
     b.   There exists an i such that $x_i = x^r$, $y_{i-1} > y^r$.
7.   Forcing a peak to be as wide as possible (also preventing multiple occurrences of x^l and x^r and hence the nesting of two positive peaks):
     a.   There exists i,j such that $x_i = x^l$, $x_j \leq x^l$, $y_j = y^l[(y_{j-1} > y_j) \rightarrow$ for all k: $x_j \leq x_k \leq x_i$ $(y_k \leq y^l)]$.
     b.   There exists i, j such that $x_i = x^r$, $x_j \geq x^r$, $y_j = y^r$ $[(y_j < y_{j+1}) \rightarrow$ for all k: $x_i \leq x_k \leq x_j(y_k = y^r)]$.

A set of similar constraints can be defined for negative peaks.

2.   Waveform Grammar

Let us define the primitive *p*, *n*, and 0 for denoting the slope characteristic of a waveform segment, where *p* denotes *positive* slope, *n* denotes *negative* slope, and 0 denotes *zero* slope. The slope of a segment can easily be computed as

$$\frac{y_{i+1} - y_i}{x_{i+1} - x_i} \tag{11}$$

A waveform can then be represented as a sentence

$$\omega = \omega_1\omega_2 \cdot \cdot \cdot \omega_i \cdot \cdot \cdot \omega_n$$

where $\omega_i = p$, *n*, or 0. An example is given in Fig. 10.

A regular expression for the left side of a positive peak is given by

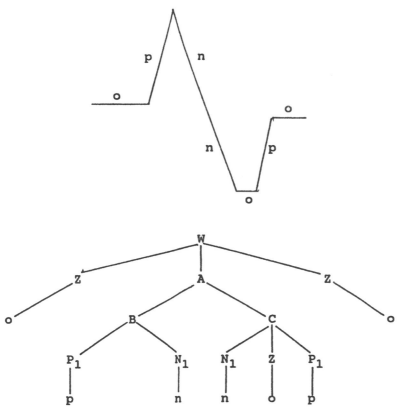

Figure 10   Example of a waveform and its syntactic representation.

$$P = p \mid p(p \mid 0)*p \tag{12}$$

where "|" represents "or" and "*" represents "reflexive-transitive closure" of the set. This expression is obtained based on the fact that the left side of a positive peak may contain any number (possibly zero) of the symbols p and 0 (by constraint 2 in Section V.A.1), and that it must start with the symbol p (constraint 6a) and end with the symbol p (constraint 1). Similarly, the regular expression for the right side of a positive peak is given by

$$Q = n \mid n \; (n \mid 0)*n \tag{13}$$

This is obtained using constraints 1, 3, and 6b.

From Eqs. (12) and (13), we can easily obtain the expression for a positive peak as

$$R = P0*Q \tag{14}$$

and for a negative peak as

$$S = Q0*P \tag{15}$$

where 0* is for the horizontal line on a flat peak. Based on these, any waveform can be considered to be an alternating sequence of positive and negative peaks:

$$W = 0* \cdots R0*S0*R0*S \cdots 0* \tag{16}$$

A deterministic context-free grammar G can then be constructed for the recognition of positive and negative peaks (if any) in a waveform:

$$G = (V_N, V_T, P, W)$$

where $V_N = \{W, A, B, C, P_1, N_1, P_2, N_2, Z\}$ is the set of nonterminals, $v_T = \{p, n, 0\}$ is the set of terminals, and the set of productions P is as follows:

$W \to ZAZ$
$W \to ZA$
$W \to AZ$
$W \to A$
$W \to Z$
$A \to B$
$A \to C$
$A \to P_1$
$A \to N_1$
$B \to CZN_1$
$B \to CN_1$
$B \to P_1ZN_1$
$B \to P_1N_1$
$C \to BZP_1$
$C \to BP_2$

$$C \rightarrow N_1 Z P_1$$
$$C \rightarrow N_1 P_1$$
$$P_1 \rightarrow P_1 p$$
$$P_1 \rightarrow P_2 p$$
$$P_1 \rightarrow p$$
$$N_1 \rightarrow N_1 n$$
$$N_1 \rightarrow N_2 n$$
$$N_1 \rightarrow n$$
$$P_2 \rightarrow P_1 \, 0$$
$$P_2 \rightarrow P_2 \, 0$$
$$N_2 \rightarrow N_1 \, 0$$
$$N_2 \rightarrow N_2 \, 0$$
$$Z \rightarrow Z \, 0$$
$$Z \rightarrow 0$$

A parser can be constructed based on the grammar G and a positive/negative peak is recognized if a sentence can be reduced to contain nonterminals B/C.

B. Other Applications

Syntactic approach has been applied successfully for texture analysis [35–40]. A texture is modeled as a repetitive pattern of texture elements. Depending on the problems, different texture elements can be defined. The repetitive pattern is described by a set of grammar rules. Zucker [41] modeled texture as an ideal texture and its transformation rules, where the ideal texture is a highly structured perfect pattern. Carlucci [37] describes a *texture language* for the description of texture made up of polygons or open polygonals. Lu and Fu [38,42] divide a texture pattern into fixed-sized windows. A tree grammar is then constructed to characterize the windowed patterns.

Anderson [43,44] formulated a simple precedence grammar [2] for the recognition of two-dimensional mathematical equations. The grammar can handle a mathematical equation consisting of letters, numbers, and the following special symbols:

$$\|- \quad -\| \quad \{ \quad \} \quad |- \quad -| \quad [$$
$$] \quad (\quad) \quad : \quad ; \quad , \quad =$$
$$\infty \quad \downarrow \quad \uparrow \quad \int \quad \sqrt{} \quad \Pi \quad \Sigma$$
$$+ \quad - \quad \cdot \quad / \quad \sin \quad \cos \quad \tan$$
$$\cot \quad \sec \quad \csc \quad \log \quad \exp$$

The syntactic approach has been used in the recognition of Chinese characters [45–52]. The primitives used are usually strokes of the characters, and rules are constructed to describe the relation between them. Chou et al. [53] presented an optical recognition system for handprinted Chinese characters based on a string-matching principle and inductive learning scheme. The system scans a character instance from four different views to obtain its peripheral segment information, which, in turn, is represented by a string. Stockman et al. [54] have applied syntactic approach to the analysis of carotid pulse waves. Giese et al. [55] applied syntactic approach to the analysis of electroencephalograms (EEGs). Other applications of syntactic approach include contour shape analysis [10,56–58], chromosome image analysis, bubble chamber picture identification [2,10], and automatic inspection [59].

VI. CONCLUDING REMARKS

We have presented the syntactic approach to image pattern recognition. This approach is used when it is desirable to describe as well as classify an input image pattern. Primitive patterns are extracted from the input image and represented by symbols. The structural relationship between primitives (or symbols) is described using a string or two-dimensional grammar. Using parsing techniques, we can classify as well as obtain a description (via a parse) of the input pattern. Using error-correcting parsing techniques, we can classify and describe a noisy or distorted pattern. When it is only necessary to classify an input pattern— without generating a description—automata or sentence-to-sentence matching (and clustering) algorithms can be used. Finally, we presented application examples of syntactic pattern recognition with references.

It should be pointed out that the basic building blocks of a syntactic pattern recognition system are analogous to those of a statistical system. In the block diagrams of Figs. 6 and 1, primitive selection and recognition correspond to feature selection and extraction, structural analysis corresponds to feature space analysis (or classification), and grammatical inference corresponds to learning. Furthermore, many techniques and algorithms developed for syntactic pattern recognition are analogous to those in the statistical approach. For example, sentence-to-sentence clustering corresponds to characteristic feature clustering, nearest-neighbor and k-nearest-neighbor classification rules for sentences correspond to nearest-neighbor and k-nearest-neighbor classification rules for features, and direct sentence-to-sentence matching corresponds to direct feature-to-feature matching.

Although we consider the statistical approach and the syntactic approach as two distinct approaches to pattern recognition, they are actually complementary to each other in practical applications. For example, in character recognition, a statistical approach may be used for the recognition of individual strokes, which in turn are used as primitive patterns in a syntactic recognizer that analyzes the structural composition of the character. On the other hand, a recognition system can be primarily a statistical system, but utilizing some simple structural information can greatly increase the classification rate. In such a case, a simple structural recognizer can be incorporated into a statistical classifier. In short, a good combination of the two approaches will, in many cases, result in an efficient and practical system.

REFERENCES

1. K. Fukunaga, *Introduction to Statistical Pattern Recognition*, Academic Press, New York, 1972.
2. K. S. Fu, *Syntactic Pattern Recognition and Applications*, Prentice Hall, Englewood Cliffs, N.J., 1982.
3. K. S. Fu, *Sequential Methods in Pattern Recognition and Machine Learning*, Academic Press, New York, 1968.
4. J. M. Mendel and K. S. Fu, eds., *Adaptive, Learning, and Pattern Recognition Systems: Theory and Applications*, Academic Press, New York, 1970.
5. E. A. Patrick, *Fundamentals of Pattern Recognition*, Prentice Hall, Englewood Cliffs, N.J., 1972.
6. R. O. Duda and P. E. Hart, *Pattern Classification and Scene Analysis*, Wiley, New York, 1973.
7. K. S. Fu, *Digital Pattern Recognition*, Springer-Verlag, New York, 1976.
8. K. S. Fu, *Syntactic Methods in Pattern Recognition*, Academic Press, New York, 1974.
9. T. Pavlidis, *Structural Pattern Recognition*, Springer-Verlag, New York, 1977.
10. K. S. Fu, ed., *Syntactic Pattern Recognition Applications*, Springer-Verlag, New York, 1977.
11. R. Z. Gonzalez and M. G. Thomason, *Syntactic Pattern Recognition: An Introduction*, Addison-Wesley, Reading, Mass., 1978.
12. E. S. Gelsema and L. N. Kanal, eds., *Pattern Recognition in Practice*, North-Holland, Amsterdam, 1980.
13. K. S. Fu, Recent developments in pattern recognition, *IEEE Trans. Comput.* C-29:845–854 (Oct. 1980).
14. R. C. Gonzalez and P. Wintz, *Digital Image Processing*, Addison-Wesley, Reading, Mass., 1987.
15. A. K. Jain, *Fundamentals of Digital Image Processing*, Prentice Hall, Englewood Cliffs, N.J., 1989.
16. J. Serra, *Image Analysis and Mathematical Morphology*, Academic Press, New York, 1982.
17. C. R. Giardina and E. R. Dougherty, *Morphological Methods in Image and Signal Processing*, Prentice Hall, Englewood Cliffs, N.J., 1988.
18. R. M. Haralick, S. R. Sternberg, and X. Zhuang, Image analysis using mathematical morphology, *IEEE Trans. Pattern Anal. Mach. Intelligence*, PAMI-9:532–550 (July 1987).
19. K. S. Fu and J. K. Mui, A survey on image segmentation, *Pattern Recognition* 13(1):3–16 (1981).
20. J. E. Hopcroft and J. D. Ullman, *Formal Languages and Their Relation to Automata*, Addison-Wesley, Reading, Mass., 1969.

21. S. Ginsburg, *The Mathematical Theory of Context-Free Languages*, McGraw-Hill, New York, 1966.

22. A. V. Aho and J. D. Ullman, *The Theory of Parsing, Translation, and Compiling*, Vol. 1: *Parsing*, Prentice Hall, Englewood Cilffs, N.J., 1972.

23. H. C. Lee and K. S. Fu, A stochastic syntax analysis procedure and its application to pattern classification, *IEEE Trans. Comput.*, C-21:660–66 (1972).

24. R. J. Nelson, *Introduction to Automata*, Wiley, New York (1968).

25. M. A. Harrison, On the relation between grammars and automata, *Adv. Inform. Systems Sci.*, 4:39–92 (1972).

26. J. Earley, An efficient context-free parsing algorithm, *Comm. ACM*, 13:94–102 (1970).

27. V. I. Levenshtein, Binary codes capable of correcting deletions, insertions and reversals, *Sov. Phys. Dokl.*, 10 (Feb. 1966).

28. R. A. Wagner and M. J. Fisher, The string to string correction problem, *J. Assoc. Comput. Mach.* 21(1):168–173 (1974).

29. S. Y. Lu and K. S. Fu, A sentence-to-sentence clustering procedure for pattern analysis, *IEEE Trans. Systems Man Cybernet.*, SMC-8(5):381–389 (1978).

30. K. S. Fu, Tree languages and syntactic pattern recognition, in *Pattern Recognition and Artificial Intelligence*, C. H. Chen, ed., Academic Press, New York, 1976.

31. J. L. Pfaltz and A. Rosenfeld, Web grammars, *Proc. First International Joint Conference on Artificial Intelligence*, Washington, D.C., 1969.

32. A. C. Shaw, The formal description and parsing of pictures, *SLAC Report 84*, Stanford Linear Accelerator Center, Stanford, Calif., 1968.

33. A. C. Shaw, The formal picture description scheme as a basis for picture processing systems, *Inform. Control*, 14:9–52 (1969).

34. T. Pavlidis and S. L. Horowitz, Segmentation of plane curves, *IEEE Trans. Comput.*, C-23:860–70 (Aug. 1974).

35. B. S. Lipkin and A. Rosenfeld, eds., *Picture Processing and Psychopictorics*, Academic Press, New York, 1970.

36. J. M. S. Prewitt, Tissues as texture: a syntactic approach to a syntactic problem, *International Conference on Pattern Recognition Cell Images*, Chicago, May 1979.

37. L. Carlucci, A formal system for texture languages, *Pattern Recognition*, 4:53–72 (1972).

38. S. Y. Lu and K. S. Fu, A syntactic approach to texture analysis. *Comput. Graphics Image Process.*, 7:303–330 (1978).

39. F. Tomita, Y. Shirat, and S. Tsuji, Classification of textures by structural analysis, *Proc. 4th International Joint Conference on Pattern Recognition*, Kyoto, Japan, Nov. 8–11, 1978.

40. S. N. Jayaramamurthy, Multilevel array grammars for generating texture scenes, *Proc. 1979 IEEE Computer Society Conference on Pattern Recognition and Image Processing*, Chicago, Aug. 6–8, 1979.

41. S. W. Zucker, Toward a model of texture, *Comput. Graphics Image Process.*, 5:190–202 (1976).

42. K. S. Fu and S. Y Lu, Computer generation of texture using a syntactic approach, *Proc. ACM SIGGRAPH Conference*, Atlanta, Ga., Aug. 21–25, 1978.

43. R. H. Anderson, Syntax-directed recognition of hand-printed two-dimensional mathematics, Ph.D. thesis, Division of Engineering and Applied Physics, Harvard University, Jan. 1968.

44. R. H. Anderson, Two-dimensional mathematical notation, in *Syntactic Pattern Recognition Applications*, K. S. Fu, ed., Springer-Verlag, New York, 1977.

45. W. W. Stallings, Recognition of printed Chinese characters by automatic pattern analysis, *Comput. Graphics Image Process.*, 1(1):47–65 (1972).

46. O. Fudimura and Y. Kagaya, Structure patterns of Chinese characters, *Annual Bulletin 3*, Research Institute of Logopedics and Phoniatrics, University of Tokyo, Japan, Apr. 1968–July 1969, pp. 131–148.

47. T. Sakai, M. Nagao, and H. Terai, A description of Chinese characters using sub-patterns, *Inform. Process.* (Jpn.), 10(5) (1969).

48. B. K. Rankin and Tan, Component combination and frame-embedding in Chinese character grammars, *NBS Technical Note 492*, Feb. 1970.

49. S. K. Chang, An interactive system for Chinese character generation and text editing, *Proc. International Conference of the Cybernetics Society*, Washington, D.C., Oct. 9–12, 1972.

50. W. W. Stallings, Chinese character recognition, in *Syntactic Pattern Recognition Applications*, K. S. Fu, ed., Springer-Verlag, New York, 1977.

51. K. S. Fu and S. Y. Lu, Applicability of pattern handling methods to Chinese language processing, *Technical Report TR-EE:79–16*, Purdue University, West Lafayette, Ind., Apr. 1979.

52. S. H. Lam, Efficient encoding of high-density Chinese character fonts by stroke composition, S.M. thesis, MIT, June 1980.

53. R. I. Chou, A. Kershenbaum, and E. K. Wong, Representation and recognition of handprinted Chinese characters by string-matching, *Inform. Sci.* (to appear).

54. G. Stockman, L. N. Kanal, and M. C. Kyle, Structural pattern recognition of carotid pulse waves using a general waveform parsing systems. *Comm. ACM*, 19:688–695 (Dec. 1976).

55. D. A. Giese, J. R. Bourne, and J. W. Ward, Syntax analysis of electroencephalogram, *IEEE Trans. Systems Man Cybernet.*, SMC-9:429–434 (Aug. 1979).

56. T. Pavlidis, Methodologies for shape analysis in *Biomedical Pattern Recognition and Image Processing*, K. S. Fu and T. Pavlidis, eds., Verlag Chemie, Weinheim, Germany, 1979.

57. T. Pavlidis and F. Ali, A hierarchical syntactic shape analyzer, *IEEE Trans. Pattern Anal. Mach. Intelligence*, PAMI-1:2–9 (Jan. 1979).

58. K. C. You and K. S. Fu, A syntactic approach to shape recognition using attributed grammars, *IEEE Trans. Systems Man Cybernet.*, SMC-9:334–345 (June 1979).

59. G. B. Porter and J. L. Mundy, Visual inspection system design, *Computer*, 13:40–49 (May 1980).

6

Heuristic Parallel Approach for 3D Articulated Line-Drawing Object Pattern Representation and Recognition

P. S. P. Wang

Northeastern University
Boston, Massachusetts

I. INTRODUCTION

Normally, one will encounter three types of objects in dealing with three-dimensional (3D) pattern recognition and computer vision problems: (1) rigid objects, which normally do not change shape, such as a piece of rock, pyramid, or dining table; (2) nonrigid (flexible, deformable) objects, which can easily change shapes, such as a handkerchief, rubber band, or piece of paper; and (3) objects whose nature and properties are between the first two types, that is, they can change their shape (appearance) partially in the sense that some portion of the object will change its relation (e.g., angle, distance, length) relative to the other portions, but each portion itself will remain rigid. Many interesting objects, known as *articulated* objects, fall into this category: boxes, swinging office chairs, refrigerators, and pairs of scissors. This class of objects is of special importance since it includes most industrial robots and man-made factory tools. Furthermore, on the many occasions in daily life when a robot tries to interact intelligently and effectively with its environment, it must not only recognize and locate objects but must understand and describe the situation or status of the objects. For example, if a robot is assigned to fetch a frozen pizza and a bottle of milk, it should be able not only to recognize and locate the refrigerator, but also tell which part is the freezer (where the frozen pizza is stored), which part is the cooling unit (where the bottle of milk is located), and whether the doors are open or closed.

Although interesting and important, the characteristics and patterns of articulated objects are more complicated than those of rigid objects. Recognizing such objects using computers is difficult and challenging, and most of the work done to date has concentrated on the recognition of rigid objects (see [4,7] for comprehensive reviews and [6,8,11,16,21] for some more recent work). To date very little work has been done on articulated object recognition. The first attempt to tackle the problem was probably by Brooks [5] in his ACRONYM system. Grimson extended the interpretation tree approach to deal with 2D objects with rotating subparts [10]. Goldberg and Lowe [9] extended Lowe's system to deal with 3D articulated objects such as staplers. Beinglass and Wolfson [3] have reviewed various factors in this area and generalized the generalized Hough transform (GHT) to recognize single-joint articulated objects such as pairs of scissors. Yet it remains to be shown how to overcome the added limitations of the GHT method, as usually happens when an existing object recognition technique is extended to handle articulated objects [3].

This research is aimed at exploring how a robot or computer can represent, recognize, and describe a 3D object, modeled by wireframe line drawings (which Sugihara calls "noble class of images" [15]) in which the phenomenon of 3D interpretations holds for many practical and interesting images in realistic world. In [12,13], Marill did a review of various approaches to 3D perception and recognition of line-drawing objects and proposed an algorithm called minimum standard deviation of angles (MSDA) for recovering 3D objects from 2D images. Recently, Baird and Wang [1] introduced a more efficient algorithm called the *gradient descent* method, which can convert a 2D line-drawing image to a 3D line-drawing object with the same effect as perceived by human beings, but reduces the time complexity from Marill's quadratic time $O(m^2)$ *to linear time $O(m)$*, where m is the number of nodes in the line-drawing objects. Therefore, it became possible to handle more complicated line-drawing 3D objects. So far, however, all these efforts were aimed at recognizing 3D nondeformable rigid line-drawing objects.

For the purpose of illustrating ideas proposed in this chapter for articulated-objects recognition, it is assumed that the input is a 3D line-drawing image with (x, y, z) coordinates of nodes and lines connecting pairs of nodes. We introduce the concept of *coordinated graph*, *layered graph representation*, and *parallel pattern matching* to tackle more complicated 3D line-drawing objects [23,24]. Such a method, although simple, is very robust for representing, recognizing, and describing not only rotations, displacements, scaling, and topological variations [14,18], but also different states for many interesting 3D objects, including *articulated* objects. In Section II we introduce some basic definitions and notations and some basic concepts of states of *articulated* objects. In Section III a 3D object recognition scheme is designed with several examples illustrated. Finally, some future research is discussed in Section IV. A preliminary version of this paper was presented in [22].

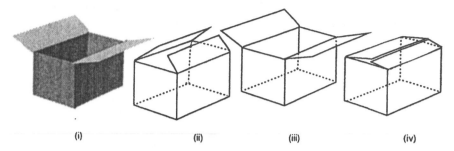

Figure 1   Example of an articulated object (i), a box, and its line-drawing wireframe model in three states: (ii) half open, (iii) wide open, and (iv) almost closed.

II. PRELIMINARIES: NOTATIONS AND DEFINITIONS

Let's take a look at the objects in Fig. 1. From what we see, they look quite different, yet they are the same object, that is, a box with three different *states*: half open, wide open, and almost closed. This is from human point of view—how about from the computer or robot point of view? How can we design a method to make a computer or robot do (learn, perceive, understand, describe, and recognize such objects) as human beings do? This has to deal with a very important concept in pattern recognition and computer vision known as *pattern representation*. Here we propose a heuristic parallel approach to tackle such problems. We adapt the definitions and notation from [19].

Definition.   A 3D *coordinated graph* (3Dcg) in 3D Cartesian space is a graph $G = (V, E)$, where V is a nonempty finite set of nodes (vertices), each denoted by a Cartesian product (x, y, z), and E is a finite nonempty set of edges (branches, lines) connecting pairs of nodes in E that are neighbors. Each line is denoted by $(\mathbf{a}, \mathbf{b})$ if $\mathbf{a}$ and $\mathbf{b}$ are two nodes and are directly connected.

   Note that each node is a vector in the 3D space, and given a pair of connected nodes (a line) (a, b), it is readily computed the vector from $\mathbf{a}$ to $\mathbf{b}$, denoted as $\mathbf{v_{ab}}$ or $\mathbf{b} - \mathbf{a}$, as shown in Fig. 2. Given two lines $\mathbf{v_1} = \mathbf{v_{ab}} = (\mathbf{a}, \mathbf{b})$ *and*

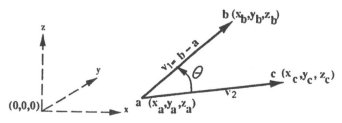

Figure 2   Nodes, vectors, lines, and angle θ between lines in 3D Cartesian space.

$\mathbf{v}_2 = \mathbf{v_{ac}} = (\mathbf{a}, \mathbf{c})$, where $\mathbf{a} = (x_a, y_a, z_a)$, $\mathbf{b} = (x_b, y_b, z_b)$, and $\mathbf{c} = (x_c, y_c, z_c)$, the angle θ between the two lines can be computed as

$$\theta = \cos^{-1} \frac{\mathbf{v}_1 \cdot \mathbf{v}_2}{|\mathbf{v}_1||\mathbf{v}_2|} =$$

$$\cos^{-1} \frac{(x_b - x_a)(x_c - x_a) + (y_b - y_a)(y_c - y_a) + (z_b - z_a)(z_c - z_a)}{\sqrt{(x_b - x_a)^2 + (y_b - y_a)^2 + (z_b - z_a)^2}\sqrt{(x_c - x_a)^2 + (y_c - y_a)^2 + (z_c - z_a)^2}} \tag{1}$$

and the distance between two nodes $\mathbf{a}$ and $\mathbf{b}$ is computed as

$$d(a, b) = d(b, a) = |\mathbf{ab}| = |\mathbf{ba}| = |\mathbf{v}_1| = \sqrt{(x_b - x_a)^2 + (y_b - y_a)^2 + (z_b - z_a)^2} \tag{2}$$

A 2D image can be converted into 3D line drawings by the gradient descent minimum standard deviation of angles (GDMSDA) as follows. The standard deviation of angles (SDA) for all the angles in the entire image is

$$SDA = \frac{1}{n} \sqrt{n \sum_{\theta} \theta^2 - \left(\sum_{\theta} \theta\right)^2} \tag{3a}$$

where the summations are over each angle where two lines in the image meet. The partial derivative of the SDA with respect to a given z coordinate z_i is

$$\frac{\partial SDA}{\partial z_i} = \frac{n\sum_{\theta} \theta\, (\partial\theta/\partial z_i) - (\sum_{\theta} \theta) \sum_{\theta} (\partial\theta/\partial z_i)}{n} \left[\sqrt{n \sum_{\theta} \theta^2 - \left(\sum_{\theta} \theta\right)^2}\right]^{-1} \tag{3b}$$

The partial derivative of each angle with respect to a given z_i is nonzero only if z_i is one of the three points forming that angle. If z_i is at point b in Fig. 2, the derivative is

$$\frac{\partial\theta}{\partial z_b} = \frac{\mathbf{v}_1 \cdot \mathbf{v}_2(z_b - z_a) - |\mathbf{v}_1|^2(z_c - z_a)}{|\mathbf{v}_1|^2\sqrt{|\mathbf{v}_1|^2 |\mathbf{v}_2|^2 - (\mathbf{v}_1 \cdot \mathbf{v}_2)^2}} \tag{3c}$$

and if z_i is at point a in Fig. 2, the derivative is

$$\frac{\partial\theta}{\partial z_a} =$$

$$\frac{\mathbf{v}_1 \cdot \mathbf{v}_2|\mathbf{v}_1|^2(z_a - z_c) + \mathbf{v}_1 \cdot \mathbf{v}_2|\mathbf{v}_2|^2(z_a - z_b) + |\mathbf{v}_1|^2|\mathbf{v}_2|^2(z_b - z_a + z_c - z_a)}{|\mathbf{v}_1|^2|\mathbf{v}_2|^2\sqrt{|\mathbf{v}_1|^2|\mathbf{v}_2|^2 - (\mathbf{v}_1 \cdot \mathbf{v}_2)^2}} \tag{3d}$$

Each x and y coordinate is given by the 2D line drawing. The goal is to find z values for which the SDA is at a local minimum. To find these z values by gradient descent, the z's should be repeatedly updated according to

$$z_i \leftarrow z_i - \sigma \frac{\partial SDA}{\partial z_i} \tag{3e}$$

where σ is a small positive number that regulates the step size of the gradient descent. The partial derivative in (3e) can be implemented by substituting (3c and 3d) into 3b). As indicated in [1], this gradient descent method is much faster than [13] (from quadratic time to linear time) with the same effect as perceived by human beings.

Once the 3D coordinates of nodes and lines connecting pairs of nodes are obtained, we call this 3D image an *coordinated graph* (3Dcg) since it is basically a graph with 3D coordinates of nodes and lines connecting pairs of nodes in 3D space. An object denoted by its 3Dcg can then be described by a string representation by the following algorithm:

Heuristic Parsing Algorithm for 3D object by 3Dcg (CGPA)
  Input: A 3Dcg
  Output: A parsing sequence representing the object
  Procedure

0. Unmark all input lines. Select a node **a** with maximum number of neighbors (if there are more than one such node, select the one with a longest line, else select anyone of them to start).
1. If **a** has n nonmarked lines connecting to b_1, b_2, . . . , b_n, record (a $-$ $b_1 b_2 \cdot \cdot \cdot b_n$). Mark all lines that have already been parsed.
2. Check all b_i nodes *simultaneously*, where $i = 1, 2, . . . , n$, to see if any of these b_i nodes have any connecting lines that are not marked yet. If no, go to step 3, else, set **a** $\leftarrow b_i$, $\forall i = 1, . . . , n$ such that b_i has nonmarked lines, and go to step 1. Repeat this process until all b_i's have been done, and no more unmarked lines left, then go to step 3.
3. Produce the parsing sequence recorded from steps 1 and 2 and halt.

Example. The three different states of a box in Fig. 1 can be described by the following linear string pattern representation:

$$(a - bdel)(b - cfk, d - c, e - hif, l - ik)(f - gj, k - j, i - j, h - g) \qquad (4)$$

which is equivalent to the layered graph representation (model syntax) with different measurements (interpretations, semantics) in Fig. 3. It is called a *layered graph* because it can be laid out in different levels from the start node downward, and if there is any node repeating itself, it repeats in at most two consecutive levels. For instance, in Fig. 3(iii), node j appears at level 2 and repeats only at level 3. So does node i, which appears at level 2 and repeats only at level 3. Such repetition occurs because the graph was generated *in parallel* layer by layer, and does not cost any extra time but makes the representation from each node *unique* and reduces search and pattern matching time [19].

Suppose that in the *learning* process it is determined that θ_1 *and* θ_2 have a range between $-\pi/2$ *and* $3\pi/2$. Suppose that the learning also shows that when θ_1 *and* θ_2 are between $\pi/4$ *and* $\pi/2$, it is called half open; less than $\pi/4$, almost closed; or if greater than $\pi/2$, wide open. The three situations (states) are shown

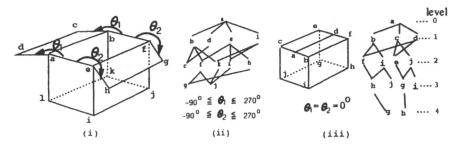

Figure 3 (i) Box, (ii) its graph representation and measurements (range in part), and (iii) completely closed.

as in Fig. 1. Note that this only needs very few *learning samples*, as few as four or five. Note also that when $\theta_1 = \theta_2 = 0°$, the box is completely closed, and its graph representation is slightly different, as shown in Fig. 3(iii).

Note the robustness of this representation method. Take the box, for instance. Its representation not only represents a variety of views of the box, its different rotations and sizes, but also different states (e.g., wide open, half open, almost closed, etc.), all in one representation (except when completely closed), with different measurements and range, of course. These measurements and range for interpretation of the model object can be obtained from heuristic learning experience. This is shown in Fig. 4. Also note that it is possible that the box can be half cover completely closed and the other half open. In this case a little different graph representation can easily be derived from the algorithm CGPA.

This method works for many interesting classes of line-drawing semideformable rigid objects. Figure 5 shows some figures for pairs of scissors, swinging office chairs and crutches, among others. Note that the three pairs of scissors in

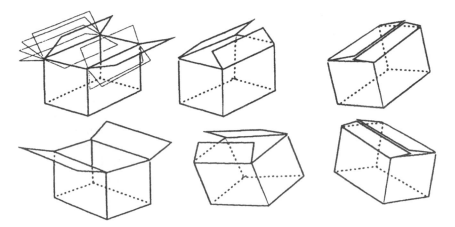

Figure 4 Some more examples of a box object, its rotations, and various states.

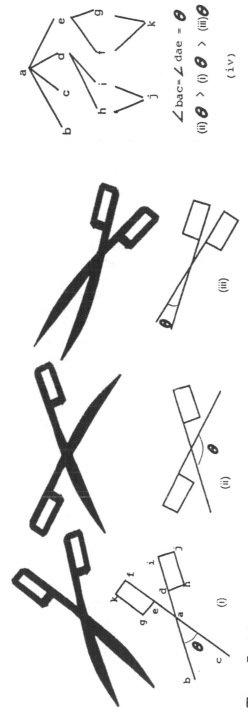

Figure 5 Scissors (type I): its coordinated graph (line-drawing skeleton [20]), three different states by angle θ, and layered graph representation.

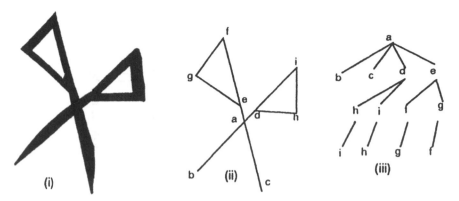

Figure 6   Scissors (type II).

Fig. 5 look quite different but are actually the same object with three different states represented by the layered graph and measurements (range) in Fig. 5(iv). The different states can be distinguished by the angle θ. Note also that the difference between two types of scissors is reflected by the representations shown in Figs. 5(iv) and 6(iii). In general, each layered graph is actually a *finite representation* of an infinite class of objects' views, where different states are characterized by a certain range of measurements from learning.

Figure 7 show different views of an office chair from different angles of θ *and* ω. The nine views, although they look quite different, actually show the same

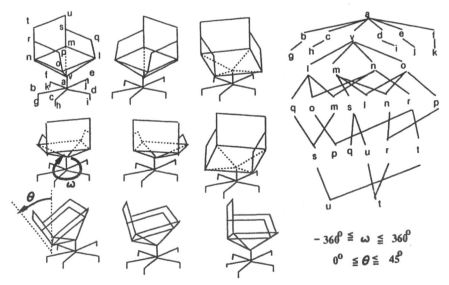

Figure 7   Swinging office chairs.

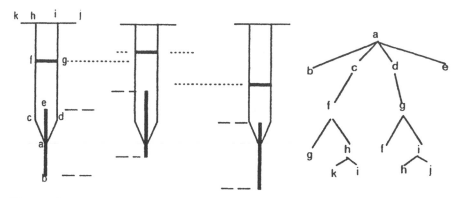

Figure 8 Adjustable crutches.

chair. Figure 8 shows a crutch and its layered graph representation for handicapped people. This example was inspired when the author had a minor knee accident and needed a crutch for temporary support. Additional examples are shown in Figs. 9 to 11. Note that a round clock is approximated by an octagon, and the three hands (hour, minute, and second) are connected to the octagon by a pseudo (dashed) link in the graph.

An additional example, a playground swing for children, is shown in Fig. 12. Figures 13 and 14 show two more examples, a compass and a stretchable blackboard. Note that both representations of stretchable blackboard are shown since there are two nodes that satisfy our start node criterion in the matching algo-

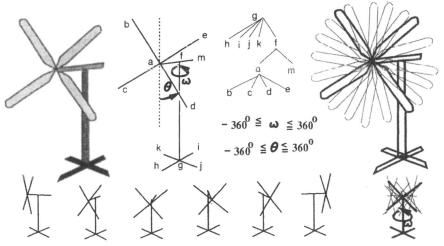

Figure 9 Windmill: its skeleton, graph, and different states (various θ and ω values).

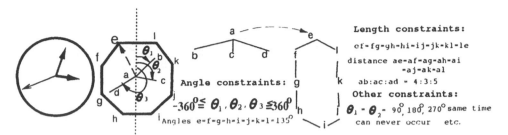

Figure 10 Clock: its approximation, graph representation, range measurements, and constraints.

rithm (i.e., it includes a connecting line with maximum length). The other two representations, starting at nodes b and j, are symmetric to a and d, respectively, and therefore are not shown in the figure.

III. IMAGE RECOGNITION AND DESCRIPTION SCHEME: ILLUSTRATIVE EXAMPLE

From [19] it was shown that from a certain start node of a 3D line-drawing object, there is a unique graph representation. This, together with the properties discussed in Section II, enables one to establish a fast parallel matching algorithm (PMA) as follows:

Parallel Matching Algorithm (PMA)

Input: Two representation graphs G1 (model object from learning) and G2 (object to be recognized)

Output: Yes, plus description (if equal); no (if not)

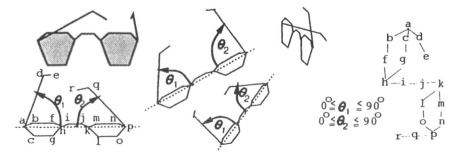

Figure 11 Pair of eyeglasses: various views, sizes, and states (range measurements).

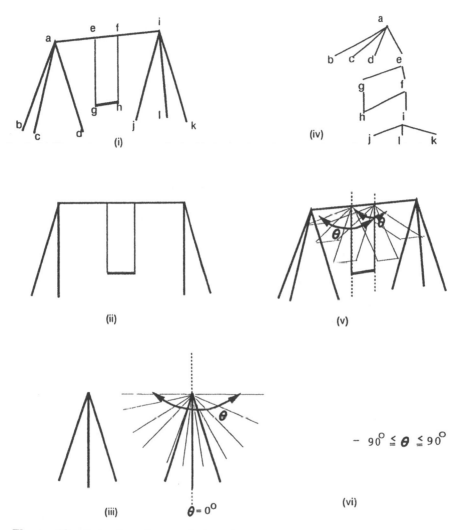

Figure 12 Swing in a playground: (i) bird's-eye view; (ii) front view; (iii) side view (with swing angles); (iv) graph representation (shown in five levels); (v) bird's-eye view of swing angles; (vi) angle constraints of θ (in part) $\pi/2 \geq \theta \geq -\pi/2$.

Procedure

1. Set level $i = 0$, compare number of neighbors of start nodes. If nonequal, go to exit 1, else go to step 2.
2. $i \leftarrow i + 1$, check all nodes and their associated lines to see if there is a match between G1 and G2 at the same level. If no, go to exit 1, else repeat step 2 until all levels are done, go to exit 2.

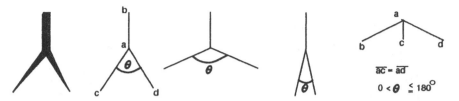

Figure 13 Example of compass.

a. *Exit 1*: No (reject)
b. *Exit 2*: Yes (recognized with descriptions)

Once the representation of the input object matches with the model, we compare their corresponding range of measurements. If all of them match, the output is yes, else rejected. From the observations above and Section II, we can establish a recognition and image understanding scheme for a 3D line-drawing object, shown in Fig. 15. Note that in the scheme, whenever an input object is rejected, it can also be included in the dictionary (or library) in the learning phase, so that it can be recognized in the future.

Figure 16 is an illustrative example for recognizing, understanding, and describing a model refrigerator shown in wireframe line drawings (shown in its skeleton). We assume that there are two types of refrigerators in the example: type I, with the freezer on the top and the refrigerator section at the bottom, and type II with the reverse arrangement. Note that if the original image is a 2D wireframe line drawing, there may be an illusion (ambiguous) problem, as shown in Fig. 17. It needs multiple-view or other means to disambiguate. From

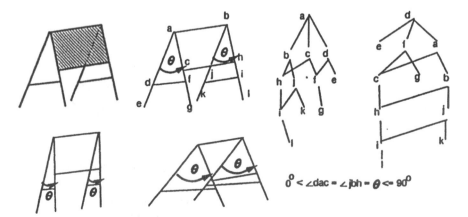

Figure 14 Example of stretchable blackboard.

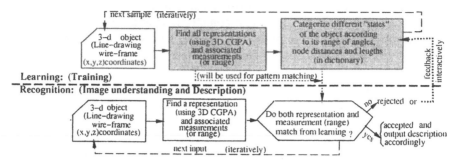

Figure 15 Three-dimensional line-drawing semirigid object recognition and description scheme.

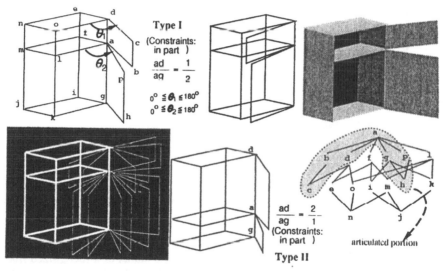

Figure 16 Refrigerators: (i) type I with freezer on top; (ii) type II with freezer at bottom.

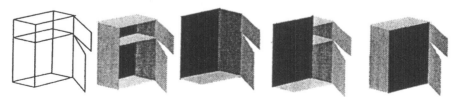

Figure 17 Refrigerator illusions: one model with four visualizations, assuming that there is only one side open, attached with doors.

learning experience that defines various states of a refrigerator, such as wide open, half open, almost closed, and completely closed for freezer door and/or refrigerator door, as shown in Fig. 18, the method can recognize (or reject) an input object and if recognized, can determine what states it is in, as shown in Fig. 19.

A robot with vision capability can use this technique to learn, understand, recognize, and describe 3D wireframe line-drawing objects. This can be demonstrated by a question–answer format. One example of such dialogue is as follows:

Input: Object F
Question (Q):   What is it?
Answer (A):   It is a refrigerator.

Q: What type is it?
A: Type I.

Q: Can you describe its status?
A: Yes, its freezer door is wide open and the door to the refrigerator section is almost closed.

Input: Object B
Q: What is it?
A: It is a refrigerator.

A: What type is it?
A: Type II.

Q: Can you describe its status?
A: Yes, its freezer door is almost closed and the refrigerator door is completely closed.

A summary of eight input objects and their testing results appears in Table 1. Not only can this method determine a certain static status of a recognized line-drawing wireframe semideformable rigid object, but using it, we can also describe the motion of a sequence of 3D images. This is illustrated in Figs. 20 to 22 for a refrigerator rotating, approaching, and opening/closing door(s). For example. Fig. 22 could be demonstrated by the following dialogue:

Q: What do you see?
A: I see a refrigerator.

Q: What type is it?
A: Type I.

Q: Can you describe its motion status?
A: Yes, its freezer door is opening, opening, opening, until about 50°, then closing, closing, closing, until completely closed.

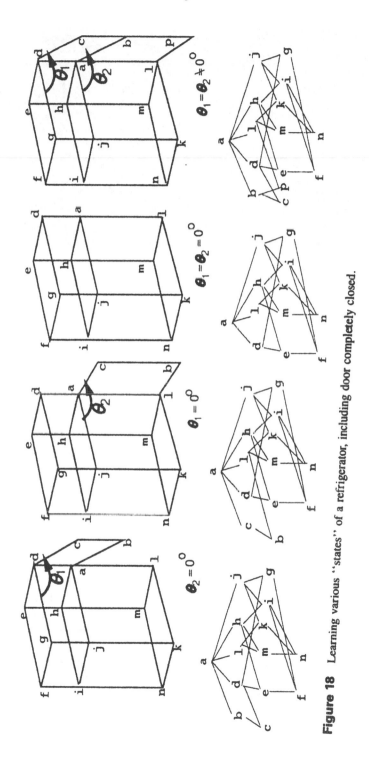

Figure 18 Learning various "states" of a refrigerator, including door completely closed.

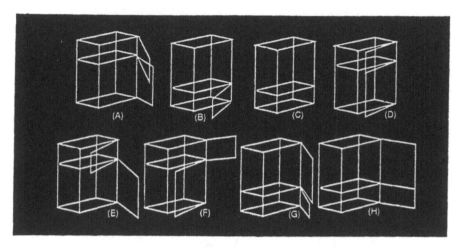

Figure 19   Recognizing and describing various types and states of a refrigerator.

Table 1   Testing Results of Eight Refrigerators

Input objects	Output measurements, ranges, and descriptions[a]					
	θ_1	θ_2	ad:al	Type	Freezer	Refrigerator
A	80°	100°	1:2	I	HO	WO
B	0°	40°	2:1	II	AC	CC
C	0°	0°	2:1	II	CC	CC
D	30°	30°	1:2	I	AC	AC
E	25°	105°	1:2	I	AC	WO
F	160°	20°	1:2	I	WO	AC
G	85°	80°	2:1	II	HO	HO
H	160°	160°	2:1	II	WO	WO

[a]HO, half open; WO, wide open; AC, almost closed; CC, completely closed.

IV.   DISCUSSIONS, CONCLUSIONS, AND FUTURE RESEARCH

We have introduced a mechanism to recognize, understand, and describe a 3D semideformable rigid wireframed line-drawing geometric object modeled by a coordinated graph and layered graph representation. A recognition scheme using parallel matching is established. This method is rather simple and robust in that not only can it recognize a 3D line-drawing wireframe object in its various rotations, locations, and topological transformation (scalings, elongations, etc.) but it can also describe different states of the same semideformable rigid object.

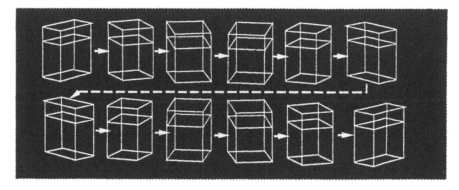

Figure 20   Type I refrigerator rotating clockwise, then counterclockwise (with respect to the observer).

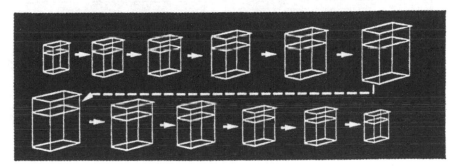

Figure 21   Type I refrigerator approaching, then departing (with respect to the observer).

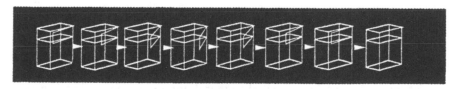

Figure 22   Type I refrigerator's freezer door is opening, then closing until completely closed.

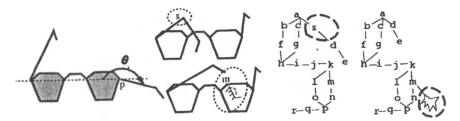

Figure 23 Examples of distorted and broken eyeglasses and their representations (node labels adapted from Fig. 11).

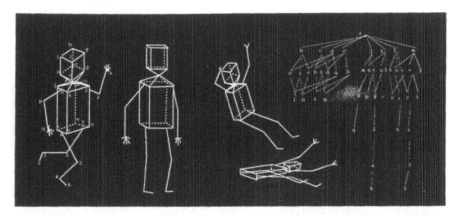

Figure 24 Model robot shown in line-drawing skeleton simulating human body movements.

In addition, it needs very few learning samples and can distinguish similar objects, such as even and uneven stairs [19], that other method cannot distinguish [12].

For distorted or noisy objects, this method is also very convenient for description, analysis, and recognition. Take Fig. 23, for instance. Suppose that the angle, θ_1 is 190°, which is larger than the limit of 180°; then it shows a broken bar (at joint p) of eyeglass. Similarly, other distorted versions and broken parts (of lens) of an eyeglass can also be reflected from their representation graphs in Fig. 23. In fact, this method can also work for more complicated objects, such as modeling a robot or recognizing (understand, describing) a class of water-pouring situations, as shown in Figs. 24 and 25. Note that there four different representations are needed for this purpose: namely, Fig. 25(i) and (vi) share the same representation, (ii) and (iii) share the same graph representation, with dif-

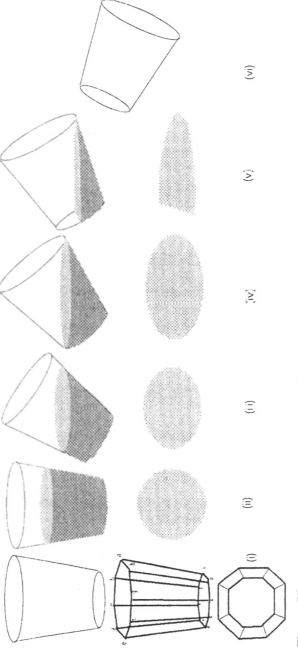

Figure 25 Understanding (describing, recognizing) water pouring out of a glass.

ferent measurements and constraints, (iv) has its own slightly different representation, and (v) has another representation and range of constraints and measurements.

As mentioned in Section III, for curved-line objects, we can use line-segment approximations. A circle can be approximated by an octagon as shown in Fig. 25. Note that different thresholding values may give different results, which is an interesting future research topic. In the future it would be interesting to extend our approach from its essentially object-centered orientation in this chapter to viewer-centered or combined objected- and viewer-centered recognition.

APPENDIX: A COMPARISON

As pointed out by Winston in Chapter 26 of his book *Artificial Intelligence* [23], there are objects with different parts in which each part is rigid but rotatable. Coincidentally, this interesting class of objects falls into the same category as my *semideformable rigid objects*, and his Fig. 26.7 (p. 543) can be considered as a simplified version of the tank model introduced in [19], which will be discussed in this appendix as a semideformable rigid object. Figure 26 shows a model tank with four components, in which components (i) and (iii) stick together as one rigid piece, but which is rotatable around its z axis relative to component (iv). Further, component (ii) (i.e., the canon), can rotate [with respect to component (iii)] in a geometric cone shape along the xy plane ($-30°$ to $+30°$) and the xz plane ($0°$ to $45°$). Several bird's-eye views of the model tank with different sizes, angles, and gray levels are shown, together with graph representation and some angle measurements. In Fig. 27 some top and side views of different "states" of the model tank are shown. This method is robust in handling the structural similarities and differences in such semideformable rigid objects.

Another example of an object [an obelisk (rigid)] from [23] is analyzed and represented by this method as shown in Fig. 28(i) in comparison with similar objects, such as a jukebox or sofa. This example also shows how robustly we can represent, describe, and distinguish similar objects in different views, orientations, sizes, and topological transformations by use of this method. In Fig. 28(ii) it shows that with a bit more effort, such as rotation to find the hidden edges and node, one can construct a (slightly different) representation that is better and more capable of covering all angles of view. In this case we need only one representation for the object for all its rotations, displacements, and topological transformations. In comparison with the LC (linear combination) method introduced in [2,17], at least 18 views (two views in each of the eight views) must be needed in Fig. 28(iii) for recognition of the object.

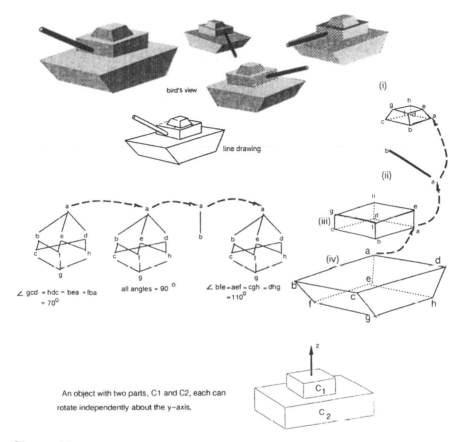

bird's view

line drawing

(i)

(ii)

(iii)

(iv)

∠ gcd = hdc = bea = fba
= 70°

all angles = 90°

∠ bfe = aef = cgh = dhg
= 110°

An object with two parts, C1 and C2, each can
rotate independently about the y-axis.

Figure 26 Model tank: its four components, different view, line drawing, representation, and angle measurements.

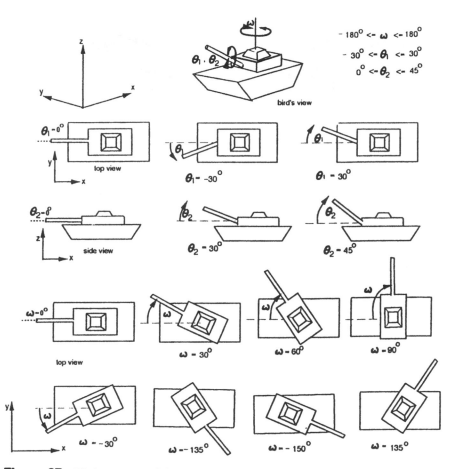

Figure 27   Various states of the model tank in Fig. 26.

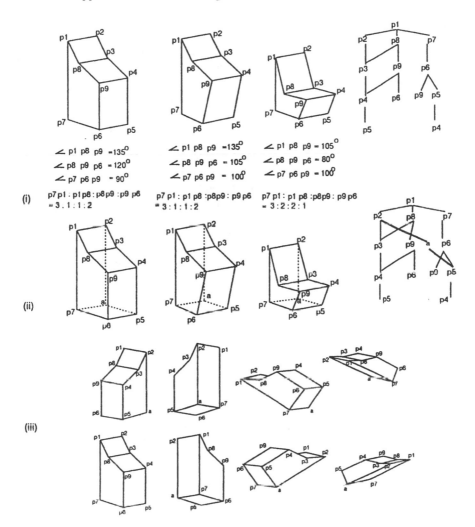

Figure 28 Obelisk, jukebox, and sofa: their representations and measurements. (From Ref. 23.)

REFERENCES

1. L. Baird and P. S. P. Wang, 3-d object recognition using gradient descent and the universal 3-d array grammar, *Proc. SPIE*, Vol. 1607, *Intelligent Robots and Computer Vision*, 1992, 711–719.

2. R. Basri, Viewer-centered representations in object recognition: a computational approach, in *Handbook of Pattern Recognition and Computer Vision*, (C. H. Chen, L. Pau, and P. Wang, eds., World Scientific Publications, Singapore, 1993 (to appear).

3. A. Beinglass and H. Wolfson, Articulated object recognition, or how to generalize the generalized Hough transform, *Proc. International Conference on Computer Vision and Pattern Recognition*, 1991, pp. 461–466.

4. P. Besl and R. Jain, 3-d object recognition, *ACM Comput. Survey*, 17(1):75–154 (1985).

5. R. Brooks, Symbolic reasoning around 3-d models and 2-d images, *Artificial Intelligence*, 17:285–348 (1981).

6. C. Chien and J. Aggarwal, Shape recognition from single silhouette, *Proc. International Conference on Computer Vision*, 1987, pp. 481–490.

7. R. T. Chin and C. R. Dyer, Model-based recognition in robot vision, *ACM Comput. Survey*, 18(1):67–108 (1988).

8. B. Girod and S. Scherock, Depth from defocus of structured light, *Proc. SPIE*, Vol. 1194, *Optics, Illumination and Image Sensing for Machine Vision IV*, 1989.

9. R. Goldberg and D. Lowe, Verification of 3-d parametric models in 2-d image data, *Proc. IEEE Workshop on Computer Vision*, 1987, pp. 255–267.

10. W. E. L. Grimson and T. Lozano-Perez, Localizing overlapping parts by searching the interpretation tree, *IEEE Trans. Pattern Anal. Mach. Intelligence*, PAMI-9(4):469–482 (1987).

11. T. S. Huang and C. H. Lee, Motion and structure from orthographic projections, *IEEE Trans. Pattern Anal. Mach. Intelligence*, PAMI-2(5):536–540 (1989).

12. T. Marill, Recognizing three-dimensional objects without the use of models, *Memo. 1157* Artificial Intelligence Laboratory, MIT Cambridge, Mass., Sept. 1989.

13. T. Marill. Emulating the human interpretation of line-drawings as 3-d objects, *Internat. J. Comput. Vision*, 6(2):147–161 (1991).

14. R. N. Shepard and J. Metzler, Mental rotation: effects of dimensionality of objects and type of task, *J. Exp. Psychol. Human Percept. Performance*, 14:3–11 (1988).

15. K. Sugihara, *Machine Interpretation of Line Drawings*, MIT Press, Cambridge, Mass., 1986.

16. M. Tuceryan and A. K. Jain, Texture analysis, in *Handbook of Pattern Recognition and Computer Vision*, C. H. Chen, L. Pau, and P. Wang, eds., World Scientific Publications, Singapore, 1993 (to appear).

17. S. Ullman and R. Basri, Recognition by linear combinations of models, *IEEE Trans. Pattern Anal. Mach. Intelligence*, PAMI-13(10):992–1006 (1991).

18. S. Ullman, Aligning pictorial descriptions: an approach to object recognition, *Cognition*, 32(3):193–254 (1989).

19. P. S. P. Wang, A parallel heuristic algorithm for 3d object pattern representation and recognition, *Proc. SSPR (Structural and Syntactic Pattern Recognition) '92*, 411–420, World Scientific Publishing Co., Bern, Switzerland, (1992).

20. P. S. P. Wang and Y. Y. Zhang, A fast and flexible thinning algorithm, *IEEE Trans. Comput.*, C-38(5):741–745 (1989).
21. P. S. P. Wang, A formal parallel model for 3-d object pattern analysis, in *Handbook of Pattern Recognition and Computer Vision*, C. H. Chen, L. Pau, and P. Wang, eds., World Scientific Publications, Singapore, 1993 (to appear).
22. P. S. P. Wang, 3D articulated object recognition: a parallel approach, *Proc. SPIE*: Vol. 1825, *Robotics and Computer Vision*, 11–12, SPIE Press, Bellingham, Wash., 1992.
23. P. Winston, *Artificial Intelligence*, 3rd ed., Addison-Wesley, Reading, Mass., 1992, Chap. 26.
24. E. K. Wong, Model matching in robot vision by subgraph isomorphism, *Pattern Recognition*, 25(3):287–303 (1992).

7

Handwritten Character Recognition

Paul Gader

University of Missouri at Columbia
Columbia, Missouri

Andrew Gillies and Daniel Hepp

Environmental Research Institute of Michigan
Ann Arbor, Michigan

I. INTRODUCTION

A. What Is Character Recognition?

In broad terms, character recognition refers to a computer reading some part of some type of document. An easy character recognition problem is that of designing computer systems to scan and read stylized characters on the bottom of bank checks. A difficult one is designing computer systems to scan and read pages of handwriting about arbitrary topics. The first problem has been solved for a long time. The latter is not.

A number of intermediate problems are solved or are close to being solved. Systems are available for personal computers that scan and recognize machine-printed documents. The wide variety of font styles available makes it difficult to ensure reliable operation of these systems. The existence of figures and other nontextual entities on a page cause difficulties as well. Analyzing pages filled with text of various styles, graphics, and images is an active research area. These capabilities are important as researchers try to develop automated systems for scanning and reading structured documents such as maps and tax returns.

Automated systems for reading handwritten characters are also being designed. Mail-order forms instruct us to write our characters using a certain style so that a machine can read them. Many banks and credit card firms are seeking to implement automated systems for reading amounts from checks and credit card receipts. Many postal services in various countries use machine-printed

223

character recognition systems. They are also actively pursuing the design and implementation of improved machine-printed and handwritten systems for automating the sorting of mail.

Finally, on-line systems are becoming feasible commercial products. In an on-line system, a specialized pen and pad is used to allow the computer to measure the velocity and pressure information used in creating the characters. This added information makes handwritten character recognition more tractable.

Character recognition involves more than just recognizing characters. Consider the problem of reading a handwritten letter from a friend. Unless your friend has very good penmanship, there are probably several characters on the page that are illegible or ambiguous when viewed in isolation. Similarly, many of us of have had the experence of reading a page of text with misspelled words without noticing or caring that the words were misspelled. For example, did you notice that the word *experience* was misspelled back there? If so, did it stop you from reading it correctly?

Character recognition systems require a variety of strategies. We need algorithms that isolate and recognize individual characters, that read words with illegible and ambiguous characters, that correct for spelling errors, and that make use of dictionaries and/or rules for forming legal words. Furthermore, these algorithms should be able to process documents quickly. In this chapter we discuss isolating and recognizing individual characters. These processes are referred to as *segmentation* and *recognition* respectively.

We show why segmentation is a very difficult problem and argue that segmentation cannot be performed accurately without recognition. A major point of the chapter is that the only reliable way to decide where the characters are in a word is to read the word. Since we are only interested in isolating the characters so that we can read the word, the last statement may seem to be a paradox. In fact, it is a special case of the recognition/segmentation paradox found throughout computer vision: you cannot recognize without segmenting and you cannot segment without recognizing. This paradox must be overcome with an approach that combines segmentation and recognition in a tightly coupled fashion.

We also demonstrate the use of some state-of-the-art neural network classifiers for individual character recognition. We discuss the processing steps in detail. Results are provided on real data sets. These in-depth examples give the reader insight into the current tools used to perform individual character recognition.

B. Character Recognition: Two Perspectives

The standard view of character recognition is limited and does not accurately reflect the nature of the problem. We provide the standard view and a more expanded view. We wish to convince the reader that the more expanded view represents the correct approach to developing realistic solutions.

The standard view of character recognition is that a character recognizer is a function f with domain the set of all two-dimensional binary images and with range a set of class labels, $C_1, C_2, \ldots, C_n$. The function should have this domain because we do not know in advance where the input image will come from. It is often true that a character recognizer will be given noncharacters as input. Therefore, one of the class labels should be a noncharacter label. Let us call that class G (for "garbage"). The function f should have the property that $f(l) = C_i$ when l represents a character of class C_i and $f(l) = G$ when l does not represent any character (e.g., l could be an image of a flower or a face, or more dangerously for f, a piece of a character).

An expanded view is: given an image S of a segment (a possible character), an image W of the word from which the segment is taken, and a character class C, let

$$f(S, W, C) = \text{confidence that segment } S \text{ represents class } C \text{ within word } W$$

The confidence values assigned by f should accurately reflect ambiguity of characters and should give low confidence to segments that are illegible and to those that do not represent characters at all. The function f should take into account the relative size, shape, and position of S within W and with respect to the other segments in the segmentation from which S is taken. It should also take into account any other subtle style factors that could influence the resemblance of S to class C. Since this recognizer f uses word-level information, we say that it uses contextual analysis to perform recognition.

Take the foregoing reasoning one step further. The function f could use contextual analysis based on information from the entire document. For example, the system could model writing style and perform parallel recognition based on combining initial tentative results on all segments in the document and the style analysis. A major research problem is to investigate if a handwritten document analysis system can be designed that uses document-level contextual analysis. This design criterion is much more vague than the standard view and much more difficult to work with. It moves closer to capturing the true problems that need to be overcome in designing high-performance character recognizers for handwritten word recognition applications.

The remainder of the chapter is organized as follows: First, the problem of segmentation is discussed. Background on feature measurement and neural network classifiers is then provided. Finally, in-depth examples of actual character recognizers are described.

II. SEGMENTATION: ISOLATING INDIVIDUAL CHARACTERS

A. The Problem of Segmentation

Character segmentation is the process of isolating the individual characters in an image of handwriting. Ideally, we would like to take the point of view that a character segmentation algorithm should take as input an image of a set of characters (such as a word or a ZIP code) and should produce as output a set of images. Each image in the output set should represent one of the characters in the original image. In practice, however, several guesses concerning the locations of the characters in the image are needed to ensure that they are found correctly.

Thus character segmentation algorithms often produce a collection of sets of images as output. Each set in the collection represents a *segmentation*. The elements of a segmentation are called *segments*. We say that a character segmentation algorithm with this property produces multiple segmentation hypotheses. An example is shown in Fig. 1.

The difficulty with the character segmentation problem is reduced substantially if the writer is constrained to write characters in boxes that are well separated. This is often done in mail-order forms and credit card slips. In many applications we are faced with unconstrained handwriting; that is, the writer is not constrained to write inside a box. We describe existing segmentation techniques and the difficulties they encounter. We illustrate why a character segmentation algorithm must use information from recognition by producing multiple segmentation hypotheses or using a tightly coupled approach. We argue that contextual analysis is necessary due to ambiguity and illegibility of characters and uncertainty in the accuracy of the recognition results.

1. Handwritten Numeric Field Segmentation

In this section we consider the problem of segmenting images of unconstrained numeric fields into isolated digits. Several methods have been described in the literature [2,11,18,35]. We describe some current work and refer the reader to the references for more details.

Numeric field segmentation often starts by finding connected components in the image. If the number of digits in the field is known, such as in a ZIP code, we can try to decide if the connected components are in a one-to-one correspondence with the digits in the field. If we expect five digits in the field and we find five connected components, we may assume that each connected component is a digit.

This approach has potential pitfalls. A small connected component can represent either noise or a small character. Some digits could be broken into multiple connected components. This can happen naturally, as is the case with many handwritten "5"s or as a result of previous processing. In the latter case, some

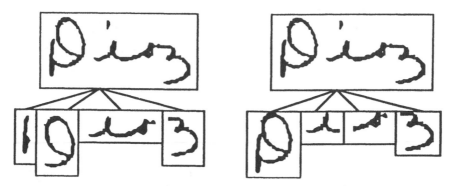

Figure 1 Handwritten word and two segmentations.

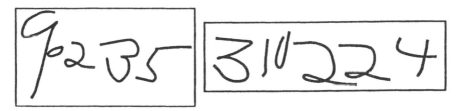

Figure 2 It can be difficult to distinguish connected components that are not digits from those that are based on size and position. In the left-hand image, the "0" is very small. In the right-hand image, a piece of a descender (such as a "y" or a "g") from the line can cause difficulties.

of the pixels that should have been included in the foreground of the binary image were to because of poor thresholding or image acquisition. Finally, if the input image were located automatically using another algorithm, the input image could contain more than the desired numeric field. For example, in a handwritten ZIP-code recognition system, a location algorithm takes an image of a handwritten address and seeks to produce a subimage of the original image that contains only the ZIP code. The algorithm may find the ZIP code together with the last letter of the state. This can obviously cause difficulties. Examples of such problems are shown in Figs. 2 to 4.

One potential solution to this problem is to find the best way to read an entire line. We can assume that the line contains a city, state, and ZIP code and nothing else. We then try to find the best way to match the line to legal combinations of cities, states, and ZIPs. Thus we see that reading can be thought of as a hierarchy of matching processes: the character level, word level, line level, and so on.

Figure 3 Digits can be broken into multiple connected components, some of which are touching. This can lead to ambiguity. The left-hand-side image could be read as "67322" or "51322." The right-hand-side image could be read as "21105" or "4105" or "465."

Figure 4 It is difficult to determine whether the "0" in the state word "MO" is part of the state or the ZIP without taking the entire city–state–ZIP line into consideration.

A segmentation algorithm must consider the possibility that a connected component consists of multiple digits. Most numeric field segmentation algorithms contain submodules to perform splitting. Splitting submodules often take an image of a connected component as input and produce one or multiple segmentations as output. Several splitting algorithms have been suggested in the literature, none of which is entirely satisfactory.

Kimura uses a clustering approach to splitting based on the vertical projection of the image of the connected component [18]. Otsu's thresholding algorithm is used to detect the best point at which to separate the vertical projection, and therefore the image, into two parts. Kimura modifies Otsu's method to perform linear splits in nonvertical directions. These techniques result in splits with spurious cuts of characters. Kimura refines the splits using connected component analysis.

Gillies et al. used mathematical morphology to identify regions in the connected component that were likely locations for splits based on shape [11]. Several hypothetical splits were constructed and initially ranked based on heuristic criteria, including relative size and shape. The final decision is made using digit recognition. The top-ranked hypothetical splits are given to a model-based digit recognition algorithm which then attempts to recognize the components of each split.

Fenrich uses a hierarchical approach that uses a popular philosophy of "getting the easy ones first" [2]. In this approach, simple computational techniques are used to perform easy splits. More computationally intensive techniques are used to perform more difficult splits. This requires the use of multiple splitting algorithms that Fenrich has developed. A similar use of multiple techniques for

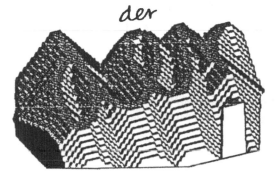

Figure 5   Distance transform-based segmentation. Multiple segmentations of an image of the word "Boulder" are shown. The distance transform on the connected component representing "der" is also shown.

classification is described below and has been popular in the character recognition field.

Whalen et al. describe an approach for segmenting handprinted words that is also applicable (with some modification) to numeric fields [35]. This approach is based on the concept of the distance transform. In this approach, the distance of each point in the background to the stroke is computed. A search technique is then used to find paths from the top of the image to the bottom of the image that maximize the distance from the stroke, as shown in Fig. 5.

From these works one concludes that multiple segmentation hypotheses should be generated. This strategy results in another problem however: How do we decide which hypothesis is best? Many people have come to believe that we should use recognition itself, as in the cases of [20,26], to make the decision. That is, we cannot decide which of the hypothetical segmentations is the correct one without using recognition. This leads to the principle of having tightly coupled segmentation and recognition. This is our recognition/segmentation paradox: "You cannot read without segmenting but you cannot segment without reading". This paradox is not unique to character recognition but is found in various forms in many computer vision problems.

The strategy of resolving multiple segmentations using recognition can lead to problems also. Consider the ZIP code and two hypothetical segmentations

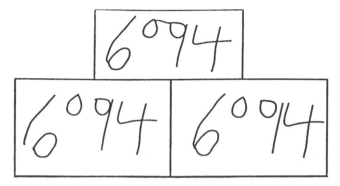

Figure 6   Handwritten numeric field and two possible segmentations into five digits. The left-hand segmentation could be read as "10094." The right-hand segmentation could be read as "60014." Each segmentation could easily be equally likely based on recognition alone.

Figure 7   Ambiguous characters often occur in handwritten words. These patterns were taken from words in which they were used as "u" and "tta." They could be used equally as well as "J" and "Ha."

shown in Fig. 6. Each segmentation is plausible using recognition. That is, if we show each individual segment to a recognizer, they will all get high scores. A human being would probably reject the right-hand segmentation but it is a challenge to program a computer to reject that segmentation and accept other correct segmentations, such as the "90235" shown in Fig. 2. We refer to the process of integrating this type of information into the recognition process as using global shape consistency. This difficult problem is still in the research stage.

2.   Handwritten Characters in Words

In word recognition, the segmentation problem becomes even more difficult. This is partly because there is more variation in the sizes and shapes of alphabetic characters since there are now 52 classes (upper and lower case) instead of 10. This wider variety of styles makes it more likely that ambiguity can occur. The same pixel pattern can represent different individual characters or multiple characters, as shown in Fig. 7.

Another difficulty is that context is much more important in word recognition than in numeric field recognition. This allows writers to write readable words with illegible characters.

When we read words, we have a dictionary of possibilities that allows us to constrain the number of possibilities for the word. Even if we are attempting to read an unknown word, we have rules for forming legal words in a language such as English. For example, many people are not familiar with the word "pergola" (which is a structure consisting of posts supporting an open roof in the form of a trellis) but would agree that it is a plausible word in the English language. However, very few English speakers would believe that "wwwdsd" represents a word in the English language.

The combination of these constraints allows us to read words that include illegible and ambiguous characters and that are misspelled. The same constraints can be applied when developing character recognition systems. For example, if a program is trying to read a street word in an address and the ZIP code and city are known, we may be able to reduce the number of possibilities for the street word to a manageable number, say, 200 words.

Programming a computer to read the image of the street word then becomes a matching problem. We give the program a set of possible candidates (represented as character strings) for the identity of the word in the image and ask which candidate is most likely to be represented in the image. More generally, the program can be designed to assign a confidence value to each candidate. These confidence values can then be combined with confidence values for other words in the document being analyzed to produce a final result. The development of methodologies for combining these confidence values is an important area of research.

We can use the notion of segmentation to match a word image against a character string. For example, if we want to match an image of a word against the string "Iowa," we can match each segmentation of the image of length 4 against the string. We match a segmentation, say {S1, S2, S3, S4} against "Iowa" by matching S1 to "1," S2 to "O," and so on. In this model of word recognition, we are asking a somewhat different question than in classical character recognition. In classical character recognition, and in digit recognition, we ask the question: Which class does the input pattern most closely represent? In this model of word recognition, we ask: How much does this pattern look like a particular class? The difference between these questions is subtle and different researchers have different opinions concerning the impact of these observations on the design of character classifiers. At the very least, we see that to be useful for word recognition, a character recognition algorithm must assign a confidence value for each class.

Segmentation algorithms proposed for handwritten words do not differ much from those proposed for numeric fields. There are some differences, however. There is a wider range in sizes and shapes; lowercase characters can be much smaller than uppercase characters. Furthermore, there are many more multiple-stroke characters, such as E, F, G, H, I, J, K, P, Q, R, and T.

Figure 8   The fact that the same pixel pattern can represent one or multiple characters implies that multiple segmentations are important for word recognition.

Multiple segmentations seem necessary for a segmentation-based word recognition algorithm. This can be seen by considering that the connected component shown in Fig. 8 could be part of the word "Marietta" or the word "Hart." It seems unlikely (although perhaps not impossible) that a segmentation-based technique would be able to read both words correctly without considering multiple hypotheses.

The requirement of using multiple segmentation hypotheses carries with it the problem of garbage. That is, the fact the many pieces of characters and unions of pieces of characters with other characters can look like characters. The potential for false matches becomes very high. There is still a need for development of techniques that perform recognition well on characters and that also recognize when a noncharacter is used as input. This also implies a need for integrating global shape consistency information, as indicated previously.

In summary, segmentation is a difficult problem due to ambiguity and illegibility of characters, imperfect inputs, and uncertainty in processing results. One approach to segmentation is to generate multiple segmentation hypotheses that can be ranked using recognition. Problems with this approach include unreliable recognition performance on noncharacters and difficulties in integrating global shape consistency information.

III. FEATURES

In pattern recognition, a feature is usually some measurement or property of a class of objects. A good feature is one that provides discrimination capability between classes. A standard approach of using features to represent an object is by making a set of numerical measurements and collecting them in a feature vector. For example, one could try to describe a character using four features: height, width, area, and number of holes. This could be represented using a four-dimensional feature vector (h, w, a, n). For a given character, this vector is the information that is used as input to a pattern recognition algorithm.

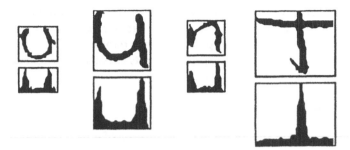

Figure 9 A "v," "u," "n," and "t" and their vertical projections.

In one very different and interesting view of features used by Gillies and Mitchell a feature is viewed not as a numerical value but as an object in an object-oriented language [12]. In their model-based approach, characters are viewed as objects and features as associated objects. Intrinsic properties of the features as well as spatial relationships between them define models of characters.

The choice of features, or representation, of a character to be used in a recognition algorithm is crucial. The features function as a window from which we view the character. The window can allow us to see certain things but be blind to others. Consider the case of the vertical projection of a character as a feature. The vertical projection allows us to see the difference between a "t" and a "u" but not an "n" and a "u," as can be seen in Fig. 9. Let us call the inside of the "u" a *north cavity*. By measuring the relative area of the north cavity we can distinguish between an "n" and a "t" and a "u" and an "n" but not between a "u" and a "v." Different features provide different windows, but no single feature is capable of discriminating all character classes.

The search for good features for character recognition is an ongoing activity. Researchers often use handcrafted features, that is, features designed by a human expert. The features presented in the detailed examples in this chapter are handcrafted. Another very interesting approach is to attempt to program a computer to "learn" features. This is a particular example of a field of study known as machine learning. Several researchers have performed experiments in machine learning of features for character recognition [9,10,21,28,36]. This is a fertile area of research.

The features we use can all be considered as being derived, at least partially, using mathematical morphology. We provide some basic definitions here for completeness. Much more information concerning mathematical morphology can be found in other chapters of this book. Mathematical morphology can be thought of as a mathematical system that can be used to describe and manipulate shape information. It is used in this chapter to compute features of shape that are then used in the character recognition process. There are two basic operations in mathematical morphology: erosion and dilation. We provide definitions here only for the restricted case of binary morphology in two dimensions.

Let A and B be subsets of $R \times R$. Thus A and B are sets of points in the plane. The erosion of A by B is defined as the set

$$C = A - B = \{a \in A : a + b \in A \; \forall \; b \in B\}$$

where $a + b$ denotes the vector sum of a and b. The dilation of A by B is defined as the set

$$C = A \oplus B = \{c : c = a + b, a \in A, b \in B\}$$

If we think of B as representing a shape, the erosion of a set A by the shape B consists of all points at which the shape fits into A. Thus erosion performs a kind of template matching. On the other hand, dilation makes a copy of the shape B at every point of A. There are a variety of other interpretations of these operations and a surprisingly large number of ways that they can usefully be applied. We use them to compute features for character recognition.

IV. CLASSIFIERS

A wide variety of classifiers have been applied to the problem of handwritten character recognition, including statistical classifiers, template matchers, model-based classifiers, decision trees, linear and polynomial discriminant functions, and neural networks. Most of these classifiers are fully described in pattern recognition books and papers. In this section we discuss two pertinent aspects of pattern classification: a specific type of neural network classifier and combining multiple classifiers. We use both in the character recognition examples that follow.

A. Pattern Classification with Neural Networks

The term *neural networks* refers to a broad class of techniques that have become very popular as classifiers and seem to perform quite well. Neural networks derive their name from possible relationships to biological systems. We take the point of view that neural networks are mathematical models of pattern classifiers that can be implemented in software or hardware. They are tools that can aid in the solution of problems involving pattern recognition.

The examples provided in this chapter use neural networks as classifiers. The type of neural network used here is called a *multilayer feedforward network* (MLFN). It is "trained" using an algorithm called *backpropagation*, an algorithm that was popularized in the mid-1980s and demonstrated convincingly to be useful by the PDP research group [29]. In this chapter we discuss the MLFN and backpropagation algorithm.

We mentioned above that the isolated character recognition problem can be viewed as a problem of estimating a function f that assigns a class to an image of a character. Backpropagation is an optimization algorithm for estimating such a function with a MLFN. The MLFN is then used to assign a value between 0 and 1 for each character class that indicates the degree to which the input pattern represents that class.

Suppose that we want to train a MLFN to perform digit recognition. We start with a training set consisting of some number of samples of each digit class, 0, 1, . . . , 9. We then try to estimate a function g that has 10 real-valued outputs, $g_0, g_1, \ldots, g_9$. That is, $g(x) = (g_0(x), g_1(x), \ldots, g_9(x))$. If the input x is a sample from class i, we want $g_i(x) = 1$ and $g_j(x) = 0$ if j is different from i. These values are referred to as desired output values. The function f can then be defined as

$$f(x) := \max\{g_0(x), g_1(x), \ldots, g_9(x)\}$$

It is very difficult to create such a function g explicitly, but neural network techniques are some of the most successful to date.

A MLFN can be thought of as a parametrized functional form. We use the backpropagation algorithm to estimate the parameters of a MLFN. The parameter estimation process is called *training*. It involves using some set of known input and output values of the function to arrive at an estimate of the parameters of the function. The input and output values are the input patterns and their desired outputs.

A simple analog that may be familiar is the process of fitting a straight line to a set of points. In this case, the functional form is

$$f(x; a, b) = ax + b$$

The task is to find the values of the parameters a and b that yield the best line to fit a given set of points. Thus if we have points $(x_1, y_1), (x_2, y_2), \ldots, (x_n, y_n)$, we would like to have

$$y_i = ax_i + b \qquad \text{for} \quad i = 1, 2, \ldots, n$$

If $n > 2$, this is not very likely. Therefore, we often settle for making the sum of the squares of the differences between the left-hand side and the right-hand side of the equations as small as possible. That is, we try to minimize

$$E(a, b) = \sum_{i=1}^{n} [y_i - (ax_i + b)]^2$$

If we set the partial derivatives of E to 0, we obtain

$$\frac{\partial E}{\partial a} = -2 \sum_{i=1}^{n} (y_i - ax_i - b) x_i = 0$$

and

$$\frac{\partial E}{\partial b} = -2 \sum_{i=1}^{n} (y_i - ax_i - b) = 0$$

or

$$\left(\sum_{i=1}^{n} x_i^2 \right) a + \left(\sum_{i=1}^{n} x_i \right) b = \sum_{i=1}^{n} x_i y_i$$

and

$$\left(\sum_{i=1}^{n} x_i \right) a + nb = \sum_{i=1}^{n} y_i$$

These are linear equations in a and b and are therefore easy to solve. We can represent this solution technique in matrix form. Let

$$X = \begin{bmatrix} x_1 & 1 \\ x_2 & 1 \\ \cdot & \cdot \\ \cdot & \cdot \\ \cdot & \cdot \\ x_n & 1 \end{bmatrix}$$

$$P = \begin{bmatrix} a \\ b \end{bmatrix}$$

and

$$Y = \begin{bmatrix} y_1 \\ y_2 \\ \cdot \\ \cdot \\ \cdot \\ y_n \end{bmatrix}$$

The ideal situation would be that we could solve the linear system of equations $XP = Y$ but we cannot in general. Therefore, we solve

$$X^t XP = X^t Y$$

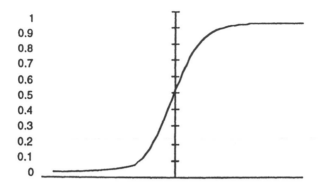

Figure 10 Sigmoid functions are used to "squash" the values in neural networks between 0 and 1.

instead. This system of linear equations is identical to the system derived above using partial derivatives; it is just rewritten in matrix form. If $X'X$ is invertible, the solution is given by

$$P = (X'X)^{-1}X'Y$$

Thus in the case of a linear functional form, there is a closed-form solution for finding the parameters that give the best instance of that form to fit the data. This is not the case for MLFNs except in a very special case. We now discuss the parameter estimation problem for MLFNs.

A MLFN consists of a set of layers; that is, it can be expressed as a composition of functional forms. Let

$$s(x) = \frac{1}{1 + e^{-x}}$$

The function s, called a *sigmoid function*, has the form shown in Fig. 10. A sigmoid function maps real numbers into the interval $(0, 1)$. We apply s to a vector by applying it to each component of the vector. If $\mathbf{x} = (x_1, x_2, \ldots, x_n)$, then $s(x) = (s(x_1), s(x_2), \ldots, s(x_n))$. Let $k > 2$ be a positive integer. Let W_1, $W_2, \ldots, W_{k-1}$ be a sequence of matrices of parameters of size $n_i \times m_i$, where $n_i = m_{i+1}$ for $i = 1, 2, \ldots, k - 2$. Given values for the parameters, each matrix W_i represents a linear transformation from R^{m_i} to R^{n_i}. Define $f_i = sW_i$, that is f_i is the composition of W_i and s. A MLFN with k layers is the composition $f = f_{k-1}f_{k-2} \cdots f_1$. If $\mathbf{x} \in R^{m_1}$, then $f(x;W_1,W_2, \ldots, W_{k-1}) = f_{k-1}(f_{k-2}(\cdots(f_1(x)) \cdots))$.

Note that f is a nonlinear functional form because of the sigmoid functions. The parameters to be estimated are the elements of the matrices. There are called the *weights* of the network. If $k = 2$, then f is essentially a linear func-

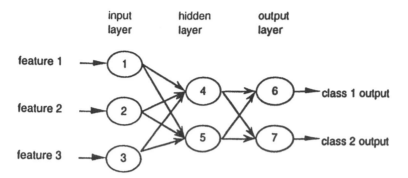

Figure 11   Three-layer neural network. The arcs connecting units i and j are weighted by weight w_{ij}. Feature values are used as inputs to the network. Values are computed at noninput units using sigmoid functions of weighted summations of output values from units in the previous layer.

tional form. A nonlinear functional form of this type can perform classification much than a linear one. Unfortunately, the parameter estimation problem is much more difficult.

The layers of a MLFN can be thought of as the domains and ranges of the mappings sW_i, for $i = 1, 2, \ldots, k - 1$. The first layer is the domain of sW_1 and is called the *input layer*. The second layer is the range of sW_1 and the range of sW_2, and so on. The final layer is called the *output layer*. The number of elements in the layer is the dimension of the associated domain or range. The coordinates of each layer are generally referred to as *units*. There are also layers of weights. Each parameter matrix is sometimes referred to as a layer also. It should be stated explicitly or be clear from the context whether "layer" refers to a layer of units or weights.

An example of a three-layer network is shown in Fig. 11. A MLFN that is being used for pattern recognition generally has as many input units as features and one output unit for each class. The layers of units between the input and output layers are called *hidden units*. It is very difficult to decide how many hidden layers to use and how many units to use in each hidden layer for a given pattern recognition problem. This is a fundamental problem in the use of MLFNs for pattern recognition.

A pattern, x, is "presented" to the network by setting the values of each of the input units to the corresponding feature value. The values are "propagated forward" using the parameter matrices and the sigmoid activation function. The output values should represent the degree to which the input pattern "looks like" each class.

In the character recognition domain, the dimensionality of the weight matrices can be quite large. The inputs to a MLFN are generally either a binary image of a character or feature vector computed from a character. If characters of size 16 × 16 are used, the dimensionality of the input layer is 256. In the examples presented in this chapter, the dimension of the feature vectors ranges from 60 to 120 elements. For example, one network has 120 input units, 69 units in the first hidden layer, 35 units in the second hidden layer, and 27 output units (one for each character class and one for a noncharacter class). Thus there are $(120)(69) + (69)(35) + (35)(27) = 11,640$ weight parameters in this single network. This parameter estimation problem involves performing optimization in extremely high-dimensional spaces, a very difficult mathematical feat.

The training problem for MLFNs can be expressed as follows: We choose a number of layers, k, and dimensionalities, $n_1 \times m_1$, $n_2 \times m_2$, . . . , $n_{k-1} \times m_{k-1}$, for each layer. We have pairs of input and output vectors $(\mathbf{x}_1, \mathbf{t}_1)$, $(\mathbf{x}_2, \mathbf{t}_2)$, . . . , $(\mathbf{x}_N, \mathbf{t}_N)$ of the appropriate dimensions. The input vectors $\mathbf{x}_i$ represent patterns. The ith output vector, $\mathbf{t}_i$, represents the desired output of the MLFN given the input $\mathbf{x}_i$. That is, we would like to have

$$\mathbf{t}_i = f(\mathbf{x}_i, W_1, W_2, \ldots, W_{k-1})$$

For example, assume there is one output unit for each class. If $\mathbf{x}_i$ is in class j, then the jth coordinate of $\mathbf{t}_i$ could have value 1 and all other coordinates could have value 0. This means that we desire a high activation value on the output associated with the correct class and a low activation value on all other output units. As in the case of fitting a line to data points, we wish to find parameter matrices $W_1, W_2, \ldots, W_{k-1}$ that minimize

$$E = 1/2 \sum_{p=1}^{N} \| \mathbf{t}_p - f(\mathbf{x}_p, W_1, W_2, \ldots, W_{k-1}) \|_2^2$$

where $\|z\|_2^2$ represents the sum of the squares of the components of z.

There is no closed-form solution to this problem unless $k = 2$, in which case we have a linear problem, as discussed previously. The backpropagation algorithm is used as an attempt to find the minimum value of E. The backpropagation algorithm uses a technique called *gradient descent*, illustrated in Fig. 12. Gradient descent is sometimes referred to as hill climbing. A value is chosen for the parameters and the gradient is computed at a point. The gradient "points" in the direction of maximum rate of change and so can be used to find the direction to change the parameters. The algorithm is a "greedy" algorithm in that it tries to take the fastest route to the minimum. As can be seen in Fig. 12, if the starting point is not chosen in a lucky spot, gradient descent will not find the global minimum value of the function. This is because the gradient uses only local information.

There are a large number of important details that are involved in actually implementing a gradient descent algorithm on a computer. Discrete sampling can cause the search algorithm to "jump" across valleys and humps, derivatives may not exist, the gradient can be 0, and so on. In particular, if we move from

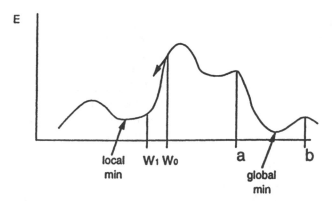

Figure 12 Gradient descent. If the algorithm starts at W_0, it will follow the gradient to a local minimum. If the algorithm starts between a and b, it will find the global minimum. Unfortunately, it is extremely difficult to know where to start.

W_0 in the direction given by the gradient to a new point W_1, we must decide how far to move. If we move too far, we may jump over the valley containing the global minimum. If we move too little, it may take an impractically long time to find any kind of minimum. The decision is made much more difficult by the fact that we are in very high-dimensional spaces and that the functions are nonlinear. Implementers of backpropagation should take such details into account.

The derivation of the backpropagation algorithm can be found in several places, including [29]. It is provided here for completeness. Let us write

$$E_p = \frac{1}{2} \| t_p - f(x_p, W_1, W_2, \ldots, W_{k-1}) \|_2^2$$

$$= \frac{1}{2} \sum_{u=1}^{n_k-1} (t_{pu} - f_{(x_p, W_1, W_2, \ldots, W_{k-1})}^{(u)})^2$$

where $f^{(u)}(x_p, W_1, W_2, \ldots, W_{k-1})$ denotes the uth component of $f(x_p, W_1, W_2, \ldots, W_{k-1})$ and t_{pu} the uth component of t_p. Then

$$E = \sum_{p=1}^{N} E_p$$

Backpropagation is an optimization procedure for changing the weights in the direction of the gradient. Let W denote the vector containing all the parameters in a given MLFN. Then the weight updating rule is given by

$$\mathbf{W}_{new} = \mathbf{W}_{old} + \Delta \mathbf{W}$$

where

$$\Delta \mathbf{W} = -\eta \, \text{grad}(E)$$

where η is the parameter, called the learning rate, that determines how far to move each time the weights are updated. As mentioned earlier, η is a very important parameter.

The gradient is the vector of partial derivatives. Thus to update the weights, we must derive formulas for the partial derivatives of E that can be implemented on a computer. The partial derivative can be computed by cleverly applying the chain rule to E_p.

Assume that the pth pattern has been presented to the network. Let a_{pi} denote the net input to the ith node. That is, a_{pi} is the input value if unit i is in the input layer and

$$a_{pi} = \sum_j w_{ij} s(a_{pj})$$

where the sum runs over the indices of all units feeding into unit i if unit i is not in the input layer. Furthermore, let o_{pj} denote the output of unit j. Thus

$$a_{pi} = \sum_j w_{ij} o_{pj}$$

where

$$o_{pj} = s(a_{pj})$$

Note that

$$\frac{\partial E_p}{\partial w_{ij}} = \frac{\partial E_p}{\partial a_{pi}} \frac{\partial a_{pi}}{\partial w_{ij}}$$

and

$$\frac{\partial a_{pi}}{\partial w_{ij}} = o_{pj}$$

if unit i is not an input unit. Let us define the delta value by

$$\delta_{pi} := -\frac{\partial E_p}{\partial a_{pi}}$$

Then

$$\Delta_p w_{ij} = \eta \delta_{pi} o_{pj}$$

That is, the change in the ij weight (the weight that is multiplied by the output of unit j resulting in one term of the input to unit i) is a product of the learning rate, the delta value for unit i, and the output value of unit j. Let us refer to unit i as the resultant unit and unit j as the source unit associated with weight w_{ij}.

As we will see below, we can interpret the delta value as a measure of the error at a unit for a given training sample. The change in the weight between two units is proportional to the product of the error at the resultant unit and the strength of the output of the source unit. The change in the weight will be large if there is a significant error at the resultant unit and if the output of the source unit is large. This rule for changing weights is sometimes referred to as the *generalized delta rule*.

We still need to develop an implementable formula for the delta value. We again use the chain rule to write

$$\delta_{pi} = -\frac{\partial E_p}{\partial a_{pi}} = -\frac{\partial E_p}{\partial o_{pi}} \frac{\partial o_{pi}}{\partial a_{pi}}$$

The second factor is easy to compute:

$$\frac{\partial o_{pi}}{\partial a_{pi}} = s'(a_{pi})$$

The derivative of the sigmoid function has a pleasantly simple form [29]:

$$s'(a_{pi}) = s(a_{pi})[1 - s(a_{pi})]$$

Two cases need to be considered to compute the derivative of the first factor: (1) unit i is an output unit and (2) unit i is not an output unit.

Case 1. In this case,

$$\frac{\partial E_p}{\partial o_{pi}} = \frac{1}{2} \frac{\partial}{\partial o_{pi}} \sum_{u=1}^{n_k-1} [t_{pu} - f^{(u)}(x_p, W_1, W_2, \ldots, W_{k-1})^2]$$

$$= -(t_{pi} - o_{pi})$$

since $f^{(u)}(x_p, W_1, W_2, \ldots, W_{k-1}) = o_{pi}$ in this case. This expression is easy to implement since both terms are known after presentation of the pth pattern. This expression provides the basis for the interpretation of the delta value as a measure of error.

Case 2. In this case the chain rule is applied yet again. It is used to express the partial derivative with respect to the output at the resultant unit i in terms of the weights and delta values of all the units that unit i feeds into. This allows us to "backpropagate" the error from the output units back through the entire network. We write

$$\frac{\partial E_p}{\partial o_{pi}} = \sum_h \frac{\partial E_p}{\partial a_{ph}} \frac{\partial a_{ph}}{\partial o_{pi}}$$

where h runs over the indices of all units connected to unit i in the layer following unit i. Thus unit i is a source unit for each weight w_{hi}.

By definition of δ,

$$\frac{\partial E_p}{\partial a_{ph}} = \delta_{ph}$$

also

$$\frac{\partial a_{ph}}{\partial o_{pi}} = \frac{\partial}{\partial o_{pi}}\left(\sum_j w_{hj}o_{pj}\right) = w_{hi}$$

Hence

$$\frac{\partial E_p}{\partial o_{pi}} = -\sum_h \delta_{ph}w_{hi}$$

At first glance, this may not seem to have solved the problem because there is still a delta in the formula. This is where the notion of backpropagation comes in. The value δ_{ph} comes from the layer after the layer that unit i is in. Thus we can compute all the delta values for the output layer first. Having done so, we can update the weights for that layer and then we can compute the delta values for the second-to-last layer. Continuing in this fashion, we can update all the weights in the network. This is the backpropagation algorithm, which can be stated as follows:

```
BACKPROPAGATION
Input pth pattern from the training set
Feed  pattern forward through the current network
FOR layer = k - 1 down to 2 DO
    Compute delta values for layer
    Compute ΔW values for layer
    Update weights for layer
ENDFOR
```

There are actually many ways to implement backpropagation and there are some subtle notions that we do not consider here. For example, the algorithm provided here updates the weights after each presentation. A different updating scheme is one that averages the weight changes computed for an entire set of patterns and updates the weights using the average value. There are many more aspects of backpropagation that the interested reader can investigate by delving into the vast number of articles and books that have been written about neural networks.

B. Character Recognition Using Multiple Classifiers

Many researchers support the idea of using multiple classifiers for character recognition. If more than one classifier is used, methods for combining the outputs must be used. One way of categorizing these methods is into parallel or pipeline classifiers as shown in Fig. 13. For example, Duerr et al. use a combination of statistical and syntactic classifiers and Gader et al. use a combination of tem-

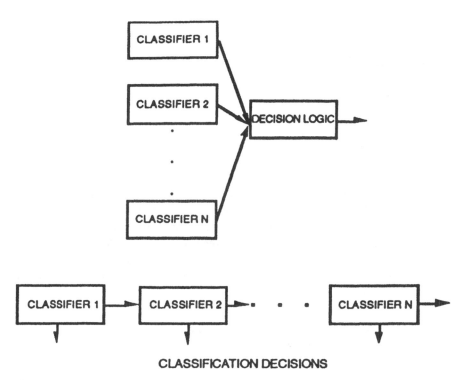

Figure 13   Parallel and pipline classification modes.

plate matching and model-based techniques; these groups combine their results in pipeline fashion [1,5–7]. That is, the classifiers are applied to characters sequentially. If the first classifier can perform classification, no other classifiers are applied to the character. If not, the second classifier is applied to the character. Again, if the second classifier can perform classification (perhaps using information from the first classifier as well as its own measurements), it does so. Otherwise, the character is sent to the next classifier. The process continues until there are no classifiers left in the pipeline. If a decision has not yet been made, the character could be rejected, or if rejection is not applicable, a confidence can be assigned.

On the parallel combination side, Hull et al. use a combination of template matching, a Bayesian classifier based on structural features, and a syntactic approach based on coding the contour of a character [16]. Suen et al. use a combination of a model-based classifier, decision tree classifiers, and a inference-based classifier [32]. In these algorithms, all classifiers are applied to all characters. Some decision logic is then applied to decide on the final result (whether that be classification decision or confidence assignment). This method is more general than the pipeline method. In fact, the pipeline method can be

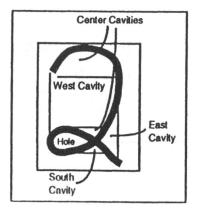

Figure 14 Cavity features.

realized as one particular type of decision logic for the parallel method. A very important area of research is in developing meaningful ways of combining the results of multiple classifiers.

V. EXAMPLES

We present examples of successful character recognizers. These recognizers use handcrafted feature vectors. The feature vectors are inputs to multilayer feed-forward neural networks trained using backpropagation. The networks perform classification and confidence assignment. We describe the training and testing data sets and results for handwritten digits and characters.

A. Features

1. Cavity Features

The notion of cavities involves the spaces that surround the actual character stroke. Figure 14 illustrates the cavity features. A *cavity* is a region of points bounded by the stroke on at least three sides. The cavities are named by the direction in which they open, that is, the side on which they are not bounded. A west cavity is a region open to the west but not open to the north, south, or east. A hole is a region completely enclosed by the stroke. A center cavity is a region that is surrounded on all four sides, but is not a hole. There are six cavity feature types: east, west, north, south, center, and hole.

The cavity features are computed using mathematical morphology. The algorithm uses combinations of dilations, or smears, in different directions and intersections. The morphological algorithm generates six feature images as shown in Fig. 15. These images, together with the image of the character itself, are used to generate a 105-element feature vector. The precise formulation is given in the next paragraph.

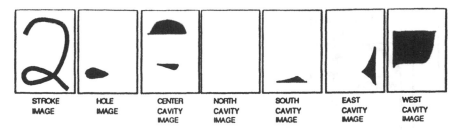

| STROKE
IMAGE | HOLE
IMAGE | CENTER
CAVITY
IMAGE | NORTH
CAVITY
IMAGE | SOUTH
CAVITY
IMAGE | EAST
CAVITY
IMAGE | WEST
CAVITY
IMAGE |

Figure 15 Feature images used for cavity-based zone features.

Let N, S, E, and W denote structuring elements that are rays in the directions north, south, east, and west. These structuring elements are depicted in Fig. 16. The feature images are denoted I, HF, CF, NF, SF, EF, and WF in the order given in Fig. 15. In particular, I denotes the image of the character itself. Furthermore, let B denote the background, or complement, of the image I. The feature images are computed according to the following morphological algorithm:

$$NF = I \oplus N \cap \overline{I \oplus S} \cap I \oplus E \cap I \oplus W \cap B$$

$$SF = \overline{I \oplus N} \cap I \oplus S \cap I \oplus E \cap I \oplus W \cap B$$

$$EF = I \oplus N \cap I \oplus S \cap I \oplus E \cap \overline{I \oplus W} \cap B$$

$$WF = I \oplus N \cap I \oplus S \cap \overline{I \oplus E} \cap I \oplus W \cap B$$

$$CF = I \oplus N \cap I \oplus S \cap I \oplus E \cap I \oplus W \cap B$$

$$HF = \overline{\text{span-until}(\text{BORDER}, B, T)} \cup I$$

The expression for the hole feature image HF needs some explanation. BORDER denotes the image that consists of the one-pixel-wide border around the edge of the image, which is assumed to be completely contained in the background. T represents the 3×3 binary structuring element. The function span-

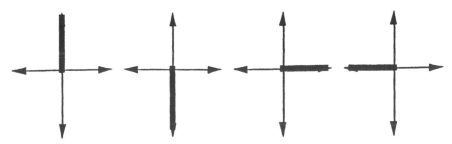

Figure 16 N, S, E, and W structuring elements used to compute the cavity feature images.

until represents the iteration of the conditional dilation operation. If I_1 and I_2 are two binary images and S is any structuring element, the conditional dilation is defined as the operation $(I_1 \oplus S) \cap I_2$. The span-until operation is defined by

DO
$$I_{old} \quad = I_1$$
$$I_1 \qquad = (I_1 \oplus S) \cap I_2$$
UNTIL $I_1 == I_{old}$

It is interesting to note that the equations defining the cavity images can be interpreted as rules. For example, the equation defining the north cavity image can be interpreted as the rule "A point is in a north cavity IF it is in the background (B) AND there are points below it $(I \oplus N)$ AND no points above it $(\overline{I \oplus S})$ AND there are points to the left of it $(I \oplus E)$ AND there are points to the right of it $(I \oplus W)$.

Feature vectors can be computed as follows: The image of the character is first normalized to the size 24×16.* Each feature image is coarse-coded into a 3×5 "blurred" image. The 15 values from each blurred image are combined to form a 105-dimensional feature vector. The coarse coding is performed as follows: An 8×8 mask is placed over each feature image at 15 different locations, 3 from left to right by 5 from top to bottom, to produce a 15-element feature vector for each feature image. The value of each element is the ratio of "on" pixels to total number of pixels in the 8×8 area (64). The resulting feature vector for each digit is 15×7 images = 105 elements long. The component values of the feature vector lie between 0 and 1.

2. Direction Values

The direction-value features seek to encode the information concerning the number of pixels contributing to a tendency of the stroke to point in a certain direction in a zone. Thus the direction values capture information from within the stroke, as opposed to regions surrounding the stroke, as in the case of the cavity features. For this reason, one would hope that the features are "complementary."

The basic procedure in computing the direction values is similar to that of the cavity features. Morphological operations are used to generate a set of feature images. A coarse coding of the feature images is used to generate a feature vector. In this case the feature vector is of dimension 60.

As before, the image of the character is normalized to a size of 24×16. The boundaries of the character image are computed and the character is thinned to

---

*The question of whether to normalize to a fixed size is another interesting aspect of character recognition that we do not address in this chapter.

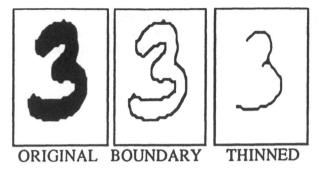

ORIGINAL BOUNDARY   THINNED

Figure 17 To compute the direction value features, the boundary of the image is computed and the image is thinned.

a one-pixel-wide stroke, creating two binary images, a boundary image and a thinned image, as shown in Fig. 17. Let I be the image of the original character, T be the 3×3 binary structuring element, and B be the background of I. The boundary is computed using the following morphological operation:

$$\text{boundary} = \text{THIN}([(I \oplus T) \cap B] \cup [(B \oplus T) \cap I])$$

That is, we first span the stroke over the background one time, span the background over the stroke one time, and take the union of the two. The two span operations generate all the pixels that are either in the foreground touching the background or in the background touching the foreground. We then thin the result. The thinning is performed using the morphological thinning algorithm described in [30].

The boundary and thinned images are eroded by the structuring elements shown in Fig. 18. This process results in a set of eight feature images. Each feature image corresponds to one direction (east, northeast, north, and northwest). The number of pixels in each feature image is a measure of how "much" the corresponding direction exists in the original image.

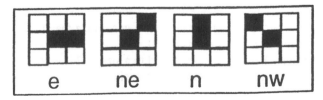

Figure 18 Directional structuring elements used to erode the boundary and thinned images to create the direction-value-feature images.

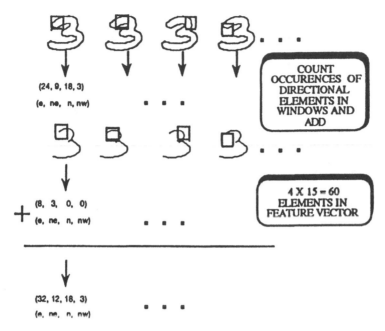

Figure 19 Coarse coding the direction-value features.

We coarse-code each feature image in the same fashion as for the cavity feature images. In this case, however, we add the counts for each zone and for each associated direction. For example, we add the counts for the upper left zone and the east direction for both the boundary and the thinned image. This process is illustrated in Fig. 19. The result is a 60-element feature vector with components having values normalized between 0 and 1.

3. Bar Features

The bar features use directional information on both the foreground and background [8]. The foreground features are based on those used by Ho [14] for machine-printed word recognition. The bar features can be thought of as an attempt to incorporate much the same information as the cave and direction value features but in a more unified framework.

The character image is normalized to 24 × 16. Eight feature images are generated. Each feature image corresponds to one of the directions east, northeast, north, and northwest in either the foreground or the background. Each feature image has an integer value at each location that represents the length of the longest bar that fits at that point in that direction. Examples of the background and foreground feature images corresponding to the east-west directions for a "B" are shown in Fig. 20.

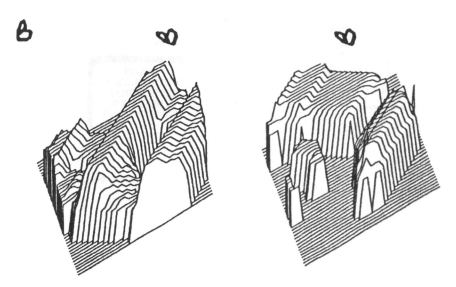

Figure 20 Uppercase "B" and foreground and background bar-feature images corresponding to the east/west directions.

A morphological algorithm can be developed to express and implement the generation of the feature images. For example, to generate the feature image corresponding to the east direction in the background, we erode successively by longer horizontal bars. At each point we keep track of the number of iterations until the point disappears from the eroded image. This is the desired value. Readers may note the similarity of this algorithm to skeletonization.

A more efficient algorithm exists for computing the bar feature images. We illustrate assuming that the bar features are computed only on the foreground. The background is similar. The algorithm requires two passes over the original image array and is an example of a recursive algorithm. The image is assumed to be surrounded by at least one row and one column of zeros at the top, bottom, and both sides. The neighborhood shown in Fig. 21a is used. Four arrays are

Figure 21 Neighborhoods used in pass 1 (a) and pass 2 (b) of the computation of the bar features.

used; each array keeps track of the length of bars at the current stage of processing.

The first pass, the forward pass, scans the image from left to right, top to bottom. At each point in the foreground, the direction array for each neighbor is updated by adding one to the value in the direction for the appropriate neighbor. For example, neighbor 1 is used for the east-west direction, neighbor 2 for the northwest/southeast direction,. and so on. On the second, or backward, pass, the maximum is propagated back up through the array from bottom to top, right to left. The neighborhood in Fig. 21b is used. Pseudocode for computing the bar features on the foreground is as follows:

```
/* FORWARD PASS */
FOR i = 1,2,  . . . , nrows DO
    FOR j = 1,2,  . . . , ncols DO
        e(i,j)  = e(i,j-1) + 1
        ne(i,j) = ne(i-1,j+1) + 1
        n(i,j)  = n(i-1,j) + 1
        nw(i,j) = nw(i-1,j-1) + 1

/* BACKWARD PASS */
FOR i = nrows, nrows - 1,  . . . , 1 DO
    FOR j = ncols, ncols - 1,  . . . , 1 DO
        e(i,j)  = max( e(i,j), e(i,j+1) )
        ne(i,j) = max( ne(i,j), ne(i+1,j-1) )
        n(i,j)  = max( n(i,j), n(i+1,j) )
        nw(i,j) = max( nw(i,j), nw(i+1,j+1) )
```

The reader may be able to see that the distance transform mentioned in the segmentation section can be computed during this algorithm.

The next stage of processing consists of generating feature vectors from the feature images using the technique of overlapping zones. In this case we do not have binary features images as is the case for the previous two features. Therefore, the values in each zone in each feature image are summed. The resulting sums are then normalized between 0 and 1 by dividing by the maximum possible sum in a zone. Thus the resulting feature vector is of dimension $15 \times 8 = 120$ and each component has values between 0 and 1.

B. Network Structures

The networks used to perform classification using these feature sets for both handwritten digits and characters are all four-layer feedforward networks. Only the number of nodes in each layer change. The network structures are illustrated

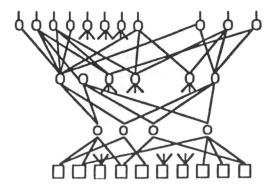

Figure 22 The neural network configuration used in these experiments are four-layer backpropagation networks. There are various numbers of input features and output nodes.

in Fig. 22. For example, the cavity feature network for digit recognition has 105 inputs, 25 units in the first hidden layer, 15 units in the second hidden layer, and 11 units in the output layer (one output unit for each digit class and one output unit for garbage). We denote this structure by (105–25–15–11). The structures of the networks reported on here are:

Handwritten digits:	cavity features (105–25–15–11)
	direction values (60–25–15–11)
	bar features (120–25–15–11)
Handwritten alphas:	cavity features (105–69–35–27)
	direction values (60–45–35–27)
	bar features (120–69–35–27)

In the case of the handwritten alphabetic characters, there is one network for uppercase and one network for lowercase characters.

C. Training and Testing Results

We describe results using a variety of data sets. Most of the data used were taken from images of mailpieces from the U.S. Postal Service. These images were collected by researchers at SUNY–Buffalo. Isolated digits and alphabetic characters were removed from the images by a combination of automatic and manual methods by a number of people at SUNY, the Environmental Research Institute of Michigan, and the University of Missouri–Columbia.

We first describe the digit data sets and results. One thousand digits from each class were used as training digits. Networks were trained for the bar features, the direction-value features, and the cavity features. Rejection parameters

Table 1 Results of Testing the Handwritten-Digit Neural Network Classifiers on the 6000 Suen Digits

Feature type	Percent incorrect	Percent correct
Cavity features	0.96	93.87
Direction-value features	0.46	86.80
Combined	0.67	91.06
Bar features	0.97	96.27

for each network were tuned on a different set of 500 digits from each class. The parameters were set at a level that resulted in an error rate of 1%. These parameters were then used to test the networks on an independent test set obtained from C. Y. Suen [6,19]. There are 600 digits in each class of Suen digits.

The results of the cavity-feature and direction-value networks were combined by averaging the outputs. Thus if for a given digit class (0 to 9) the cavity network produced a confidence value of 0.75 and the direction-value networks produced a confidence value of 0.85, the combined output for that class would be 0.8. The bar-feature networks were trained at a different location and time and with different software and hardware as those used for the other networks. Thus it was not combined with the other networks. The results of the tests are shown in Table 1. Examples of the Suen data are shown in Fig. 23.

Two different training sets were used to train the character networks: the NIST data and the HWAB data. The NIST data consist of characters taken from a database created and sold by the National Institute of Standards and Technology. The HWAB data consist of isolated characters extracted from images of handwritten address blocks. Examples from these data sets are shown in Figs. 24 and 25. The lowercase characters are more difficult than the uppercase characters. Both cases exhibit a considerable amount of ambiguity and illegibility. The data sets were divided into training, threshold, and test subsets. The training sets were used to train neural networks, the threshold sets to set thresholds in the decision logic, and the test sets to generate results.

Figure 23 Some examples of handwritten digits from the Suen data set. These examples represent difficult cases. The true identities are 2, 0, 6, 7, 9, 5, 8, 5, 3, 0, 2, and 6.

Figure 24 Some lowercase characters from the HWAB data set. These images represent the character classes "a," "f," "f," "y," "z," "d," "i," "l," "e," "p," "i," "g," and "q."

The NIST data set is uniformly distributed. We therefore were able to generate 250 characters from each class for training, parameter setting, and testing. The distribution of characters in the HWAB data set is much more skewed. For example, there are only seven lowercase "j"s but there are over 1000 uppercase "A"s. We therefore created the training and testing HWAB data sets by, first, dividing the characters in each class into two equal groups, group 1 and group 2. Group 2 was used for testing. Group 1 was further subdivided into two equal groups, one for training and the other for threshold selection. The training set for the HWAB data consisted of either of the following: the first 250 characters from each class of the HWAB data, if there were that many in the training set, or as many characters as possible from each class, with the difference between the number available and 250 being made up using NIST data. A threshold set consisting of 250 characters from each class was constructed in the same fashion as the training sets. All characters not included in either the training or threshold selection sets were included in the test sets. The training rates for the cavity and direction-value feature-based networks are shown in Table 2.

For the cavity and direction-value networks, we performed the same type of testing as for the handwritten digits. That is, we found rejection parameters that yielded about 1% error and then tested the networks on the test sets. The results for the HWAB and NIST data sets are shown in Tables 3 and 4. As before, the bar-feature networks were trained at a different time and location and were not tested in the same way. In this case the networks were run on a test set with no rejection criteria. Thus the error percentage is just (1 − classification percentage). The results, shown in Tables 5 and 6, indicate the greater level of difficulty involved in recognizing alphabetic characters as opposed to digits. Furthermore,

Figure 25 Some uppercase characters from the HWAB data set. These images represent the character classes "D," "O," "P," "S," "J," "U," "W," "M," "W," "B," "C," and "G."

Table 2 Training Rates for Cavity- and Direction-Value-Based Networks

		Training rate (%)
CAVE	Lowercase NIST	93
	Uppercase NIST	96
DIR-VALS	Lowercase NIST	88
	Uppercase NIST	92
CAVE	Lowercase HWAB*	90
	Uppercase HWAB*	92
DIR-VALS	Lowercase HWAB*	85
	Uppercase HWAB*	86

Table 3 Testing Results for the Cavity- and Direction-Value-Based Networks on the HWAB Data

	Error (%)	Correct (%)
Uppercase		
Cavity features	0.89	54.60
Direction values	0.81	45.38
Combined	0.27	57.18
Lowercase		
Cavity features	1.61	42.15
Direction values	0.91	33.71
Combined	1.16	52.75

Table 4 Testing Results for the Cavity- and Direction-Value-Based Networks on the NIST Data

	Error (%)	Correct (%)
Uppercase		
Cavity features	1.07	60.77
Direction values	0.83	66.14
Combined	0.94	80.90
Lowercase		
Cavity features	0.73	47.35
Direction values	0.83	40.82
Combined	1.02	60.54

Table 5 HWAB Character Data Correct Classification Rates for the Bar Features

	Training set (%)	Test set (%)	RMS error	Learning cycles
Uppercase	92.98	85.91	0.063959	2158
Lowercase	88.43	82.15	0.071576	3255

Table 6 NIST Character Data Correct Classification Rates for the Bar Features

Uppercase	Training set	Test set	RMS error	Learning cycles
Crisp	97.5%	95.6%	0.045	2911

the effect of constraining handwriting can also be seen. Recall that the NIST data were constrained to lie in boxes of a fixed size. The NIST data are much easier to recognize than data taken from the unconstrained realm of handwritten address blocks, which is where the HWAB data were taken.

To conclude our examples section, we provide a further indication of the complementarity of the direction value and cavity neural network classifiers. This is part of the justification for the use of multiple-character classifiers, as discussed earlier. As shown in Tables 1, 3, and 4, the error rates for both the character classifiers and the digit classifiers can be lowered by combining the direction-value and cavity-feature outputs.

In Table 7, the recognition and error rates of each of the HWAB uppercase classifiers is given by class. The complementarity is evident in the X and Y classes. The direction-value network had an error rate of 5.3% on the X class, whereas the cavity network had 0% error on the X class. Furthermore, the direction-value network had an error rate of 0% on the Y class and the cavity network had an error rate of 9.2% on the Y class. The combined network had an error rate of 0% on both classes. That is, the different networks did not make errors on the same characters, so the "bad" characters could not receive high overall confidence.

VI. CONCLUSIONS

In the latter part of this chapter we presented results on recognition of isolated characters. The results on digits approach those of human beings. The results on alphabetic characters are mixed; further work is needed. However, as we pointed out in the beginning of the chapter, isolated character recognition has only limited utility in the problem of reading handwritten words and, beyond that, constrained and unconstrained handwritten documents. Research needs to be

Table 7 Results (%) by Combining the Direction-Value and Cavity-Feature-Based Neural Network Classifiers

	Direction values		Cave		Combined	
	Error	Recognition	Error	Recognition	Error	Recognition
A	0.50	62.80	0.20	62.60	0.00	70.20
B	0.80	70.60	1.60	34.90	0.40	59.50
C	0.00	70.00	0.90	48.40	0.50	61.80
D	1.30	48.70	0.70	39.50	1.30	50.70
E	1.60	59.30	0.80	49.20	0.40	58.90
F	1.50	62.70	0.00	64.20	0.00	64.20
G	3.90	37.30	2.00	45.10	2.00	54.90
H	0.00	48.90	0.00	30.00	0.00	45.60
I	1.30	49.40	0.00	75.30	0.00	79.20
J	0.00	27.30	0.00	50.00	0.00	59.10
K	0.00	57.00	0.00	25.60	0.00	50.00
L	0.00	59.30	0.00	65.70	0.00	75.70
M	2.10	45.70	0.70	22.70	0.40	42.20
N	1.00	24.00	0.30	32.30	0.00	34.10
O	0.00	51.60	0.30	36.80	0.90	51.60
P	1.50	59.10	1.20	47.90	0.20	59.40
Q	0.00	60.00	0.00	20.00	0.00	20.00
R	1.50	53.60	2.10	31.40	0.30	48.50
S	0.30	71.30	2.00	68.70	0.00	76.00
T	0.00	56.30	0.00	52.30	0.00	72.20
U	3.60	10.70	0.00	35.70	0.00	28.60
V	0.00	40.60	5.80	40.60	0.00	46.40
W	0.00	62.00	0.00	41.30	0.00	58.70
X	5.30	15.80	0.00	47.40	0.00	31.60
Y	0.00	51.50	9.10	27.30	0.00	51.50
Z	0.00	50.00	0.00	37.50	0.00	62.50

performed in the area of combining segmentation and recognition to perform handwritten recognition of both words and numeric fields.

The general area of character recognition and document analysis, particularly in the handwritten domain, requires use of a variety of computer vision, pattern recognition, and artificial intelligence techniques. We have presented a small portion of that here. In addition, we have not addressed the problem of computational complexity at all.

There are still many challenges ahead for developers of handwritten document readers. We hope that we have conveyed a sense of the richness of the area and of the challenges in store for developers of handwritten document analysis systems.

ACKNOWLEDGMENTS

The examples reported on in this chapter were developed on research projects supported by the U.S. Postal Service, Office of Advanced Technology. Most of the work reported on in this chapter was performed at the Environmental Research Institute of Michigan (ERIM). Some of the work was performed at the University of Missouri–Columbia under funding from ERIM. We would like to thank Carl O'Connor of the USPS, and John Tan, Peter Heldt, and Gilles Houle of A. D. Little, Inc. for their support of our research. We would also like to acknowledge the many people who participated in various experiments reported on here, including Brian Forester, Margaret Ganzberger, Mike Whalen, Todd Yocum, Alex Popodich, Magdi Mohamed, and Jung-Hsien Chiang. I would also like to thank Professor Harry Tyrer for reviewing parts of this chapter.

REFERENCES

1. B. Duerr, W. Haettich, H. Tropf, and G. Winkler, A combination of statistical and syntactical pattern recognition applied to classification of unconstrained handwritten numerals, *Pattern Recognition*, 12:189–199 (1980).
2. R. Fenrich, Segmentation of automatically located handwritten words, *Proc. International Workshop on Frontiers of Handwriting Recognition 2*, Bonas, France, Sept 23–27, 1991, pp. 33–45.
3. P. D. Gader and M. P. Whalen, Advanced research in handwritten ZIP code recognition, *Final Technical Report to United States Postal Service Office of Advanced Technology from ERIM*, Dec. 1990.
4. P. D. Gader, D. Hepp, D. Fay, and A. Popodich, Handprinted word recognition, I: character recognition, *Mach. Vision Its Appl.* (submitted).
5. P. D. Gader, D. Hepp, B. Forester, T. Peurach, and B. Mitchell, Pipelined systems for recognition of handwritten digits in USPS ZIP codes, *Proc. United States Postal Service Advanced Technology Conference*, Washington D.C., Nov. 1990.
6. P. D. Gader, B. Forester, M. Ganzberger, A. Gillies, B. Mitchell, M. Whalen, and T. Yocum, Recognition of handwritten digits using template and model matching, *J. Pattern Recognition*, 24(5):421–431 (1991).
7. P. D. Gader and B. Forester, Integrating template and model matching for unconstrained handwritten numeral recognition, *SPSE's 43rd Annual Conference Summaries*, Rochester, N.Y., May 1990.
8. P. D. Gader, M. Mohamed, and J.-H. Chiang, Fuzzy and crisp handwritten alphabetic character recognition using neural networks, *Proc. Artificial Neural Networks in Engineering*, St. Louis Mo., Nov. 1992 (to appear).
9. A. M. Gillies, Machine learning procedures for generating image domain feature detectors, doctoral dissertation, University of Michigan, 1985.
10. A. M. Gillies, Automatic generation of morphological template features, *Proc. SPIE*, Vol. 1350, *Image Algebra and Morphological Image Processing I*, P. D. Gader, ed.), San Diego, Calif., July 1990, pp. 252–262.

11. A. M. Gillies, P. D. Gader, M. Whalen, and B. Mitchell, Application of mathematical morphology to handwritten ZIP code recognition, *proc. SPIE*, Vol. 1199, *Visual Communications and Image Processing IV*, Philadelphia, Nov. 1989, pp. 380–390.

12. A. M. Gillies and B. T. Mitchell, A model-based approach to handwritten digit recognition, *Mach. Vision Appl.*, 2:231–243 (1989).

13. D. Hepp, An application of backpropagation to the recognition of handwritten digits using morphologically derived features, *Proc. 1991 SPIE/SPSE Symposium on Electronic Imaging Science and Technology, Nonlinear Image Processing II Conference*, San Jose, Calif., Feb. 1991.

14. T. K. Ho, J. J. Hull, and S. N. Srihari, Word recognition with multi-level contextual knowledge, *Proc. First International Conference on Document Analysis and Recognition*, Saint-Malo, France, Oct. 1991, pp. 905–915.

15. J. J. Hull, T. K. Ho, J. Favata, V. Govindaraju and S. N. Srihari, Combination of segmentation and wholistic handwritten word recognition algorithms, *Proc. International Workshop on Frontiers of Handwriting Recognition 2*, Bonas, France, Sept. 23–27, 1991, pp. 229–240.

16. J. J. Hull, and T. K. Ho, Multiple algorithms for handwritten character recognition, *Proc. International Workshop on Frontiers of Handwriting Recognition 2*, Bonas, France, Sept 23–27, 1991, pp. 117–131.

17. H. Joo, R. Haralick, and L. Shapiro, Toward the automatic generation of mathematical morphology procedures using predicate logic, *Proc. 3rd International Conference on Computer Vision*, 1990, pp. 156–165.

18. F. Kimura, and M. Shridhar, Recognition of connected numeral strings, *Proc. First International Conference on Document Analysis and Recognition*, Saint-Malo, France, Oct. 1991, pp. 731–740.

19. L. Lam and C. Y. Suen, Structural classification and relaxation matching of totally unconstrained handwritten zip code numbers. *Pattren Recognition*, 21:19 (1988).

20. Y. Le Cun et al. Handwritten ZIP code recognition with multilayer networks, *Proc. 10th International Conference on Pattern Recognition*, IEEE CS Press, Los Alamitos, Calif., 1990, pp. 35–40.

21. Y. Le Cun et al., Constrained neural network for unconstrained handwritten digit recognition, *Proc. International Workshop on Frontiers of Handwriting Recognition 1*, Montreal, Quebec, Canada, Apr. 1990, pp. 145–155.

22. G. Lorette, ed., *Proc. First International Conference of Document Analysis and Recognition*, Saint-Malo, France, Oct. 1991.

23. O. Matan, H. Baird, J. Bromley, C. Burges, J. Denker, L. Jackel, Y. LeCun, E. Pednault, W. Satterfield, C. Stenard, T. Thompson, Reading handwritten digits: a ZIP code recognition system, *Computer*, 25(7):59–63 (1992).

24. B. T. Mitchell and A. G. Gillies, A system for feature-model-based image understanding, *VISION '87* June 1987, pp. 81–90.

25. B. T. Mitchell, and A. G. Gillies, Advanced research in recognizing handwritten ZIP codes, *United States Postal Service Advanced Technology Conference*, May 1988.

26. B. T. Mitchell, A. G. Gillies, M. Whalen, and P. D. Gader, A model-based computer vision system for the recognition of handwritten address ZIP codes, *Final*

Technical Report to United States Postal Service Office of Advanced Technology from ERIM, June 1989.

27. B. T. Mitchell, A. G. Gillies, M. Whalen, and P. D. Gader, Handwritten ZIP code recognition mid-term system performance, *Technical Report to United States Postal Service Office of Advanced Technology from ERIM.*, March 1989.

28. M. Rizki, L. Tamburino, and M. Zmuda, Adaptive search for morphological feature detectors, *Proc. SPIE*, Vol. 1350, *Image Algebra and Morphological Image Processing I*, P. D. Gader, ed., San Diego, Calif., July 1990, pp. 150–159.

29. D. Rumelhart and J. McClelland, *Parallel Distributed Processing*, MIT Press, Cambridge Mass., 1986.

30. J. Serra, *Image Analysis and Mathematical Morphology*, Academic Press, New York, 1982.

31. M. Shridhar and A. Badreldin, Recognition of isolated and simply connected handwritten numerals, *Pattern Recognition*, 19(1):1–12 (1986).

32. F. M. W. Stentiford, Automatic feature design for optical character recognition using evolutionary search procedures, *IEEE Trans. Pattern Anal. Mach. Intelligence*, PAMI-7(3): 349–355 (1985).

33. C. Y. Suen, C. Nadal, T. Mai, R. Legault, and L. Lam, Recognition of totally unconstrained handwritten numerals based on the concept of multiple experts, *Proc. International Workshop on Frontiers of Handwriting Recognition 2*, Bonas, France, Sept. 23–27, 1991, pp. 131–145.

34. R. C. Vogt, *Automatic Generation of Morphological Set Recognition Algorithms*, Springer-Verlag, New York, 1989.

35. M. P. Whalen, M. J. Ganzberger, and P. D. Gader, Handprinted word recognition, II: Segmentation and verification, *Mach. Vision Its Appl.* (submitted).

36. S. S. Wilson, Unsupervised training of structuring elements, *Proc. SPIE*, Vol. 1350, *Image Algebra and Morphological Image Processing I*, P. D. Gader, ed., San Diego, Calif., July 1990, pp. 188–199.

37. M. A. Zmuda, L. A. Tamburino, and M. M. Rizki, Automatic generation of morphological sequences, *Proc. SPIE*, Vol. 1769, *Image Algebra and Morphological Image Processing III*, P. D. Gader, E. R. Dougherty, and J. C. Serra, eds., San Diego, Calif., July 1992, pp. 106–199.

8

Digital Image Compression

Paul W. Jones and Majid Rabbani

Eastman Kodak Company
Rochester, New York

I. INTRODUCTION

The use of digital images has increased at a rapid pace over the past decade. Photographs, printed text, and other hard-copy media are now routinely converted into digital form, and the direct acquisition of digital images is becoming more common as sensors and associated electronics improve. Many recent imaging modalities in medicine, such as magnetic resonance imaging (MRI) and computed tomography (CT), also generate images directly in digital form. Computer-generated (synthetic) images are becoming an additional source of digital data, particularly for special effects in advertising and entertainment. The reason for this interest in digital images is clear: representing images in digital form allows visual information to be easily manipulated in useful and novel ways.

The downside to representing images in digital form is that a large number of bits is required to represent even a single digital image, and with the rapid advances in sensor technology and digital electronics, this number grows larger with each new generation of products. Furthermore, the total volume of digital images produced each day increases as more applications are found. As a result, it becomes necessary to find efficient representations for digital images in order to (1) reduce the memory required for storage, (2) improve the data access rate from storage devices, and (3) reduce the bandwidth and/or the time required for transfer across communication channels. The branch of digital image processing that deals with this problem is called *image compression* (or sometimes, *picture coding*). A few books are devoted entirely to the subject of image compression [1–7], and numerous textbooks on image processing also contain detailed coverage [8–12]. Several excellent review articles on image compression have been published in the literature [13–18], and there have also been several special journal issues dedicated to image coding and visual communications [19–27].

A. Need for Image Compression: Examples

The need for image compression can best be illustrated by considering a few storage and/or transmission examples based on current or proposed technology.

  1. *Storage*. An electronic still camera using a single-chip CCD sensor (with a color filter array) might required 484 × 768 pixels × 8 bits/pixel, resulting in about 3×10^6 bits/image. A 1-megabyte solid-state memory card used on-board the camera could store less than three images in uncompressed form. Obviously, this is neither convenient nor cost-effective. If an 8:1 *compression ratio* (defined as the ratio of the number of bits used for the original image to the number of bits for the compressed image) can be achieved, the memory card can then hold 24 images, making the system competitive in terms of storage with a conventional film-based camera using a 24-exposure roll.

  2. *Transmission*. A low-resolution facsimile (bilevel) image (e.g., an 8½- by 11-in. document scanned at 200 pixels/in. horizontally and 100 pixels/in. vertically) transmitted using a 4800-bit/s modem over ordinary telephone lines requires approximately 6.5 min to transmit in uncompressed form. In comparison, current fax compression standards allow the same document to be sent usually in under 30 s. For a multipage document, the savings in total transmission time are significant.

  3. *Storage/transmission*. A full-motion color video signal (consumer TV quality) requires 512 × 480 pixels/frame × 24 bits/pixel × 30 frames/s, resulting in about 180×10^6 bits for 1 s of video. A CD-ROM with a 650-megabyte storage capacity can only hold about 30 s of full-motion video. Furthermore, the CD-ROM data transfer rate is about 150 kilobytes/s, which means that it would take 1.2 h to play back the 30 s of uncompressed video. Appropriate compression techniques would not only allow a longer video sequence to be stored, but it would also allow more video data (in compressed form) to be transferred per second from the CD-ROM. If the compression ratio is high enough (around 150:1), the sequence could potentially be played back in real time.

B. Achieving Compression: Redundancy and Irrelevancy

Given the need for image compression, the question arises as to why images can be compressed at all. If each pixel value represents a unique and perceptually important piece of information, it would be difficult indeed to compress an image. Fortunately (at least from the standpoint of compression), the data comprising a digital image or sequence of images are often redundant and/or irrelevant. *Redundancy* relates to the statistical properties of images, while *irrelevancy* relates to the observer viewing an image [6]. Redundancy can be classified into three types: (1) spatial (due to the correlation between neighboring pixels in an image), (2) spectral (due to the correlation between color planes or spectral bands), and (3) temporal (due to the correlation between neighboring frames in a sequence of images). Similarly, irrelevancy can be classified as spatial, spectral, and/or temporal in nature, but the key issues in this case are the limitations and variations of the human visual system (HVS) when presented with certain stimuli under various viewing conditions. Ideally, an image compression technique removes redundant and irrelevant information and then efficiently encodes what remains. Practically, it is often necessary to throw away both nonredundant and relevant information to achieve the necessary degree of compression.

The preceding comments point to a fundamental dichotomy in the classification of image compression techniques, namely, whether the compression is lossless or lossy. In *lossless compression* (also known as bit-preserving or reversible compression), the reconstructed image after compression is numerically identical to the original image on a pixel-by-pixel basis. In *lossy compression* (also known as irreversible compression), the reconstructed image contains degradations relative to the original image. However, under certain conditions, these degradations may not be visually apparent (sometimes called *visually lossless compression*). Obviously, lossless compression is ideally desired since no information is compromised. Unfortunately, only modest compression ratios (an average of 2:1 for a single-band image) are possible with lossless compression. Much higher compression ratios can be obtained with lossy techniques in exchange for *potentially* visible degradations.

C. Compression System Components

Figure 1 depicts the three basic components of a general compression scheme: (1) image decomposition or transformation, (2) quantization, and (3) symbol encoding. The *image decomposition* or *transformation* is usually a reversible operation and is performed to eliminate the redundant information, or more generally, to provide a representation that is more amenable to efficient coding. This stage is used in both lossless and lossy techniques. Examples include the prediction error signal formation in differential pulse code modulation (DPCM), the discrete cosine transform (DCT), and subband/wavelet decompositions. The next stage, *quantization*, is a many-to-one mapping found only in lossy techniques, and it is the point at which errors are introduced. The type and degree of

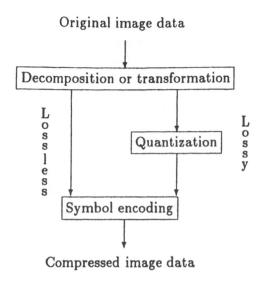

Figure 1 General compression framework.

quantization has a large impact on the bit rate and the reconstructed picture quality of a lossy scheme. In essence, quantization can be viewed as a control knob that trades off image quality for bit rate. Examples of quantization strategies include uniform or nonuniform scalar quantization or vector quantization. It is also desirable to quantize in such a way that the resulting output can be efficiently encoded by the last stage. This final stage, *symbol encoding*, is a means for mapping the symbols (values) resulting from the decomposition and/ or quantization stages into strings of 0's and 1's, which can then be transmitted or stored. This mapping may be as simple as using fixed-length binary codewords to represent the symbols, or it might use a variable-length code, such as a Huffman code or an arithmetic code, as a means of achieving rates close to the fundamental information-theoretic limits. It should be noted that the three components shown in Fig. 1 often mutually interact, and their *joint* optimization is a complicated task. As a result, they are often optimized individually based on assumed inputs.

An important aspect of compression techniques is that one or more of the components shown in Fig. 1 can be implemented in either an adaptive or a nonadaptive mode. A component is adaptive if its structure (or parameters) changes within an image to take advantage of locally varying image characteristics. Since most images vary significantly from one region to another, adaptivity offers the potential for improved performance, in exchange for an increase in complexity. In systems with *causal adaptivity*, the adaptivity is inferred only from the previously reconstructed pixel values (e.g., reconstructed in a raster scan fashion), and as such, no overhead information needs to be transmitted to the decoder. On the other hand, in systems with *noncausal adaptivity*, the encoder parameters are based on previous pixel values (reconstructed or actual) in addition to future input values. Although this may result in a higher bit rate due to the required overhead information, it also usually results in superior performance and lower decoder complexity.

The remainder of this chapter is organized as follows. Symbol encoding (Section II) and quantization strategies (Section III) are discussed first, since these two components are rather general in nature and common approaches are used in all compression techniques. The primary difference between most techniques is in the decomposition applied to the original data, and various approaches are discussed in Sections IV to VII. In Section IV we focus on several common techniques (primarily lossy) for *intraframe coding* (removing spatial redundancy in single images), and in Section V we outline techniques for hierarchical coding (coding images in such a way that different resolutions or quality levels are easily accessible). In Sections VI and VII we discuss extending these techniques to the compression of color images (removing spectral redundancy) and image sequences (removing temporal redundancy in image sequences). In Section VIII, various image compression standards that have been adopted or are being proposed are discussed; such standards are critical in allowing compressed image data to be transferred across application boundaries. Finally, in Section IX we outline some of the factors that should be considered in selecting a compression scheme for a particular application.

II. SYMBOL ENCODING

In this section we consider the third component of Fig. 1: the mapping of output symbols (values) resulting from the decomposition and/or quantization stages into channel symbols, which typically are 0's and 1's. This mapping operation is referred to as *symbol encoding, noiseless coding, lossless coding*, or *data compaction coding*. Symbol encoding may be as simple as using fixed-length binary codewords to represent the symbols, or it might use variable-length codewords for better efficiency.

A. Fixed-Length Codes

In some applications it is desirable to use fixed-length codewords to minimize implementation complexity or to satisfy certain channel constraints, such as the need for a constant data rate. As an example, consider a case in which one needs to encode the values of a discrete ternary source S. Let's assume that the symbol A means that the signal value equals zero, the symbol B means that the value is positive, and the symbol C means that the value is negative. Further, assume

Table 1 Fixed-Length and Variable-Length Codes for the Ternary Source *S*

Source symbol	Symbol probability	Fixed-length code	Variable-length code
A	0.70	00	0
B	0.15	01	10
C	0.15	10	11

that the source is *memoryless* (i.e., the probability of a particular symbol occurring does not depend on any previous symbols). A fixed-length code for this example is given in Table 1, where each signal value is encoded with 2 bits.

The most important criterion in designing an efficient code is minimization of the *average codeword length*, $\bar{L}$, since it translates directly to the storage space or the transmission time required by the application. Obviously, for the fixed-length code in Table 1, $\bar{L} = 2$ bits/symbol. In this example, notice that there is an inefficiency since there are four possible 2-bit codewords but only three are used. When the number of source symbols is not a power of 2, the value of $\bar{L}$ can usually be reduced by assigning fixed-length codewords to blocks of source symbols (*block coding*). For example, if we encode blocks of three symbols from the ternary source, there are $3^3 = 27$ possible combinations, which requires 5 bits/block. The average codeword length per original source symbol has now been reduced from 2 bits to 1.67 bits. It can be shown that by making the block size arbitrarily large, $\bar{L}$ asymptotically approaches $\log_2 n$, where *n* is the number of source symbols. For our example, $\log_2 3 = 1.59$ bits per source symbol.

B. Variable-Length Codes

1. Huffman Coding

If all source symbols were equally probable, fixed-length block coding would be the optimal strategy. In practice, the various source symbols typically occur with different probabilities, and in such cases, a smaller average codeword length can be achieved by assigning shorter codewords to the more probable symbols. Returning to the example for a ternary source, assume that the symbols occur with the following probabilities: $p(A) = 0.70$, $p(B) = 0.15$, and $p(C) = 0.15$. An example of a variable-length code for this example is given in Table 1, where the codeword for the most probable symbol, *A*, consists of only 1 bit, while the less probable symbols, *B* and *C*, are each assigned 2-bit codewords. For this variable-length code, $\bar{L}$ is only 1.30 bits/symbol.

Note that this code is *uniquely decodable*: that is, there is a one-to-one correspondence between any sequence of input symbols and its corresponding output sequence. A necessary and sufficient condition for constructing such codes

Table 2 Variable-Length Code for the Second Extension of S

Symbol	Probability	Code length	Codeword
AA	0.4900	1	0
AB	0.1050	3	100
AC	0.1050	3	101
BA	0.1050	3	110
BB	0.0225	6	111100
BC	0.0225	6	111101
CA	0.1050	4	1110
CB	0.0225	6	111110
CC	0.0225	6	111111

is that no codeword be a prefix of some other codeword. Any code satisfying this condition is called a *prefix condition code*. A code is *compact* (for a given source) if its average length is less than or equal to the average length of any other prefix condition code for the same source and the same code alphabet. A general method for constructing compact codes is due to Huffman [28] and is based on the following two principles:

1. Consider a source with an alphabet of size α. By combining the two least probable symbols of this source, a new source with $\alpha - 1$ symbols is formed. Suppose that the codewords for this reduced source are known. It can be shown that the codewords for the original source are identical to the reduced source codewords for all symbols that have not been combined. Furthermore, the codewords for the two least probable symbols of the original source are formed by appending (to the right) "0" or "1" to the codeword corresponding to the combined symbol in the reduced source.
2. The Huffman code for a source with only two symbols consists of the trivial codewords "0" and "1."

Thus, to construct the Huffman code for a given source, the original source alphabet is repeatedly reduced by combining the two least probable symbols at each stage until a source with only two symbols is obtained. The Huffman code for this reduced source is known ("0" and "1"). Then the codewords for the previous reduced stage are found by appending a "0" or "1" to the codeword corresponding to the two least probable symbols. This process is continued until the Huffman code for the original source is found.

As with fixed-length coding, the average length of a variable-length code can usually be reduced by encoding the source symbols in blocks. Table 2 shows the Huffman code for blocks of two symbols of the ternary source S. The average length of this code is found to be 1.20 bits/original source symbol, a savings of about 8% compared to the code given in Table 1 for single source symbols.

Shannon, in his celebrated *noiseless source coding theorem* [34,35], showed that by using block coding with variable-length codes, a source can be losslessly encoded to an average bit rate arbitrarily close to, but not less than, the source *entropy*. The source entropy, $H(S)$, is defined as

$$H(S) = -\sum_{i}^{n} p(s_i) \log_2 p(s_i) \qquad \text{bits/symbol} \tag{1}$$

where $p(s_i)$ denotes the probability of occurrence of the symbol s_i. Based on Shannon's theorem, there is no variable-length code that can code our example source at a rate below its entropy of 1.18 bits/symbol. The code in Table 2 gets within 2% of the source entropy, which is adequate for most practical applications. Because the entropy provides a lower bound to the average length of lossless codes and because most coding strategies that we consider in this section perform near this bound, these strategies are also referred to as *entropy coding*.

2.   Modified Huffman Codes

For a source with a large symbol set, many of the symbols will often have very small probabilities. The corresponding Huffman codewords will be long, which can affect the implementation efficiency of certain fast Huffman decoding algorithms. In these cases it is advantageous to lump these less probable symbols under a single symbol, ingeniously called "ELSE,", and design a Huffman code for the reduced symbol set, including the ELSE symbol [29]. This procedure is known as the *modified Huffman code*, and it results in a substantial reduction in the storage requirements and decoding complexity over the traditional Huffman code. Whenever a symbol belonging to the ELSE category needs to be encoded, the encoder transmits the Huffman codeword for ELSE followed by extra bits needed to identify the actual message within the ELSE category. If the symbols in the ELSE group have a small probability, the loss in coding efficiency (the increase in average bit rate) will be very small.

As a simple example, consider the two-symbol blocks given in Table 2. The block *AA* occurs with a probability of 0.49, while the other blocks occur much less frequently. A simple modified Huffman coding strategy is to assign the codeword "0" to the *AA* block and the codeword "1" to the ELSE symbol, which represents the remaining eight possible blocks. Every time the ELSE symbol is employed, it is followed by three extra bits to identify uniquely the particular block that occurred. The average codeword length for this modified Huffman code is $(0.49 \times 1 + 0.51 \times 4)/2 = 1.26$ bits/symbol, which is only slightly higher than the bit rate for the full Huffman code given in Table 2. Also note that for this example, a simple modified Huffman coding of source symbol *pairs* outperforms the full Huffman coding of *individual* source symbols given in Table 1.

3. Limitations on Huffman Coding

To design a Huffman code, it is necessary to know the symbol probabilities. In many practical situations, these may not be known a priori, and it becomes necessary either to make assumptions about the probabilities or to estimate them. Assuming a set of probabilities can lead to poor coding efficiency if there is a mismatch with the actual values, so it is common to implement the Huffman coding as a two-pass algorithm. The first pass gathers the symbol statistics and generates the codebook, and the second pass encodes the data. The end result is a fixed (static) Huffman code for the entire data set, which works well if the symbol probabilities are relatively constant. Unfortunately, symbol probabilities often change significantly within a given data set (e.g., within an image), and a fixed Huffman code, though tuned for the particular data, can still be suboptimal. *Dynamic Huffman coding* schemes [30,31], where the codewords are adaptively adjusted during the encoding and decoding processes, will perform better with nonstationary statistics, but their implementations are complex. The UNIX utility "compact" is an adaptive Huffman code based on the ideas presented in [32].

Another limitation of Huffman coding is in situations where the source symbol probabilities deviate significantly from negative powers of 2. Based on Shannon's theorem, the *ideal binary codeword length* for the encoding of the source symbol s_i is $- \log_2 p(s_i)$. Since codeword lengths must be integers, this condition is met exactly only when the source symbol probabilities are negative powers of 2 (e.g., $\frac{1}{2}$, $\frac{1}{4}$, $\frac{1}{8}$, etc.). Furthermore, since the minimum codeword length is at least 1 bit/symbol, there is no way to achieve average codeword length of less than 1 bit (at least for the coding of single symbols). To illustrate this point, consider a binary image used in facsimile transmission where the white pixels are represented by "0" and the black pixels are represented by "1." Although the symbol "0" occurs much more frequently than the symbol "1" in typical documents, direct Huffman coding of these symbols would result in no compression at all, regardless of the symbol probabilities. We can use block coding, but convergence to the source entropy may be slow and the implementation complexity grows exponentially with the block size. A better strategy is to redefine the symbol set to be nonbinary, as is done in international facsimile standards, where the run lengths of 0's and 1's are encoded instead of the actual binary value [33]. However, even this strategy is inadequate if the redefined symbol probabilities vary throughout the image, as is typically the case.

In the next two sections we consider two alternative entropy coding strategies, arithmetic coding and LZW coding. These techniques can often circumvent some of the limitations of Huffman coding, resulting in improved coding efficiency.

4. Arithmetic Coding

In Huffman coding there is a one-to-one correspondence between the codewords and the source sequence blocks. In comparison, arithmetic coding is a nonblock code (also known as tree code), where a codeword is assigned to the entire input sequence s_m of length m symbols. In arithmetic coding, slightly different source sequences can result in dramatically different code sequences. The credit for the idea of arithmetic coding is usually given to Elias [34,35]. We first describe the fundamental concepts of arithmetic coding by examining Elias' code and then briefly discuss its practical implementations. For simplicity, we consider a binary source, but the discussion applies to multisymbol sources as well.

Consider encoding a sequence of m binary symbols, s_m, that has a probability of occurrence $p(s_m)$. Since the sum of $p(s_m)$ over all of the 2^m possible source sequences of length m must be 1, it is possible to assign a subinterval within the half-open interval $[0, 1)$, to each source sequence s_m, such that the length of the subinterval is equal to $p(s_m)$ and the subintervals are nonoverlapping. In the Elias code, this subinterval assignment is accomplished in the following manner. Let p_1 denote the probability of the first binary symbol being a "0" and $(1 - p_1)$ denote the probability of that symbol being a "1." The interval $[0,1)$ is first partitioned into two subintervals $[0, p_1)$ and $[p_1, 1)$, as shown in Fig. 2a.. The subinterval $[0, p_1)$ is chosen if the first symbol in the sequence is a "0," and the subinterval $[p_1, 1)$ is chosen otherwise. Let p_2 denote the probability of the second binary symbol being a zero. Each subinterval is further partitioned into two subintervals in a similar fashion. For example, for the subinterval $[0, p_1)$, this partitioning would result in two subintervals, $[0, p_1 p_2), [p_1 p_2, p_1)$, as shown in Fig. 2b. Together with the first symbol, the second symbol in the sequence is used to specify one of these subintervals. For example, the source sequence '01' corresponds to the subinterval $[p_1 p_2, p_1)$. Note that in general, p_1 and p_2 need not be the same, and in fact the value of p_j might depend on the binary symbols encoded previously. Let us denote the subinterval specified after $j - 1$ binary source symbols by its beginning point $C^{(j-1)}$ and its length $A^{(j-1)}$. According

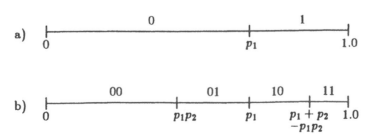

Figure 2   Example of interval partitioning for arithmetic coding.

to the procedure above, a new subinterval $(C^{(j)}, A^{(j)})$ is specified, where if the next symbol is a "0",

$$C^{(j)} = C^{(j-1)} \tag{2}$$

$$A^{(j+1)} = p_j A^{(j)} \tag{3}$$

and if it is a "1,"

$$C^{(j)} = C^{(j-1)} + p_j A^{(j-1)} \tag{4}$$

$$A^{(j)} = (1 - p_j) A^{(j-1)} \tag{5}$$

For any sequence it can be shown that the subinterval generated by this method has a width equal to the probability of that sequence. Furthermore, the subintervals produced by all the possible sequences of length m are nonoverlapping, and their union completely covers the interval $[0,1)$. Once the subinterval corresponding to a certain source sequence s_m has been identified, a codeword for s_m can be constructed by the binary expansion of the subinterval beginning point, $C^{(m)}$ (e.g., $C^{(m)} = 0.8125 = \frac{1}{2} + \frac{1}{4} + \frac{1}{16} \rightarrow 0.1101$). Since the beginning point of each subinterval is separated from the beginning point of its nearest right-hand neighbor by the subinterval width $p(s_m)$, it can be shown that it is only necessary to retain $l(s_m) = \lceil -\log_2 p(s_m) \rceil$ bits after the decimal point to uniquely specify the subinterval ($\lceil x \rceil$ is the smallest integer that is larger than x). This method of encoding results in a single codeword for the entire sequence s_m that has a length within 1 bit of the sequence's ideal codeword length, $-\log_2 p(s_m)$.

The major implementation problem of the Elias code is the precision required to carry out the subinterval computations. As the length of the source sequence increases, the length of the subinterval specified by the sequence decreases, and more bits are required to identify the subinterval precisely. Practical implementations of arithmetic coding address this problem by introducing a scaling (renormalization) strategy and a rounding strategy [36–42]. The scaling strategy magnifies each subinterval prior to partitioning so that its length is always close to 1. The rounding strategy uses b-bit (e.g., 12-bit) finite-precision arithmetic to measure the subinterval and perform the partitioning. Also, in several implementations, the multiply operations in Eqs. (2)–(5) have been approximated by add and shift operations [36,37]. Despite the inefficiencies introduced by modifications, practical implementations of arithmetic coding generally produce a bit rate that is only a few percent above the entropy of the source model.

An important point is that arithmetic coding creates a clear separation between the model for representing the data and the actual encoding of the data with respect to that model. In many practical situations, this flexibility can result in higher compression. Consider our previous example of encoding a binary image for facsimile transmission. Instead of coding in blocks or run lengths, each binary pixel value can be fed individually into the arithmetic coder along with its probability of occurrence, which can vary from pixel to pixel. These varying probabilities can be estimated based on fairly sophisticated models that utilize the values of the pixel's nearest neighbors [41]. In fact, some implementations of arithmetic coding, such as the IBM Q-coder [37], provide adaptive estimation of the binary symbol probabilities as an integral part of the encoding process. It is important to note that because arithmetic coding is a nonblock code, the encoding of a single pixel value may result in no output bits or several output bits, depending on the likelihood of that value.

5.   Lempel–Ziv–Welch Coding

The final coding strategy that we consider is a *universal* algorithm called Lempel–Ziv (LZ) coding after its inventors [43]. By "universal" is meant an adaptive code that does not require a priori knowledge of the source statistics. A practical implementation of this algorithm is due to Welch [44], and not surprisingly, is known as Lempel–Ziv–Welch (LZW) coding. It is based on progressively building up a dictionary (table) containing the symbol strings of various lengths that are encountered in the source sequence. Because the dictionary is built from the actual sequence, it accurately reflects the source statistics. The UNIX utility "compress" is a variation of LZW coding. Although the LZ class of algorithms are particularly suitable for the encoding of one-dimensional data (such as text or computer source files) where certain patterns or strings of source symbols occur frequently, they have also found use in coding two-dimensional images [45]. Following is a brief description of the encoder and the decoder operations of the LZW algorithm.

Encoder.   The string table is initialized to contain each individual source symbol. The input sequence is then examined one symbol at a time, and the longest input string for which an entry in the table currently exists is parsed. The codeword (table address) for this recognized string is transmitted to the receiver. The parsed input string is extended by the next symbol in the input sequence to form a new string, which is added to the table. This added string is identified by a unique codeword, namely, its table address, making it available for subsequent use by the encoder. The encoding process continues by parsing the longest recognizable string in the remaining portion of the sequence (beginning with the symbol immediately after the previously parsed string), extending this string by the next symbol, adding it to the table, etc.

Decoder.   The decoder is initialized to the same string table as the encoder. Each received codeword is translated by way of the string table into a source string. Except for the first symbol, every time a codeword is received, an update to the string table is made in the following way. When a codeword has been translated, its first source symbol is appended to the prior string to add a new

Table 3 LZW Code

Input string	Index
A	0
B	1
C	2
AB	3
BA	4
AA	5
AAA	6
AAAC	7
CA	8
AAB	9
BAA	10
AAAA	11
AC	12
CAB	13
BAAA	14
AAB· · ·	15

string to the table. In this way the decoder incrementally reconstructs the same table used at the encoder. A potential problem arises when an input source string contains the sequence $K\omega K\omega K$, where K denotes a single source symbol, ω denotes a string of source symbols, and the string $K\omega$ already appears in the encoder string table. The compressor will parse off $K\omega$, and send the codeword for $K\omega K$, and add $K\omega K$ to the table. Next, it will parse out $K\omega K$ and send the (*just generated*) codeword for $K\omega K$. However, the decoder, on receiving the codeword for $K\omega K$, will not yet have added that string to its table because it does not yet know an extension symbol for the prior string. In such cases, the decoder should continue decoding based on the fact that this codeword's string should be an extension of the prior string. Thus the first symbol in the new string is the same as the first symbol of the prior string, which now also becomes the extension symbol of the prior string.

The concepts described above are best illustrated by an example. Consider the encoding of the following terriary source sequence:

ABAAAAAACAABAAAAACABAAAAB· · ·

The initial dictionary contains the three symbols A, B, and C. For simplicity, let's assume that the maximum size of the table is 16 (i.e., a 4-bit word is used to address each entry in the table). After completion of the encoding operation, the dictionary in Table 3 has been created. The parsed sequence and the generated codewords are

$$\underset{A}{\overset{0}{}} \; / \; \underset{B}{\overset{1}{}} \; / \; \underset{A}{\overset{0}{}} \; / \; \underset{AA}{\overset{5}{}} \; / \; \underset{AAA}{\overset{6}{}} \; / \; \underset{C}{\overset{2}{}} \; / \; \underset{AA}{\overset{5}{}} \; / \; \underset{BA}{\overset{4}{}} \; / \; \underset{AAA}{\overset{6}{}} \; / \; \underset{A}{\overset{0}{}} \; / \; \underset{CA}{\overset{8}{}} \; / \; \underset{BAA}{\overset{10}{}} \; / \; AAB \cdot \cdot \cdot$$

At the beginning of the encoding process, the LZW method is inefficient, but once the table reaches a reasonable size, many symbols can be represented as a single entry. Because of the way the entries are formed in the dictionary, LZW coding naturally encodes run lengths of frequent symbols, as can be seen in Table 3. It can be shown that LZW coding becomes asymptotically optimal in the limit of a large dictionary; in most practical applications, a 12-bit table is chosen. Once the dictionary is filled, it can be fixed and used to encode the remaining symbols, or it can be updated adaptively to reflect the more recent source statistics. A study of LZ dictionary adaptation schemes is included in [46].

III. QUANTIZATION

A quantizer is essentially a staircase function that maps the possible input values into a smaller number of output levels. In this way, the number of symbols that need to be encoded are reduced at the expense of introducing error in the reconstructed image. The type and degree of quantization has a large impact on the final bit rate and the reconstructed picture quality of a lossy scheme. The individual quantization of each signal value is called *scalar quantization*(SQ), and the joint quantization of a block of signal values is called block or *vector quantization* (VQ). For the same encoding strategy, VQ can always be made to outperform SQ, but in many cases the gain is so minimal that it is not worth the additional complexity. In this section we discuss various techniques for both SQ and VQ.

The quantizer design problem is usually formulated as the minimization of some distortion measure for a given average output bit rate. The distortion measure may be *context-free* (i.e., independent of the neighboring signal values) or *context-dependent* (e.g., a perceptual measure based on the properties of the human visual system). An important aspect of quantizer design is to determine, for a desired level of distortion, the theoretical lower bound on the bit rate of any quantizer. This knowledge would enable a system designer to compare the performance of various quantizers to that bound and quantify the potential gains of adopting a more complicated quantization strategy (e.g., VQ instead of SQ). Rate-distortion theory [47], a branch of information theory, deals with obtaining such performance bounds. Due to the sophisticated mathematics involved, key results have been established only for context-free distortion measures such as mean-squared error (MSE) and for a few well-known signal probability distributions such as the Gaussian distribution. Despite these limitations, the results still provide useful insights into quantizer performance.

Several approaches to designing quantizers based on visual criteria have been suggested [48–51], but a justifiable debate continues as to the best criterion to use. As a result we restrict ourselves to the design of context-free quantizers based on the minimization of the MSE. Although it is well known that MSE does not always correlate well with perceived image quality, it does provide a measure of relative quality for the same algorithm at different bit rates, and its mathematical tractability has led to its widespread use. We consider several different quantization strategies and compare their relative complexity and performance.

A. Scalar Quantization

As mentioned previously, scalar quantization (SQ) refers to the independent quantization of each signal value. The main advantage of SQ is its implementation simplicity, but as we shall see, in many situations its performance is close to optimal in a rate-distortion sense. In the following we consider two basic types of scalar quantization: one where the output levels of the quantizer are fixed-length encoded, and one where the output levels are entropy encoded. Also discussed is the problem of bit allocation, where a block of signal values is scalar quantized with the aim of minimizing the overall distortion subject to a fixed quota of bits.

1. Lloyd–Max Quantizer

In many transmission or storage applications, the communication channel is fixed-rate, and the complexity of a rate-equalizer buffer to accommodate variable-length codes cannot be justified. In such cases it is desirable to find the quantizer that minimizes the distortion for a given number, N, of quantizer output levels. Such a quantizer is referred to as the Lloyd–Max [52,53] quantizer, after the two researchers who arrived independently at its solution. Note that since the quantizer output bit rate is $\log_2 N$, the value of N is usually chosen to be a power of 2 for efficiency, although this is not strictly necessary.

The mathematical development of the Lloyd–Max quantizer is straightforward. Let x be a continuous random variable with a probability density function (pdf) $p(x)$. A scalar quantizer maps the signal x into a discrete variable x^* that belongs to a finite set $[r_i, i = 0, \ldots, N - 1]$ of real numbers referred to as *reconstruction levels*. The range of values of x that map to a particular x^* are defined by a set of points $[d_i, i = 0, \ldots, N]$, referred to as *decision levels*. The quantization rule states that if x lies in the interval $(d_i, d_{i+1}]$, it is mapped (quantized) to r_i, which lies in the same interval. Given the signal pdf $p(x)$, the number of quantizer levels N, and the MSE distortion criterion, the Lloyd–Max quantizer is the solution that minimizes the expression

$$D = \sum_{i=0}^{N-1} \int_{d_i}^{d_{i+1}} (x - r_i)^2 p(x) \, dx \qquad (6)$$

with respect to $[d_i, i = 0, 1, \ldots, N]$ and $[r_i, i = 0, \ldots, N - 1]$. The solution results in decision levels that are halfway between the neighboring reconstruction levels and in reconstruction levels that lie at the center of the mass of the probability density enclosed by the two adjacent decision levels (i.e., at the conditional mean of the signal distribution in that interval). Mathematically, the decision and reconstruction levels are solutions to the following set of nonlinear equations:

$$d_i = \frac{r_{i-1} + r_i}{2} \tag{7}$$

$$r_i = \frac{\int_{d_i}^{d_{i+1}} x\, p(x)\, dx}{\int_{d_i}^{d_{i+1}} p(x)\, dx} \tag{8}$$

The Lloyd–Max quantizer has nonuniform decision regions. Since its objective is to minimize the *average* distortion, it tends to allocate more levels to those regions where the signal pdf is large. In general, except for a few special cases [54], Eqs. (7) and (8) do not yield closed-form solutions, and they need to be solved by numerical techniques [55]. When a numerical solution is necessary, the following iterative algorithm due to Lloyd can be used [52]. First, an arbitrary initial set of values for $[d_i]$ is chosen, and the optimal $[r_i]$ for that set are found by using Eq. (8). For the calculated $[r_i]$, the optimal $[d_i]$ are then determined using Eq. (7). This process is iterated until the differences between two successive approximations is below a threshold. In most cases, rapid convergence is achieved for a wide range of initial values.

2. Entropy-Constrained Quantizers

In applications where entropy coding of the quantizer output levels is allowed, the final bit rate is determined by the entropy of the quantizer output levels rather than strictly by the number of levels, N. This leads to another criterion for quantizer design, known as *entropy-constrained quantization*, where the quantization error is minimized subject to the constraint that the entropy of the quantizer output levels has a prescribed value. This means finding the quantizer that achieves the smallest MSE among all scalar quantizers with the same output entropy (but not necessarily the same number of output levels, N). The optimal entropy-constrained quantizer can be found numerically by an iterative technique similar to that used for the Lloyd–Max quantizer.

The problem of constrained-entropy quantization has been studied in detail. The reader is referred to [1,56] for comprehensive reviews. It has been established theoretically [57] that for memoryless sources and at high output bit rates, the optimal entropy-constrained SQ is a quantizer with uniformly spaced decision levels (i.e., a *uniform quantizer*). With some mild restrictions, these results are applicable to all pdf's and distortion measures of practical interest at high bit rates. Furthermore, it has been shown [57] that for the MSE distortion, the output bit rate of this optimal quantizer is within 0.255 bit/sample of the theoretical rate-distortion bound achievable by vector quantizers only in the limit of very large block sizes. Similarly, for other difference-based distortion measures, the entropy-constrained uniform scalar quantizer is close to the rate-distortion bound, but the difference from the bound may be more or less than 0.255 bit/sample, depending on the distortion measure. More recently, it was shown experimentally [58] that even at low bit rates, for the MSE distortion criterion and for a variety of memoryless sources, the performance of the uniform quantizer is virtually indistinguishable from that of the optimal entropy-constrained scalar quantizer. Interestingly, for the MSE distortion, the entropy-constrained quantizer at low bit rates is even closer to the rate-distortion bound than 0.255 bit, especially for more peaked pdf's such as Laplacian and gamma densities.

Based on the results above, a uniform scalar quantizer followed by entropy coding (e.g., Huffman or arithmetic coding) should be considered under the following conditions:

1. *The source is memoryless* (i.e., the sample values are independent). In such cases, VQ can at best result in a decrease of 0.255 bit/sample in quantizing the samples to the same distortion level. While the original pixel values in most images are not independent, many coding schemes first apply a transformation to the image with the aim of creating uncorrelated sample values that are subsequently subjected to quantization. Data that are uncorrelated are not necessarily independent (except for Gaussian random variables, where uncorrelated implies independence), but in many practical situations, decorrelating the data can remove many of the dependencies, and the performance advantage of VQ over SQ is reduced. However, for sources with memory (e.g., when the original image values are directly quantized), VQ can result in significant gains.

2. *The bit rate is moderate to large.* In such cases, the performance gap of 0.255 bit between SQ and VQ is only a small fraction of the quantizer output bit rate, and the gain in adopting a more complicated strategy is insignificant.

3. *Variable-length coding of the quantizer output levels is permitted.* As mentioned before, in applications where entropy coding is not allowed, the Lloyd–Max quantizer is the best choice for SQ. As we shall see, a vector quantizer can achieve optimal performance arbitrarily close to the rate-distortion curve without the need for entropy coding. However, this result is usually achieved for very large vector (block) sizes, which can be impractical.

3. Bit Allocation in Scalar Quantization

Consider a set of n random variables (rv's) x_k, $k = 1, 2, \ldots, n$, that have identical pdf shapes but unequal variances σ_k^2. Suppose that it is desired to allocate a fixed total number of bits, B, among all these rv's, using scalar quantizers to quantize each component independently, with the aim of minimizing the overall distortion. This situation often arises in a compression scheme where blocks (e.g., 8×8) of original image data are subjected to a decorrelating transform and are then quantized subject to a fixed total number of bits per block. Denote by R_k the number of bits allocated to an individual component x_k, and by R the average bits per component (i.e., $R = B/n$). It can be shown [59] that optimal bit allocation is achieved when all components are quantized to the same distortion. In particular, for the squared-error distortion measure and using a high-bit-rate approximation, the bit allocation is given by

$$R_k = R + \frac{1}{2} \log_2 \frac{\sigma_k^2}{(\Pi_{i=1}^n \sigma_i^2)^{1/n}} \tag{9}$$

for $k = 1, 2, \ldots, n$. Equation (9) implies that the bit allocation to quantizer k exceeds the average allocation R if the variance of x_k is greater than the geometric mean of the variances. Also, the number of quantization levels $N_k = 2^{R_k}$ of the kth quantizer is proportional to the standard deviation (σ_k) of the corresponding component. Under the above-mentioned assumptions, the actual value of R_k is independent of the rv pdf, but given R_k, a Lloyd–Max quantizer is designed for each component based on its pdf in order to achieve optimal performance.

Allocating the bits according to Eq. (9) has two major practical shortcomings. First, some of the resulting R_k values may be negative. This happens when the distortion in quantizing a component is larger than its variance. Second, the R_k's are in general continuous-valued and are not necessarily integers. Several ad hoc modifications have been suggested that start with the solution of Eq. (9) and set all the negative values to zero and round off the nonnegative values to the nearest integer. Then the integer bit rates are further adjusted to satisfy the constraint of the total number of bits. Although these techniques work, they are not optimal.

The optimal bit allocation problem under the nonnegativity constraint has been solved [60], but only for continuous values of R_k. An algorithm has been developed for optimal nonnegative integer bit allocation in the general case where component pdf's may be different [61]. A similar, but suboptimal, algorithm for nonnegative integer bit allocation is the *greedy bit allocation algorithm* [1].

B. Vector Quantization

As mentioned before, the joint quantization of a block of signal values is called vector quantization (VQ). In VQ, an n-dimensional input vector $\mathbf{X} =$

$[x_1, x_2, \ldots, x_n]$, whose components represent the discrete or continuous signal values, is mapped (quantized) into one of N possible *reconstruction vectors* $\mathbf{Y}_i$, $i = 1, 2, \ldots, N$. The distortion in approximating (quantizing) $\mathbf{X}$ with $\mathbf{Y}_i$ is denoted by $d(\mathbf{X}, \mathbf{Y}_i)$ and is defined according to the application. The most common distortion measure is MSE, which corresponds to the square of the Euclidean distance between the two vectors; that is,

$$d_{MSE}(\mathbf{X}, \mathbf{Y}) = \frac{1}{n} \sum_{i=1}^{n} (x_i - y_i)^2 \tag{10}$$

The set $\mathbf{Y}$ is sometimes referred to as the reconstruction *codebook* and its members are called *codevectors* or *templates*. The *codebook design* problem is to find the optimal codebook (in the sense of minimizing average distortion) for given input signal statistics, distortion measure, and codebook size, N. Once the codebook has been determined, the quantization process is straightforward and is based on the *minimum distortion rule*: a given input vector is compared to all the entries in the codebook and is quantized to the codevector that results in the smallest distortion; that is, choose $\mathbf{Y}_k$ such that $d(\mathbf{X}, \mathbf{Y}_k) \leq d(\mathbf{X}, \mathbf{Y}_j)$ for all $j = 1, \ldots, N$. With N possible codevectors, the output of the vector quantizer can be specified with $\log_2 N$ bits, and the resulting bit rate per vector component is $R = (\log_2 N)/n$ bits.

An important theoretical result of rate-distortion theory is that as $n \to \infty$, the quantizer output distortion can get arbitrarily close to the rate-distortion bound. Furthermore, for a given codebook size N and for very large n, the quantizer output entropy approaches $\log_2 N$, that is, all codevectors are utilized with equal probability and the need for entropy coding is eliminated. This is in direct contrast to SQ, where the quantizer output levels need to be entropy-encoded to achieve the best results. Another theoretical advantage of VQ is in the encoding of sources with memory. Unlike the SQ results that apply only to memoryless sources, the asymptotic VQ results apply to sources with any arbitrary joint pdf shapes.

Despite these theoretical advantages, it should be noted that large values of n can make designing, storing, and searching a codebook impractical, so a central issue in using VQ for a given application is determining if the desired performance can be achieved with vectors of modest dimension. In the following sections we discuss two key elements in designing a codebook: determining what vectors should be included in the codebook for best performance, and structuring the codebook to allow for efficient searching.

1. Codebook Generation

In SQ, the basic assumption is that the source is memoryless, and the source pdf is often modeled by a distribution such as Gaussian or Laplacian or, alternatively, is approximated by a histogram based on the actual input values. The optimal scalar quantizer is then found by solving Eqs. (7) and (8) using a numerical technique such as Lloyd's iterative method. In comparison, the source in VQ often has memory (since the vector components typically are correlated), and the task of finding an appropriate statistical model for the input vectors becomes difficult. As a result, most applications generate codebooks based on a *training set* of vectors that are representative of the vectors to be encoded; this is similar to the use of histograms in the SQ case. This approach eliminates the need for any a priori information about the underlying input source statistics.

This training set approach has certain implications. For example, in encoding any particular image, the generated codebook will be optimal only if the image itself is used as the training set. Despite its optimality, such a codebook, called a *local codebook*, has two disadvantages. First, a codebook must be generated for every image, which is a computationally intensive task that can hardly be performed in real time. Second, the local codebook must be transmitted to the receiver as overhead information. For example, if the vectors are formed from 4×4 blocks of the image pixels ($n = 16$), and each vector component is represented by 8 bits, a codebook of size 2^{10} results in an additional 0.5 bit/pixel of overhead for a 512×512 image. At low bit rates, this overhead may be a substantial portion of the total. The disadvantages associated with a local codebook can be overcome at the cost of introducing more distortion by using a *global codebook*, where several representative images are used as a training set. The resulting codebook may then be used to vector quantize all the images within that application. In general, a global codebook should be developed using as large and diverse a training set as possible to achieve reasonable average performance.

The algorithm generally used to generate VQ codebooks is the Linde–Buzo–Gray (LBG) algorithm [62], which is a generalization of Lloyd's iterative algorithm for SQ [52]. The steps in the LBG algorithm are as follows:

1. Start with a set of training vectors, an initial codebook $Y_i^{(1)}$, $i = 1, 2 \ldots$, N, a distortion measure d, and a fractional distortion change threshold ϵ. Initialize the iteration counter, l, to 1, and initialize the average distortion over all training vectors, $D^{(0)}$, to a large number.
2. The minimum distortion rule defines a decision region for each codevector $Y_i^{(l)}$, so that any training vector enclosed by a particular decision region is mapped to the corresponding $Y_i^{(l)}$. Quantize the training set by mapping each vector in the training set to its nearest codevector. Compute the average distortion $D^{(l)}$ resulting from the quantization process. If the fractional change in the average distortion from the previous iteration is less than or equal to the threshold, that is,

$$\frac{D^{(l-1)} - D^{(l)}}{D^{(l-1)}} \le \epsilon \tag{11}$$

convergence has been achieved and the algorithm terminates. Otherwise, continue to step 3.

3. Update the codebook by replacing each codevector $Y_i^{(l)}$ within a decision region by a new vector $Y_i^{(l+1)}$ that minimizes the quantization error within that decision region. For example, for the MSE measure, the minimum distortion codevector is simply the average (centroid) of the training vectors enclosed by that decision region (i.e., each component of the new codevector is found by computing the mean of the corresponding components of the training vectors). Set $l = l + 1$ and go to step 2. While each iteration results in a nonincreasing distortion, convergence may take many iterations if the threshold ϵ is set too low. As a result, the algorithm is usually terminated after a maximum number of iterations.

The LBG algorithm only guarantees a locally optimal codebook, and generally numerous local optima will exist. Many of these may yield poor performance, and therefore, choosing an initial codebook is a very important part of the LBG algorithm. Good performance can usually be obtained by providing a good initial codebook [62–64]. In generating an initial codebook, one can start with some simple codebook of the correct size, or one can start with a simple smaller codebook and recursively construct larger ones until the desired size is reached. Specific techniques for codebook initialization include:

1. *Random codes* [63]. One approach is to use a random initial codebook where the first N vectors (or any widely spaced N vectors) of the training set are chosen as the initial codebook.
2. *Splitting* [62]. In a splitting procedure, the centroid for the entire training set is found, and this single codeword is then split (i.e., perturbed by a small amount) to form two codewords. The LBG algorithm is applied to get an optimal codebook of size 2. The design continues in this way (i.e., the optimal code of one stage is split to form an initial code for the next stage until N codevectors are generated).
3. *Pairwise nearest neighbor (PNN) clustering* [64]. Starting with M clusters, each containing one training vector, the two closest vectors are merged to create the optimal $(M - 1)$ cluster codebook. This process is repeated until the number of clusters equals N. The centroids of these clusters are then used as the initial codebook.

2. Codebook Design: Tree-Structured Codebooks

In finding the minimum distortion codevector for a given input vector, a full search of the codebook can be performed at a computational cost of $O(nN)$. The

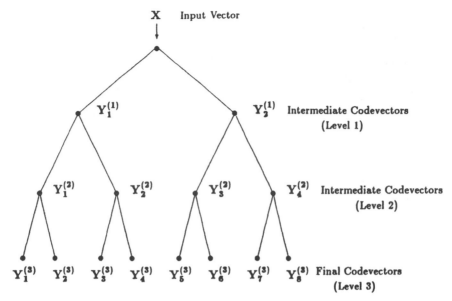

Figure 3 Tree-structured VQ codebook example.

associated storage cost is nN. For large codebooks, the search process becomes computationally intensive. To reduce the search time, a *tree structure* can be imposed on the codebook, where each node has m branches and there are $p = \log_m N$ levels to the tree [56,63]. This reduces the computational cost to $O(nm \log_m N)$ since only certain branches of the tree are examined. However, the storage cost increases to $nm(N - 1)/(m - 1)$, since vectors must be stored for every level (including the final level). The codebook is designed by using successive applications of the LBG algorithm at each node for a codebook of size m. An example is shown in Fig. 3 for a binary tree ($m = 2$) with $N = 8$. In this example, 14 codevectors must be stored (compared to 8 codevectors for a full search), but only 6 codevectors are actually searched in the tree (compared to 8 for a full search).

A tree-structured codebook can never perform better than a single-level full-search codebook, since a tree structure effectively limits the possible codevectors once a particular branch has been selected. However, by increasing the number of branches at each node, performance can be improved and storage requirements can be reduced. The trade-off is that the number of computations is increased by using more branches.

VQ tree structures can also be designed with a nonuniform number of branches at each node. Two such structures are *tapered trees* [65] and *pruned trees* [67]. In tapered trees, the number of branches per node is increased as one moves down the tree (e.g., 2 branches/node at the first level, 3 branches/node at the second level, etc.). Improved performance has been reported with such trees compared to uniform trees [65]. In pruned trees, a large initial tree is pruned by removing codevectors, so that the final tree achieves a given average length with minimum average distortion. The basic idea is to remove those codevectors that do not contribute significantly to reducing the distortion. Pruned trees can be designed for either fixed- or variable-rate schemes (possibly with entropy constraints [66]), and results reported for speech signals indicate that pruned tree structures can outperform full-search VQ at a given bit rate [67].

IV. COMMON TECHNIQUES IN IMAGE COMPRESSION

In the previous sections we discussed methods for assigning binary codewords to the symbols produced by a source (*symbol encoding*) as well as methods for reducing the number of symbols prior to encoding (through *quantization*). The techniques used in quantization and symbol encoding are common to most compression algorithms, and the distinction between various algorithms lies primarily in the decomposition or transformation process that is applied prior to these other two components. In this section we provide a brief overview of some popular compression techniques and describe how all three components can work together to provide useful and efficient compression of digital images. The focus is on lossy techniques for compressing single-channel images (intraframe coding), since they form the basis for techniques that encode multichannel (color) images and image sequences (interframe coding).

A. DPCM

The values of neighboring pixels in an image are often similar (i.e., they are highly correlated). As a result, the value of a given pixel can often be predicted quite accurately by examining the surrounding pixels. This observation is the basis for a number of *predictive coding* techniques. By far, the most common approach to predictive coding is *differential pulse code modulation* (DPCM), first described in 1952 [68].

In DPCM, an image is encoded one pixel at a time across a raster scan line. In Fig. 4, the pixel to be encoded is labeled x_m, and it is assumed that all pixels scanned prior to x_m (shown in the figure as A, B, C, etc.) have already been encoded and transmitted. Both the transmitter and receiver can then access these previous values to form a prediction for x_m. A linear prediction is typically formed, that is,

$$\hat{x}_m = \sum_{i=0}^{m-1} \alpha_i x_i \qquad (12)$$

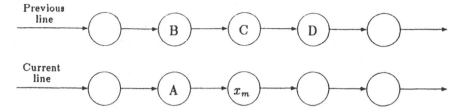

Figure 4 DPCM predictor configuration.

where $\hat{x}_m$ denotes the prediction, the x_i's are the m previous pixel values, and the α_i's are predictor coefficients (weighting factors) specified by the user. The number of pixels employed in the prediction is called the order of the predictor, and studies performed on television images [69] and radiographs [70] have shown that there is usually not much to be gained beyond a third-order predictor. An optimal set of predictor coefficients can be computed for a given image based on its autocorrelation values; as an example, for an image modeled as a first-order Markov process with correlation coefficient ρ, the optimal fourth-order predictor is

$$\hat{x}_m = \rho A - \rho^2 B + \rho C \tag{13}$$

In practice, simple fixed predictors such as

$$\hat{x}_m = A \tag{14}$$

$$\hat{x}_m = 0.5A + 0.5C \tag{15}$$

or

$$\hat{x}_m = A - B + C \tag{16}$$

are often used regardless of the actual image data, with varying degrees of success. If only pixels from the current scan line are used in forming the prediction, the predictor is referred to as one-dimensional (1D). If pixels from the previous lines are also included, the predictor is two-dimensional (2D). Two-dimensional prediction usually results in a large subjective performance improvement compared to 1D prediction, due primarily to the elimination of jaggedness around nonhorizontal edges.

After the prediction has been formed, it is subtracted from the original pixel value to form a differential (or error) signal, that is,

$$e_m = x_m - \hat{x}_m \tag{17}$$

The motivation for this simple transformation of the data is that the differential signal, e_m, has a greatly reduced variance and is significantly less correlated than x_m. Furthermore, it has a stable histogram well approximated by a Laplacian distribution [69], regardless of the histogram of the original image (although the standard deviation of e_m will vary with the image). A comparison between the histograms for an original image and its corresponding differential signal (with $\hat{X} = 0.5A + 0.5C$) is shown in Fig. 5. Because of its characteristics, the differential signal can be encoded much more efficiently and robustly than the original image.

In *lossless* DPCM, the differential signal is encoded directly, often using a Huffman code or modified Huffman code tailored to the statistics of e_m. In *lossy* DPCM, the differential signal is quantized prior to encoding in order to reduce the bit rate at the expense of errors in the reconstructed image. The optimal quantizer structure depends on the subsequent encoding method; for fixed-rate applications, a Lloyd–Max MMSE quantizer (Section III.A.1) tuned for the Laplacian distribution of e_m can be used, and for variable-rate applications that include entropy coding, a uniform quantizer (Section III.A.2) is more appropriate. A DPCM scheme using a 1-bit quantizer is also known as *delta modulation*. A block diagram for a lossy DPCM system is shown in Fig. 6. Notice how the quantizer is incorporated into the predictor loop at the transmitter, so that quantized, reconstructed pixel values, x_m^*, are used in forming the predictions. This allows the transmitter and receiver to track each other, since only quantized values are available to the receiver.

DPCM schemes can be made adaptive in terms of the predictor, the quantizer, or both. Adaptive quantizers and predictors increase performance by changing in response to local image statistics rather than being optimized for global statistics. Typical approaches to adaptive DPCM include switching between a number of different predictors or quantizers based on some criterion [71–74] (such as reconstruction error over a block of pixels) or scaling the predictor or quantizer by some predetermined factor based on previous differential signals [75–77].

B. Transform Coding

A general transform coding scheme involves subdividing an image into $n \times n$ (typically, $n = 8$ or 16) blocks of pixels and performing a reversible orthogonal transformation on each block. For each $n \times n$ block, the transformation produces an $n \times n$ set of transform coefficients. The goal of this process is to decorrelate the original signal, resulting in the energy being distributed among only a small set of coefficients. In this way, many coefficients can be discarded after quantization and prior to encoding. A block diagram of a basic transform coding scheme is shown in Fig. 7.

A transformation can be viewed as a decomposition of the original block of pixels into a set of basis images. In the case of the Fourier transform (or the general family of sinusoidal transforms), the basis images consist of sines and/or cosines with different spatial frequencies, and each transform coefficient is proportional to the fraction of energy in the original block at that particular fre-

(a)

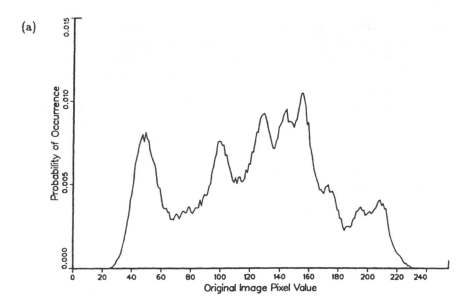

(b)

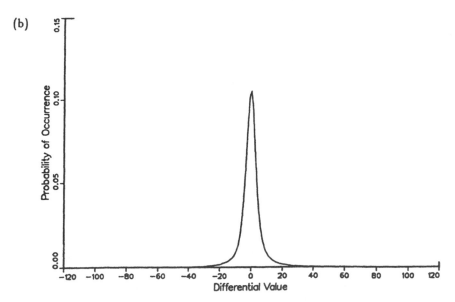

Figure 5   (a) Original image and (b) differential signal histograms.

Transmitter

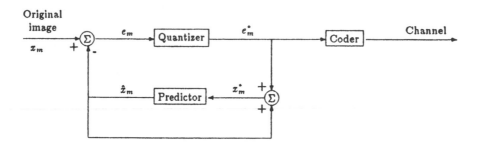

Receiver

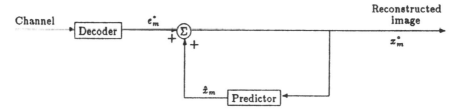

Figure 6 DPCM block diagram.

Transmitter

Receiver

Figure 7 Transform coding block diagram.

quency. Because of this relationship, it is also natural to interpret the Fourier transform as a spectral or frequency decomposition of the original block. This spectral interpretation can be extended to other transforms if the concept of frequency is generalized to include functions other than sines and cosines. For example, in the Hadamard transform, rectangular waveforms with increasing numbers of sign changes (referred to as *sequency*) are used as the basis functions. Each transform coefficient is then proportional to the fraction of energy in the original image associated with its corresponding spectral function.

A practical transform for the purpose of image compression should have a strong decorrelating effect on typical imagery, should preferably consist of image-independent basis functions, and should have a fast implementation. Since its introduction in 1974, the discrete cosine transform (DCT) has by far become the most popular transform for image compression applications [3,78]. The forward two-dimensional (2D) DCT of an $n \times n$ block of pixels can be defined in a number of ways, depending on the scaling factors that are chosen. The definition specified in the proposed JPEG compression standard [79] is

$$F(u, v) = \frac{16C(u)C(v)}{n^2} \sum_{j=0}^{n-1} \sum_{k=0}^{n-1} f(j, k) \cos \frac{(2j + 1)u\pi}{2n} \cos \frac{(2k + 1)v\pi}{2n} \quad (18)$$

where $f(j,k)$ represents the original pixel values in the block, $F(u, v)$ represents the transform coefficients at spatial frequencies u and v, and

$$C(w) = \begin{cases} \frac{1}{\sqrt{2}} & \text{for } w = 0 \\ 1 & \text{for } w \neq 0 \end{cases} \quad (19)$$

To recover the original pixel values from the transform coefficients, an inverse 2-D DCT is used:

$$f(j, k) = \frac{1}{4} \sum_{u=0}^{n-1} \sum_{v=0}^{n-1} C(u)C(v)F(u, v) \cos \frac{(2j + 1)u\pi}{2n} \cos \frac{(2k + 1)v\pi}{2n} \quad (20)$$

The DCT is *separable*; that is, the transform kernel can be decomposed into two one-dimensional (1D) kernels specifying separate horizontal and vertical operations. Thus an $n \times n$ block of pixels can be transformed by first performing a 1D transform along each row of the block, followed by another 1D transform along each column of the resulting output.

The effect of the DCT on a typical image block is best illustrated through an example. Consider the following 8×8 block of pixel values from an actual 8-bit image:

$$f(j,k) = \begin{bmatrix} 30 & 35 & 36 & 43 & 59 & 95 & 115 & 134 \\ 38 & 38 & 45 & 63 & 80 & 110 & 126 & 139 \\ 45 & 50 & 56 & 75 & 97 & 121 & 139 & 149 \\ 47 & 57 & 74 & 95 & 114 & 132 & 142 & 153 \\ 54 & 70 & 83 & 105 & 126 & 141 & 146 & 153 \\ 70 & 76 & 94 & 118 & 135 & 143 & 147 & 150 \\ 73 & 87 & 105 & 126 & 145 & 147 & 150 & 149 \\ 84 & 100 & 115 & 133 & 147 & 151 & 148 & 150 \end{bmatrix}$$

To reduce the hardware and software complexity, several implementations of Eq. (18) have been developed with the aim of minimizing the number of multiplications and additions. These practical DCT algorithms all use fixed-precision integer arithmetic. Using Eq. (18) with the integer arithmetic implementation given in [80], the transformed block for our example is

$$F(u,v) = \begin{bmatrix} 819.1 & -263.2 & -8.5 & 1.7 & 0.6 & -4.5 & 0.7 & -4.8 \\ -155.8 & -38.3 & 54.9 & 0.8 & -4.1 & 2.5 & 3.6 & -4.5 \\ -22.1 & 19.2 & 11.4 & -1.2 & -3.3 & -0.2 & 1.1 & -1.0 \\ -17.7 & 0.6 & 2.6 & -0.4 & -0.4 & 0.3 & 1.1 & -0.3 \\ -2.4 & 2.0 & 1.1 & -5.0 & -3.9 & -4.1 & 1.6 & -2.0 \\ -4.1 & -1.8 & 0.7 & -2.6 & -0.5 & -0.1 & -0.5 & 2.2 \\ -0.7 & 2.9 & 6.1 & 0.7 & 0.2 & 0.2 & -0.1 & -0.2 \\ -1.3 & -2.4 & 2.8 & 0.8 & -3.2 & -3.1 & -2.8 & 0.3 \end{bmatrix}$$

The top-left transform coefficient is called the *DC coefficient*, and because of the scaling used in Eq. (18), it is eight times the average brightness of the original image block. The other coefficients are called the *AC coefficients*, with increasing horizontal frequency from left to right and increasing vertical frequency from top to bottom. For the scaling used in Eq. (18), the AC coefficients can take on values from -1024 to $+1024$ for 8-bit input data. It can be seen that the majority of the block energy is concentrated in the low-frequency coefficients. This result is typical since most image blocks are fairly smooth, which means that there is not much contribution from the high-frequency basis functions.

It is important to realize that the transform operation by itself does not achieve any compression, but by changing the representation of the information contained in the image block, it makes the data more suitable for compression. Compression is achieved by the subsequent steps of quantization and encoding of the transform coefficients. The level of compression usually depends on the degree of quantization performed; coarser quantization results in higher compression but also introduces more degradation. There are numerous strategies for the selection, quantization, and encoding of the transform coefficients [5,18].

In *zonal sampling*, only those transform coefficients that are located in a pre-specified zone in the transformed block (usually the low frequencies) are retained; all other coefficients are set to zero. Each retained coefficient is then quantized and encoded with a fixed-length codeword. The same quantizer can be used for each retained coefficient (e.g., an 8-bit uniform quantizer), or different quantizers can be specified for various groups of coefficients. For example, an 8-bit quantizer could be used for a few lowest-frequency coefficients (because of their higher average energy), a 7-bit quantizer for the next few low-frequency coefficients, and so on. Since the selected sample locations and the quantizer(s) are both prespecified, no overhead information is required.

Given a fixed total number of bits to encode the retained coefficients, a more sophisticated approach is to minimize the quantization error through bit allocation (Section III.A.). This is done by assigning bits in proportion to the logarithm of coefficient variances, where the variance at each coefficient location is found by computing the corresponding coefficient energy in each block and then averaging over all image blocks. This is a two-pass algorithm, where the first pass computes the coefficient variances and allocates the bits, and the blocks are then encoded during the second pass. Because the bit allocation matrix will vary from image to image, it must be transmitted as overhead. In this technique the bit allocation matrix is computed based on global image statistics. Due to the nonstationary nature of images, there may be significant variations among the statistics of different blocks, and using the same bit-allocation matrix for all blocks can cause large errors when a coefficient value greatly deviates from its expected value. This problem can be addressed by classifying each block into one of several classes, and then using a different bit allocation matrix for each class [81]. The classification can be based on the total variance (energy) of the block since this is a good measure of the block activity. This scheme also requires two passes. During the first pass, the different classes are formed by computing the statistics of all image blocks. To maintain a fixed bit rate, a fixed proportion of blocks are assigned to each class. The coefficient retention zone and the bit allocation matrix for each class are then determined and transmitted to the receiver. During the second pass, this information is used to classify each block and encode it according to the corresponding bit allocation matrix.

While the zonal sampling strategies have the advantage of a fixed bit rate, a major drawback is that the reconstructed images generally do not exhibit the same subjective quality. This is because some of the coefficients outside the coefficient retention zone may contain significant energy and their omission can result in large reconstruction errors. To overcome this problem, a *threshold sampling* strategy can be used where a threshold level is chosen, and only the coefficients whose values are above the threshold are quantized and encoded [82]. Either uniform or nonuniform quantizers may be used for the coefficients above the threshold. If a uniform quantizer is used, it output levels are usually entropy encoded [82].

In threshold sampling, the addresses of the selected coefficients must be transmitted since the number and location of the coefficients that are above the threshold change from block to block. The techniques for encoding the addressing information are usually based on some variation of run length encoding of the zero-valued (i.e., subthreshold) coefficients. A variation on the threshold sampling approach is to transmit the position and value of the L largest coefficients in each block [18]. The value of L may either be fixed or may vary depending on some attribute of the block, such as its variance. To provide more adaptivity, the number of quantizer levels can also vary, depending on the activity of the block.

The quantizer structure can also be made a function of the coefficient location so as to control the amount of distortion introduced in each coefficient. Usually, a finer quantizer (producing less distortion) is used to quantize the low-frequency coefficients, while a coarser quantizer can be used for the higher frequencies. To simplify the implementation, the various quantizers can be designed so that they differ only by a scaling factor. A single quantizer can then be used for all the coefficients, provided that the coefficients are scaled (normalized) prior to quantization in order to achieve the desired step size. At the decoder, each quantized coefficient is denormalized prior to the inverse DCT operation. The DCT-based algorithm used in the JPEG system [79] is essentially a threshold sampling technique that allows a user-specified normalization array to be applied to the coefficients. There is actually no explicit threshold specified in this technique, but the use of a normalization array combined with the subsequent quantization effectively results in a thresholding process, where the threshold varies from coefficient to coefficient.

Continuing our DCT example, consider a quantization scheme like that used in the JPEG system, where each coefficient is divided by a scaling factor and then uniformly quantized by being rounded to the nearest integer. (The use of a uniform quantizer is optimal if it is followed by variable-length encoding; see Section III.A.2.) The scaling factors define an array of values that determine quantization step size; larger values correspond to larger quantization steps. Because of the frequency decomposition provided by the DCT, it is easy to include human visual system (HVS) sensitivity variations as a function of spatial frequency (described by the contrast sensitivity function) into the design of the quantization array. In this way, the error introduced in quantizing each coefficient is matched to the perceptual importance of the coefficient. To vary the overall bit rate of an encoded image while maintaining these perceptual weightings, the quantization array can be scaled by a constant factor. An example of an HVS-based normalization array is one taken from [79]:

$$Q(u,v) = \begin{bmatrix} 16 & 11 & 10 & 16 & 24 & 40 & 51 & 61 \\ 12 & 12 & 14 & 19 & 26 & 58 & 60 & 55 \\ 14 & 13 & 16 & 24 & 40 & 57 & 69 & 56 \\ 14 & 17 & 22 & 29 & 51 & 87 & 80 & 62 \\ 18 & 22 & 37 & 56 & 68 & 109 & 103 & 77 \\ 24 & 35 & 55 & 64 & 81 & 104 & 113 & 92 \\ 49 & 64 & 78 & 87 & 103 & 121 & 120 & 101 \\ 72 & 92 & 95 & 98 & 112 & 100 & 103 & 99 \end{bmatrix}$$

The resulting normalized and quantized coefficient, $F^*(u, v)$, is given by

$$F^*(u, v) = \text{nearest integer}\left(\frac{F(u, v)}{Q(u, v)}\right) \tag{21}$$

Quantizing the integer-DCT array in the example according to this rule results in

$$F^*(u, v) = \begin{bmatrix} 51 & -24 & -1 & 0 & 0 & 0 & 0 & 0 \\ -13 & -3 & 4 & 0 & 0 & 0 & 0 & 0 \\ -2 & 1 & 1 & 0 & 0 & 0 & 0 & 0 \\ -1 & 0 & 0 & 0 & 0 & 0 & 0 & 0 \\ 0 & 0 & 0 & 0 & 0 & 0 & 0 & 0 \\ 0 & 0 & 0 & 0 & 0 & 0 & 0 & 0 \\ 0 & 0 & 0 & 0 & 0 & 0 & 0 & 0 \\ 0 & 0 & 0 & 0 & 0 & 0 & 0 & 0 \end{bmatrix}$$

As expected, the normalization and quantization process produces many zero-valued coefficients. As discussed, the position of the nonzero quantized coefficients in each block is not known beforehand, and this information, along with the actual amplitude of each nonzero coefficient, has to be encoded. One approach is to use a variable-length code to encode the amplitude of each nonzero coefficient in combination with its relative position from the previously encoded nonzero coefficient. A variation of this encoding strategy is used in the JPEG system; for more details, the reader is referred to [79].

At the receiver, the quantized coefficients are recovered and are then denormalized according to

$$\hat{F}(u, v) = F^*(u, v)Q(u, v) \tag{22}$$

The denormalized block is then inverse transformed using an inverse DCT such as in [80]. For our example, the reconstructed block is

$$f_{(j,k)} = \begin{bmatrix} 31 & 31 & 33 & 44 & 64 & 91 & 118 & 134 \\ 36 & 38 & 44 & 57 & 77 & 103 & 126 & 140 \\ 43 & 49 & 60 & 77 & 97 & 119 & 137 & 147 \\ 50 & 59 & 75 & 95 & 115 & 132 & 144 & 150 \\ 56 & 68 & 87 & 109 & 127 & 140 & 147 & 150 \\ 64 & 77 & 97 & 119 & 135 & 145 & 148 & 149 \\ 74 & 86 & 106 & 126 & 141 & 148 & 150 & 149 \\ 81 & 93 & 112 & 131 & 144 & 150 & 151 & 150 \end{bmatrix}$$

The errors introduced in the code values of the original image block due to compression are

$$
e(j,k) = \begin{bmatrix}
-1 & 4 & 3 & -1 & -5 & 4 & -3 & 0 \\
2 & 0 & 1 & 6 & 3 & 7 & 0 & -1 \\
2 & 1 & -4 & -2 & 0 & 2 & 2 & 2 \\
-3 & -2 & -1 & 0 & -1 & 0 & -2 & 3 \\
-2 & 2 & -4 & -4 & -1 & 1 & -1 & 3 \\
6 & -1 & -3 & -1 & 0 & -2 & -1 & 1 \\
-1 & 1 & -1 & 0 & 4 & -1 & 0 & 0 \\
3 & 7 & 3 & 2 & 3 & 1 & -3 & 0
\end{bmatrix}
$$

where $e(j, k) = f(j, k) - \hat{f}(j, k)$. These errors are relatively small given the limited number of data that have been retained after the quantization process.

C. Subband Coding

In *subband coding*, an image is first filtered to create a number of subimages representing various spatial frequency bands of the original full-band signal. Since each of these *subband images* has a reduced bandwidth compared to the original image, they may be downsampled. This process of filtering and down-sampling (called the *analysis stage*) does not produce any compression in itself since the total number of samples remains the same. However, the subband images can be coded more efficiently than the original full-band image, typically using different bit rates or even different coding techniques for each subband, thus taking advantage of the properties of each subband. Reconstruction is achieved by upsampling the decoded subbands, applying appropriate interpolation filters, and adding the results together. This process is termed the *synthesis stage*.

The typical approach to analysis/synthesis in the subband coding of images is to use a tree structure with Cartesian-separable filters [83]. This means that filtering and downsampling (or upsampling and filtering) can be performed first along one direction, and the result then processed along the other direction. Separable filters are relatively easy to design and apply, but nonseparable filters may offer some performance advantages [84,85]. A block diagram for the analysis and synthesis stages of a 2D, four-band system using a tree structure is shown in Figs. 8 and 9. The analysis filters, $h_1(\cdot)$ and $h_2(\cdot)$, are low-pass and high-pass filters, respectively, and similarly, the synthesis filters, $g_1(\cdot)$ and $g_2(\cdot)$, are low-pass and high-pass, respectively. The subband images are denoted as $y_{11}(m, n)$, $y_{12}(m, n)$, and so on, where the subscript indicates the filters used to generate the subband [e.g., 11 indicates that filter $h_1(\cdot)$ was applied in both directions, meaning that $y_{11}(m,n)$ is a low-pass/low-pass signal]. The approximate frequency ranges produced by the four-band system are shown in Fig. 10. If more than four subbands are needed, the analysis process can be reapplied to subbands formed previously. For example, the 11-subband image can be further

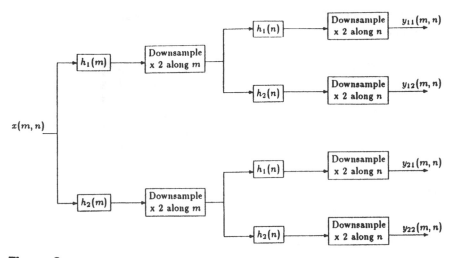

Figure 8 Two-dimensional four-band analysis bank.

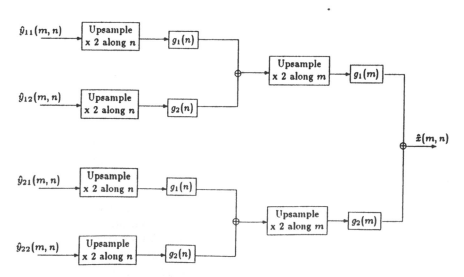

Figure 9 Two-dimensional four-band synthesis bank.

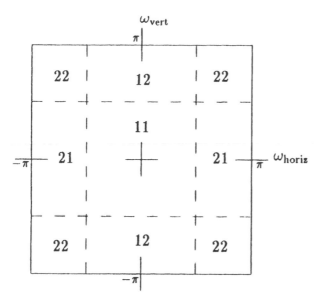

Figure 10 Approximate frequency ranges for a four band split.

decomposed into four smaller subband images to create a total of seven subbands.

The actual design of the low-pass/high-pass filter pairs used in analysis and synthesis can be done in several ways [86–88], but the typical approach is to use *quadrature mirror filters* (QMFs) [89]. Quadrature mirror filters are so named because the low-pass and high-pass filters exhibit mirror symmetry about $\pi/2$ (the quadrature point on a normalized frequency axis), as shown in Fig. 11 for an idealized QMF pair. This constraint eliminates any aliasing introduced by the

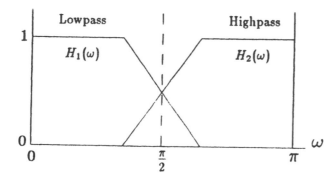

Figure 11 One-dimensional idealized QMF pair.

downsampling process. In mathematical terms, the relationships between analysis and synthesis filters are:

$$h_2(n) = (-1)^n h_1(n) \tag{23}$$

$$g_1(n) = 2h_1(n) \tag{24}$$

$$g_2(n) = -2(-1)^n h_1(n) \tag{25}$$

The QMF approach is to start with the specification of the analysis low-pass filter, $h_1(n)$, and then compute the remaining filters using Eqs. (23)–(25). Symmetric (linear phase) finite impulse response (FIR) filters are used to prevent phase distortion, and they are also designed to introduce minimal amplitude distortion. Larger filters (i.e., those with more coefficients) will generally result in improved reconstructed images, but this is quite dependent on the specific technique used to encode the subbands.

A variety of techniques have been proposed to encode subband images. A simple approach is to use DPCM encoding on each subband [90]; this works well when only a small number of subbands have been formed since each subband will still have significant pixel-to-pixel correlation. When a large number of subbands have been formed, a better approach is to use DPCM encoding on the lowest-frequency band (where the correlation is still high) and straight quantization (PCM encoding) on the higher-frequency subbands (where the correlation is much lower) [91]. In a manner similar to the bit allocation and HVS-based quantization strategies used in encoding DCT coefficients (Section IV.B), different quantizers can be used for each subband to distribute the quantization noise either equally across the bands or according to the perceptual importance of each band. Another technique is to use VQ encoding, either on blocks of pixels *within* each subband or on vectors formed *across* the subbands (i.e., by concatenating a pixel from the same spatial location in each subband) [92]. However, note that forming the vectors across the subbands is appropriate only if all subbands have the same dimensions.

There is an interesting relationship between subband coding and block transform coding that is worth discussing. Both approaches provide a localized frequency representation of the original image, with the main difference being the manner in which the output data are organized. This difference can best be explained pictorially. Consider a transform (such as the DCT) that is applied to 2×2 nonoverlapping blocks of an image. Referring to Fig. 12a, we see that each transformed block contains four transform coefficients, labeled a (the lowest frequency) to d (the highest frequency). Now consider decomposing the same image into subbands using 2×2 analysis filters and downsampling by a factor of 2 in each direction. It can be shown that by designing an *appropriate* set of analysis filters, a subband representation can be found that is entirely equivalent to the transform representation [93]. Referring to Fig. 12b, the low-pass/low-

Transformed Blocks   Subband Images

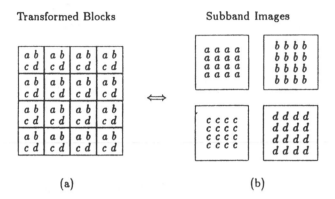

(a)   (b)

Figure 12 Equivalence between transform and subband representations.

pass subband image merely corresponds to a grouping of the lowest-frequency transform coefficients, one from each block. The other transform coefficients are similarly grouped to form the remaining subband images.

In general, the exact relationship between the transform kernel and the subband filters is more complex than the simple example suggests. This is because a subband filter of length greater than the decimation factor (e.g., an eight-tap filter and downsampling by two at each stage) corresponds to a *lapped orthogonal transform* (LOT) [93], where the transform is applied to overlapping blocks of pixels. Lapped transforms are used to reduce the blocking artifacts that can occur with disjoint block transforms. Since subband filter lengths are typically much greater than the decimation factor, blocking is generally not a problem in subband coding.

It is worthwhile to mention the term *wavelets* in conjunction with subband coding. A wavelet (literally "little wave") is basically a periodic function modulated by an envelope (such as a sinusoid multiplied by a Gaussian function). A discrete wavelet transform uses wavelets with different scalings and translations as basis functions to decompose an image into a localized frequency representation that is not unlike a nonuniform (octave) subband decomposition* (where the bandwidth of a subband is a function of its center frequency). Indeed, wavelet transforms have been implemented using a subband filtering approach [94,95]. There are some subtle aspects to the filters used in wavelet decompositions that are not always satisfied by commonly used subband filters [94], but

---

*In a nonuniform subband decomposition, lower-frequency components are split into smaller subbands than are the higher-frequency components. This results in low-frequency components with a fairly large spatial extent, and high-frequency components with a more restricted spatial extent.

from the standpoint of image compression, any practical difference between sub-bands and wavelets seems to be more one of nomenclature than anything else.

D. VQ Techniques

In Section III.B we discussed the design and performance of vector quantizers. Although it was never explicitly stated, it is obvious that a vector quantizer could be used to directly encode image data in a lossy manner. The input vector components might merely be the original pixel values, or they might be transformed values such as DCT coefficients or image subbands. Compression is achieved (and errors are introduced) by replacing each input vector with one of a limited number of codevectors. All that needs to be transmitted to the receiver is the index of the selected codevector, which can then be used as an entry into a look-up table (so decoding is extremely fast).

To improve image quality while maintaining the same bit rate R with basic VQ, it is necessary to increase the block size n and the codebook size N. Unfortunately, since $N = 2^{Rn}$, the codebook size (and encoder complexity) grows exponentially with n for a fixed bit rate. As described in Section III.B., a tree-structured codebook can be used to reduce the search time with large codebooks, but at the expense of increased storage requirements. Rather than increasing the codebook size, a different approach is to use a codebook with a product structure (i.e., a codebook that is formed as the Cartesian product of several smaller codebooks) [63]. If a vector can be characterized by certain independent features, a separate codebook can be developed to encode each feature. The final codeword is the concatenation of all the various encoder outputs. The advantage of a product code can be illustrated by considering an example where two separate codebooks of sizes N_1 and N_2 are used to encode the orientation and magnitude of a vector, respectively. The effective number of codevectors available using a produce code is $N_{eff} = N_1 N_2$, since there are N_2 different magnitudes available for each of the N_1 possible orientations. However, the storage and computational complexity is proportional to $N_1 + N_2$ rather than $N_1 N_2$, resulting in a substantial savings. Product codes can also result in more robust codebooks since the statistical distributions of certain image features are not overly sensitive to the particular input image. Product codes are suboptimal compared to full-search codebooks of the same effective size, but they can potentially outperform full-search codebooks with the same complexity and bit rate. This is due to the fact that for the same complexity, the effective size of a product codebook is much larger than a full-search codebook, and the product codebook can be used to quantize vectors of larger dimensions while maintaining the same bit rate.

One approach to product codes falls into a prediction/residual framework, where a prediction is made for the original image based on a limited set of data, and a residual is then formed by taking the difference between the prediction and the original image. The data used for the prediction are encoded using a scalar quantizer (one-dimensional VQ), and the residual image is encoded using a vector quantizer. In *mean/residual VQ* (M/RVQ) [96], the original image vectors are blocks of pixels (e.g., 4 × 4 blocks), and the prediction for each block is merely the mean value of the block. In essence, this approach creates a prediction image by replicating each mean value over its corresponding block. The motivation for M/RVQ is that many image vectors exhibit similar variations about different mean levels. By removing the mean from each vector, fewer codevectors are required to represent the residual vectors. In *interpolative/residual VQ* (I/RVQ) [97], the prediction image is formed by subsampling the original image and then interpolating back to full resolution. This process is very similar to M/RVQ except that subsampled values are used rather than mean values, and interpolation (usually bilinear) is used rather than replication. The motivation for I/RVQ over M/RVQ is the reduction in blocking artifacts achieved by using a relatively smooth prediction image rather than the replicated block means.

Another type of product code is *gain/shape VQ* (G/SVQ) [98], where separate codebooks are used to encode the shape and gain of a vector. The vectors are typically small blocks of pixels (e.g., 4 × 4 blocks), and the shape vector is defined as the original image vector normalized by the removal of a gain factor such as its energy or variance. This gain is quantized using a scalar quantizer and is transmitted to the receiver separately. The motivation for this approach is that many codevectors have similar variations (such as the presence of an edge), but the amplitude of those variations varies from one vector to another.

There are also other VQ techniques that use several smaller, separate codebooks but are not product codes by the strict definition. In *classified VQ* (CVQ) [99], a number of different codebooks are developed, each designed to encode blocks of pixels that contain a specific feature (e.g., a horizontal edge, a vertical edge, a uniform area, etc.). The codebook used for a particular block is determined by a classifier capable of differentiating among the various features. The rationale for this approach is that numerous small codebooks, each tuned to a particular class of vectors, can provide comparable image quality with lower search complexity than can a single large codebook. In *finite-state VQ* (FSVQ) [100], the VQ process is modeled as a finite-state machine where each state represents a separate VQ codebook. Rather than use a classifier as in CVQ, the codebook selection in FSVQ is done using a next-state function that maps the current state (with its associated codebook) and the current output codevector to another state (and associated codebook). FSVQ is an example of VQ with memory. The motivation for FSVQ is that adjacent blocks of pixels are often very similar, and we can take advantage of this correlation by choosing an appropriate codebook given previous encoding decisions.

E. Fractal Image Compression

A recent approach to image compression (and one that has generated much interest) involves using fractals to represent the structures in images. *Fractals* can be loosely defined as shapes that are irregular (compared to the lines, circles, etc., of standard Euclidean geometry) and have the interesting property of *self-similarity*. A self-similar object is one that looks roughly the same regardless of the scale at which it is viewed. Fractals have been found to be good models to describe natural objects [101], and their usefulness in image compression lies in the fact that they can be described very concisely by a *fractal code*. A fractal code merely consists of one or more simple transformations of a particular type that when applied iteratively to an arbitrary initial image, will converge to the desired image. However, the trick in fractal compression is finding the necessary transformations for a given image.

Much of the excitement surrounding fractal-based compression is due to Barnsley and Sloan [102–106], who are two of the principal researchers in this field. Their claims of extremely high compression ratios, while supplying somewhat limited technical descriptions of the exact process (supposedly because of patent issues), have left many with doubts regarding the validity of fractal-based compression. However, Jacquin, a former student of Barnsley, has described a compression method that applies fractals in a relatively straightforward way [107,108]. An in-depth discussion of fractal image compression is beyond the scope of this chapter, but in the following discussion, we try to give the reader a feeling for the basic approach presented in [108].

The process of fractal image compression begins by partitioning the original image into nonoverlapping blocks of size $B \times B$, where $B = 8$ in [108]. These blocks are known as *range blocks*, and each range block will be represented by its own fractal code.* As noted previously, the problem is finding the corresponding fractal code for each block. The solution is to make use of the self-similarly of the structures in an image in the following way. From the original image, all possible (overlapping) blocks of size $D \times D$, where $D > B$, are extracted. These larger blocks are known as domain blocks, and in [108], they were chosen to be 16×16. Associated with these domain blocks is a set of simple transformations that are chosen so that the application of one or more transformations will map one of the domain blocks to a selected range block. The transformations include such operations as averaging over 2×2 blocks, adding an offset to each pixel value, and scaling all pixel values by a constant (between 0 and 1), as well as transformations that shuffle the pixels within a block through rotation or reflection. These transformations can be classified as either

---

*The $B \times B$ range block could be subdivided into smaller blocks, e.g., into four $B/2 \times B/2$ blocks, to improve coding accuracy. A discussion of this aspect can be found in [108].

geometric transformations, which affect the geometry (e.g., size) of a domain block, or *massic* transformations, which change or redistribute the mass (i.e., the pixel value) within a block.

The key aspect of all these transformations is that they are *contractive*.* This means that the transformed object is never larger in spatial extent or in its range of pixel values than the original object. The importance of contractivity lies in the fact that a set of contractive transformations always has a *fixed point*, and if the transformations are iteratively applied to any arbitrary starting point, the result will converge to the fixed point. In the case of fractal image coding, the fixed point is a range block, and the convergence of the transformations means that we can start with any arbitrary values, iteratively apply a particular set of transformations, and eventually end up with the desired range block. As a result of this property, it is not necessary to know the actual domain block to reconstruct an image; the only information that must be transmitted to a receiver is the set of transformations that produced the given range block from a domain block. As noted previously, the set of transformations associated with a range block is called the fractal code for that block.

One potential disadvantage of this technique is the large number of computations needed to find the domain block and the set of transformations that will map to a desired range block. Computing every possible combination of transformations and domain blocks to find the best match is obviously inefficient. To overcome this problem, both the range and domain blocks can be classified into categories such as shade, midrange, or edge blocks. Certain transformations are also associated with each class. The end result is a decrease in the number of computations required to find the best match.

An interesting aspect of this block-based fractal coding technique is that it has some similarity to a block-based VQ technique [108]. In VQ, the nearest-neighbor match is found for each image block, and the corresponding index from the VQ codebook is transmitted. At the receiver, the index is used as an entry to an identical codebook in order to reconstruct the image. In fractal coding, the nearest-neighbor match is found using transformed domain blocks, but the domain blocks are never actually used by the receiver. Because of this, the fractal technique can be thought of as having a *virtual codebook*. Importantly, this virtual codebook is developed based on the image itself, analogous to a VQ technique in which the image is used as the training set for the VQ codebook. The ability to tune the coding process for each image is one potential advantage of fractal techniques. Additionally, the classification of range blocks prior to en-

---

*More precisely, all transformations are at least nonexpansive, and some must be contractive, so that the overall set of transformations applied to a domain block is contractive. The higher the overall contractivity, the faster the fractal code will converge to its fixed point (i.e., range block) during reconstruction.

coding is analogous to a classified VQ scheme, where an image vector is classified and then encoded using a codebook that is matched to its class.

V. HIERARCHICAL CODING

In *hierarchical coding*, image data are encoded in such a way that it is possible to access a given image at different quality levels or resolutions. Hierarchical coding is not one specific technique, but rather, it is a basic design philosophy that offers distinct advantages in certain applications. For example, in searching an image database, hierarchical coding allows the user to initially access a low-quality version of an image (at a correspondingly low bit rate) to determine if the image is the desired one. If needed, additional data can then be transmitted in stages to refine the image further. This type of scheme is termed *progressive transmission* (PT). An important aspect of progressive transmission is that the transmission can be stopped at any time, resulting in large *effective* compression even if the image data are not compressed. As another example, an image database might be used to support a number of output devices, each having a different resolution or quality requirement. A hierarchical coding scheme allows for efficient access of the appropriate version of the image by each device. This is an example of a *multiuse environment*. The common aspect of PT and multiuse environments is the need for *image hierarchies* (i.e., a way to organize the image data in order of importance). Typically, we speak of the various levels of the hierarchy, where each level corresponds to a reconstructed image at a particular resolution or level of quality. As we shall see, some of the coding techniques described in Section 4 lead naturally to image hierarchies; others do not. Hierarchies find use in other areas of image processing besides image compression, and they can be based on a variety of image attributes. For our purposes it is convenient to classify them into fixed-resolution hierarchies or variable-resolution hierarchies.

In *fixed-resolution hierarchies*, the reconstructed image is inherently the same size as the original image, and the value at any particular pixel location is refined as one moves from level to level. This type of hierarchy is used primarily for progressive transmission applications since it is poorly suited to multiuse environments, where the devices generally require different spatial resolutions. In *variable-resolution hierarchies*, the images corresponding to the levels of the hierarchy vary in spatial resolution. This approach naturally results in a *pyramid structure*, where the base of the pyramid represents the full-resolution image (i.e., the original image), and as one moves up the pyramid, the images decrease in spatial resolution and size. The variable-resolution hierarchy is particularly well suited to a multiuse environment supporting devices of varying resolutions as well as being useful for PT. For display purposes in a PT scheme, it may be desirable to interpolate or replicate the reduced-size images to a larger size for improved recognizability.

An important distinction between the fixed-resolution and variable-resolution hierarchies is the *incremental bit rate*, the number of bits required to move from one level of the hierarchy to the next (normalized to the number of pixels in the original image). In the fixed-resolution techniques, the incremental bit rate is held more or less constant as one moves through the levels. While a constant incremental increase in the number of transmitted bits does not necessarily correspond to constant incremental improvements in image quality, it does allow the reconstructed image to be refined at fixed time intervals. By comparison, the incremental bit rate for the variable-resolution techniques grows exponentially as one moves to higher levels of resolution. For example, increasing the image dimensions by a factor of 2 between each level results in an incremental bit rate that quadruples from one level to the next. Thus it takes longer to transmit and reconstruct as the progression proceeds, which may be a problem in some systems.

A. Fixed-Resolution Hierarchies

1. Bit Planes

A k-bit image can be represented by k bit planes, each of the same dimension as the original image. Progressive transmission can be easily achieved by transmitting the bit planes in a sequence, starting with the most significant bit plane and ending with the least significant bit plane. The image reconstructed from the most significant bit plane is a binary image, and additional gray levels are added as more bit planes are received. A lossless reconstruction is possible if all bit planes are used. Since each bit plane is a binary image, binary compression techniques (such as arithmetic coding) can be used to reduce the bit rate [109]. The amount of compression is largest for the most significant bit plane and decreases as one moves through the bit plane sequence. This property allows for large effective compression if the transmission is terminated after receiving the first few bit planes.

2. Tree-Structured VQ

Progressive transmission can be achieved with a tree-structured VQ codebook (Section III.B.2) by transmitting the index of the best-match intermediate codevector at each level of the tree [110]. As one progresses farther into the tree, the reconstruction quality improves as the codevector choices become more refined. Tree-structured VQ provides lossy compression at the final level unless a residual error image is transmitted. Given the index of the codevector selected at a particular level, the specification of a codevector index at the next level requires an additional $\log_2 m$ bits, where m is the number of branches. Since these additional bits are distributed as overhead over the entire vector, the progressive transmission process is relatively efficient.

3. Transform-Based Hierarchical Coding

In Section IV.B we discussed the ability of transforms to compact the image energy into a small number of coefficients. This property leads naturally to a hierarchical arrangement of the transform coefficients and allows for a recognizable reconstruction with a relatively small amount of data. Due to coefficient quantization, all of these techniques are lossy unless a residual error image is transmitted. The main distinction between various schemes is the method used to hierarchically order the coefficient information.

The most straightforward approach is to transmit the coefficients in order of approximately decreasing variance or in order of decreasing contribution to reconstructed image quality. This is generally accomplished by scanning a transformed block using a zigzag pattern (as is used in the JPEG DCT algorithm [79]), but other scanning patterns have also been examined [111,112]. The partial information transmitted for each stage may be a single coefficient or a block of coefficients, and images are reconstructed by progressively including more of the coefficients in the inverse transform (with the unknown coefficients set to zero). This points to an obvious disadvantage of all transform-based techniques: the need to perform an inverse transform at each step in the progression.

Rather than transmit only a few coefficients at each stage in the progression, another approach is to consider all of the coefficients within a given block and to allocate a limited number of bits to each one. As the progression proceeds, additional bits are added to each coefficient to improve the accuracy. One technique dynamically allocates a fixed number of bits at each stage and incorporates multistage quantization to iteratively refine any errors remaining from the preceding stage [113]. Another technique uses embedded quantizers and predefined incremental bit maps to control the reconstruction quality at each stage [114].

B. Variable-Resolution Hierarchies

1. Subsampling Pyramid

A simple spatial domain hierarchy can be generated by repeated subsampling of the original image data. For example, the original $N \times N$ image could be subsampled by a factor of 2 in both dimensions, and then the resulting $N/2 \times N/2$ image could again be subsampled, and so on, until an image consisting of a single pixel is reached. The reconstruction process at any level simply uses the subsampled points from all previous levels plus the new points from the current level. The total number of pixel values stored is equivalent to that of the original image, and the original image can be recovered exactly. Although this approach is efficient from this point of view, it suffers from several serious disadvantages. First, subsampling introduces aliasing, which becomes more pronounced at the higher levels of the pyramid. Second, the subsampled points may not be good representatives of the areas from which they are taken. Finally, it becomes difficult to apply compression techniques to the various levels because of the reduction in spatial correlation from the subsampling. However, these drawbacks may not be too severe in a system that requires only one or two hierarchical levels, and the subsampling approach may be worth considering for such a situation.

2. Mean Pyramids

Forming a hierarchy by averaging over blocks of pixels (typically, 2×2 blocks) eliminates many of the problems associated with the subsampling approach. The use of averaging reduces the aliasing at higher levels of the pyramid (since the averaging acts as a prefilter), and the mean values are generally better representatives of the block regions. Although techniques based on a mean pyramid allow for a lossless reconstruction of the original image, they may result in data expansion, depending on how the mean values are represented.

As an example, suppose that we start with a k-bit/pixel image and average repeatedly over 2×2 blocks to generate a full pyramid, as shown in Fig. 13 (i.e., we average until only a single value remains). If k bits are used to represent the means at each level, the total number of bits required to store the pyramid would be 33% more than that required for the original image [115]. Of course, the number of bits could be reduced by applying a variable-length code to the mean values, but the savings will generally be small. This is because the averaging process does not greatly affect the probability distribution of the pixel values (which is generally not highly skewed), and thus the entropy of this pyramid is approximately the same as the entropy of the original image. This type of pyramid has been termed the *truncated mean pyramid* [116].

The data expansion problem associated with averaging can be reduced by using a slightly more complex approach that increases the bit precision as one moves up the pyramid [115]. In this technique, the pyramid is generated by repeatedly forming sums, rather than averages, over 2×2 blocks. If k bits/pixel are used for the original image, then $k + 2$ bits are required to represent each sum exactly at the next level, $k + 4$ bits are required for each sum at the level above that, and so on. (The values used for display can easily be computed from the sums.) A reduction in storage is achieved by noting that given a sum from one level and three of its four components from the lower level, it is possible to recover the fourth component. Thus we can discard one-fourth of the values at each level. This approach has been termed the *reduced-sum pyramid* [116]. The resulting number of values that must be stored is equal to the number of pixels

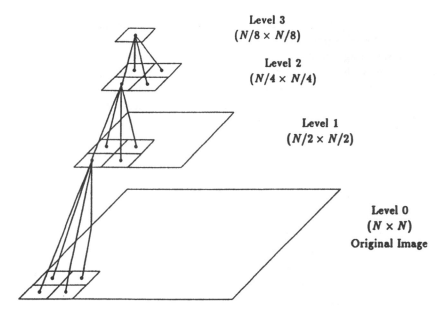

Level 3
$(N/8 \times N/8)$

Level 2
$(N/4 \times N/4)$

Level 1
$(N/2 \times N/2)$

Level 0
$(N \times N)$
Original Image

Figure 13   Image pyramid example.

in the original, but because of the increased bit precision needed for the sums, the total of bits required is approximately 8.3% more than the original for an 8-bit image [115]. As in the previous approach, variable-length coding would generally result in only a small savings.

Another approach that is more efficient than the reduced-sum pyramid in terms of coding performance is the *reduced-difference pyramid* [116]. In this technique, the truncated mean pyramid is formed as described previously (i.e., repeatedly average over 2×2 blocks and use only k bits to represent each mean value at any level). At each level, differences are then formed between neighboring values within the 2×2 blocks. That is, if the four values within a block are x_1, x_2, x_3, and x_4, we form the differences

$$
\begin{aligned}
d_1 &= x_1 - x_2 \\
d_2 &= x_2 - x_3 \\
d_3 &= x_3 - x_4 \\
d_4 &= x_4 - x_1
\end{aligned}
\tag{26}
$$

using $k + 1$ bits to represent each one. It is only necessary to retain three of the four difference values, since the other can be recovered given the three values and the corresponding mean from the next higher level. (Refer to [116] for the algorithm used to recover the data). The resulting number of values that must be stored is equal to the number of pixels in the original image, but because $k + 1$

bits are required for each difference, the total number of bits is increased. (For an 8-bit image, the increase is 12.5%.) However, the difference values generated by this method have a substantially lower entropy than the original image (and also the truncated mean pyramid or the reduced-sum pyramid), and variable-length codes can be used to provide efficient storage of the pyramid [116].

An approach that is similar to the reduced-difference pyramid is the *S-transform* [117–119]. As before, the mean values over 2 × 2 blocks are used to form the image at the next level of the pyramid. The following difference values are also computed for each 2 × 2 block:

$$
\begin{aligned}
d_1 &= \frac{1}{2}(x_1 + x_2 - x_3 - x_4) \\
d_2 &= \frac{1}{2}(x_1 - x_2 - x_3 + x_4) \\
d_3 &= x_1 - x_2 + x_3 - x_4
\end{aligned}
\tag{27}
$$

These three difference values represent the additional information needed to resolve the mean value of a block into its original four pixel values x_1, x_2, x_3, and x_4. It can be shown that by retaining only k bits for the mean, $k + 1$ bits for d_1 and d_2, and $k + 2$ bits for d_3, a lossless reconstruction of the x_i values can be obtained. As in the reduced-difference pyramid, the number of values that must be stored is equal to the number of original pixel values, while the number of stored bits is increased. (For an 8-bit image, the increase is approximately 16.7%.) However, the difference values formed using the S-transform can be encoded even more efficiently than those generated in the reduced-difference pyramid. The result is a slight gain in compression efficiency at the expense of an increased number of computations.

3. Prediction/Residual Pyramid

In the prediction/residual approach to coding, we use a limited number of data to form a prediction image and then subtract the prediction from the original image to form a residual image. The motivation for this process is that the residual image can be encoded more efficiently than the original image. To improve the efficiency further and to provide a hierarchical structure, the prediction/residual process can be iterated at several different scales to create a pyramid of residuals, often called the *Laplacian pyramid* [120]. In the following, we first describe the Laplacian pyramid and then extend it to a more general prediction/residual pyramid.

To create the Laplacian pyramid, the original image is successively low-pass filtered and subsampled to produce a pyramid consisting of reduced resolution versions of the original. This pyramid has been termed the *Gaussian pyramid* since the low-pass filters examined in [120] were approximately Gaussian in shape. Each low-pass filtered image is then expanded (by upsampling and filtering) to the dimensions of the next level to provide a prediction image for that

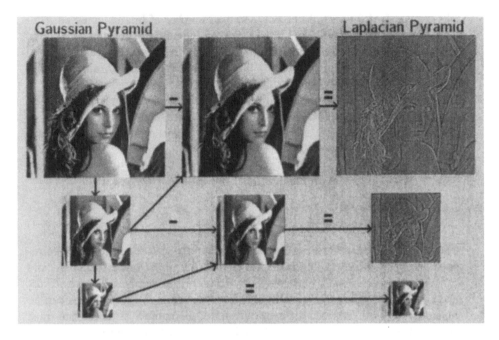

Figure 14   Gaussian and Laplacian pyramids.

level. By subtracting these prediction images from their corresponding images in the Gaussian pyramid, a pyramid of residual images is created. This pyramid is termed the Laplacian pyramid because it is similar to the output produced by filtering the original image with a Laplacian (or equivalently, a difference-of-Gaussians) weighting function. Figure 14 illustrates the generation of the Gaussian and Laplacian pyramids.*

   The information that is stored consists of the residual images as well as the lowest-resolution image in the Gaussian pyramid (the base image). The total

---

*The pyramids were generated using a 5 × 5 low-pass filter with $a = 0.3$, as described in [120]. This filter is somewhat broader and flatter than a Gaussian.

number of values that must be stored is 33% more than the number of original pixels, but the residual images have smaller variances and are less correlated than the original image, and can thus be encoded efficiently. In [120], encoding was performed by quantizing the base image and the residuals and applying variable-length codes. The quantization levels were determined via simulations so that visual degradations in the reconstructed images were minimized.

Progressive reconstruction is achieved by starting with the base image, expanding to the dimensions of the next level and adding in the corresponding residual to generate the next level in the Gaussian pyramid. This image is then expanded, added to its residual to create the next level, and so on. Each of the levels in the reconstructed Gaussian pyramid is available for display or output to an appropriate device. Since the Laplacian pyramid can be viewed as a set of bandpass-filtered versions of the original image [120], reconstruction of the Gaussian pyramid is equivalent to starting with a low-frequency representation (the base image) and then successively adding in additional higher-frequency components (the residuals).

In the method just described, there is a fundamental asymmetry between the decomposition process and the reconstruction process. During decomposition, the residuals are generated using the full Gaussian pyramid created from the original, but during reconstruction, the only information available is the base image and the quantized residuals. Reconstruction is a recursive expansion process, starting with the base image, and quantization errors introduced at one level are propagated to all higher levels. Since the residuals were originally created assuming perfect reconstruction (i.e., they were created from the full Gaussian pyramid), there is no way to correct for these errors with this approach.

To overcome this problem, the asymmetry between the decomposition and the reconstruction processes can be eliminated, resulting in a more general prediction/residual approach. The basic idea in the decomposition process is to start with the base image, expand it to the next level to create a prediction image, and subtract it from the corresponding low-pass image to create a residual. This residual is quantized, added back to the expanded base image (the prediction image), and the result is then expanded to form the prediction image for the next level. The residual at this next level can also be quantized, added back to the corresponding prediction image, the resulting image expanded, and so on. In this way, any errors introduced at one level are taken into account at the next level. If the final residual is losslessly encoded, the original image can be recovered completely.

It is worthwhile to note that the Gaussian pyramid can be generated using almost any low-pass filter (not necessarily Gaussian in shape) or even simple averaging. Since the images in the Gaussian pyramid are the ones that will be displayed, it is really up to the system designer to determine the level of quality needed at each level. Also, any interpolation method, including simple pixel replication, can be used to expand the image at one level up to the next level. Obviously, some methods will result in better prediction images, which lead to more efficient compression.

4. Hierarchical Interpolation

A variation on the prediction/residual approach is to form the variable-resolution pyramid by strict subsampling (i.e., without any prefiltering, as described in Section V.B.1.). Residuals are then generated using the standard approach of interpolation and subtraction. This method has been termed HINT (Hierarchical INTerpolation) [121,122]. Forming the pyramid in this way has the advantage that when an image at one level is interpolated to the next-higher resolution level, only the interpolated pixels need a residual in order to be reconstructed exactly since the subsampled pixels are already correct and hence need no residual. As a result, the total number of values that must be stored is equal to the number of original pixels, as compared to the 33% expansion for the Laplacian pyramid. Unfortunately, the use of subsampling with no prefiltering may yield images at the lower resolution levels that are unacceptable for display. This is because (1) aliasing artifacts are introduced by the subsampling process, and (2) the subsampled points may not be good representatives of the regions from which they have been taken. However, if a subsampling pyramid is necessary for a particular application, the HINT method does offer improved performance over the method described in Section V.B.1. since the residuals can be stored more efficiently than can the intermediate subsampled pixel values.

5. Subband Pyramid

Subband coding using quadrature mirror filters (QMFs) provides a natural hierarchical structure that is quite similar to the Laplacian pyramid. Unlike the Laplacian pyramid, the number of values stored in the subband pyramid is the same as the number of pixels in the original image. The lowest-frequency subband is a low-pass filtered and subsampled version of the original and is analogous to the base image in the Laplacian pyramid. The other subbands contain bands of higher-frequency components, and they can loosely be interpreted as a more refined representation of the Laplacian residual image information. Reconstruction proceeds by using the lowest-frequency subband as the initial approximation and then successively incorporating the higher-frequency subbands by appropriate upsampling and filtering with QMFs. Due to quantization effects, subband coding typically results in a lossy reconstruction unless a final residual error image is included.

VI. COLOR IMAGE CODING

All the compression techniques we have described so far assume single-band (i.e., monochrome) images. Extending these techniques to color or multispectral images can easily be done by encoding each channel independently, but with the same technique. Unfortunately, in many color space representations, there are significant correlations between the various channels, and such redundancies are detrimental to achieving efficient compression. These correlations are due primarily to the smooth spectral reflectances that occur in typical scenes, and partly because most multispectral sensors (the human eye included) have overlapping sensitivity functions. To remove these correlations, the images can be transformed into a different color space, where the resulting components are largely uncorrelated. The transformed color components are then separately encoded as monochrome images.

With respect to decorrelating the color components, the optimal transformation is the Karhunen–Loeve transform (KLT), but the KLT is image-dependent, making it impractical for many applications. Fortunately, image-independent color transforms that decorrelate nearly as well as the KLT can generally be found for typical imagery. Some well-known examples include the YIQ [123] and the YCrCb (Y, Y-R, Y-B) [124] transforms. These transforms produce a luminance (achromatic) component and two chromatic components, with the majority of the spectral energy being packed into the luminance component. Because of this energy compaction, the chromatic components contain a fairly small portion of the total energy and can be encoded at a lower bit rate.

Besides providing decorrelated color components, these luminance/chrominance transformations provide representations that allow sensitivity variations in the human visual system (HVS) [125] to be taken easily into account. It is not accidental that this is the case, since the HVS appears to perform a spectral decorrelation as an early part of visual processing, probably as a means of compressing information prior to transfer through the optic nerve. Much controversy exists regarding the precise nature of the transform performed by the visual system [126], but with respect to image compression, the performance differences of the various color models are relatively minor. In any event, the important point is that the HVS has a somewhat lower response (primarily because of reduced bandwidth) for the chromatic components compared to the luminance component. Accordingly, the chromatic components can be encoded with lower fidelity (and lower bit rates) than the luminance component.

An example of a commonly used color transform is that specified in the NTSC national color television standard [123]. *RGB* values are transformed into *YIQ* (luminance, in-phase, and quadrature, respectively) components according to the following transformation:

$$\begin{bmatrix} Y(i, j) \\ I(i, j) \\ Q(i, j) \end{bmatrix} = \begin{bmatrix} 0.299 & 0.587 & 0.114 \\ 0.596 & -0.274 & -0.322 \\ 0.212 & -0.523 & 0.311 \end{bmatrix} \begin{bmatrix} R(i, j) \\ G(i, j) \\ B(i, j) \end{bmatrix} \tag{28}$$

For a typical NTSC-bandlimited color picture, this spectral conversion has the desirable property of packing most of the signal energy into the Y channel (as much as 93%), and significantly less energy into the I (about 5% and Q (about 2%) channels [123]. For the purposes of image compression, the I and Q chrominance channels are often spatially averaged and subsampled by a factor of $2:1$ or $4:1$ in both the horizontal and vertical directions to take advantage of the reduced bandwidth of the HVS for these components. At the receiver, the reconstructed I and Q planes are interpolated back to their original size. The YCrCb color difference transform produces results that are similar to the YIQ transform but is much easier to implement in hardware.

In some applications, it is impractical or undesirable to include a color transformation in the system due to hardware/throughput considerations, quantization artifacts, and so on. If the final image is to be viewed by a human observer, however, additional compression can sometimes be achieved without a transform by making use of properties of the HVS. As discussed before, the reason for this is that the HVS response varies with wavelength (mainly through changes in its bandwidth). Although these variations are largest for luminance/chrominance representations, most color spaces will exhibit enough variations to warrant taking advantage of them. As before, the basic approach is to introduce errors where the HVS is the least sensitive. For example, in an RGB image, the blue color plane can often be compressed 25% more than the green color plane without creating any additional visual artifacts.

VII. IMAGE SEQUENCE CODING

It is obvious that the frames that make up an image sequence could be compressed independently using one of the techniques described previously. Unfortunately, much of the information in a sequence is often redundant, and independent processing extracts a steep price in terms of compression efficiency. As a result, *interframe coding*, where more than one frame is considered at a time, is used to take advantage of frame-to-frame correlations. A good review of interframe coding can be found in [127].

To deal with the added dimension provided by a sequence of frames, there are two basic approaches. The first is to use full 3D encoding, where several frames are encoded at once. This can be accomplished by extending intraframe (2D) techniques such as vector quantization [128], transform coding [129], or subband coding [130] to include another dimension. The second approach is to reduce the problem from 3D (encoding a sequence of frames) to 2D (encoding a single frame) by taking the difference between the current frame and a prediction based on one or more previously transmitted frames. Analogous to the way that DPCM reduces pixel-to-pixel correlation, this process reduces frame-to-frame correlation. Many of the techniques described for intraframe coding can then be used to encode the individual difference frames. The key to efficient encoding with this method is, of course, making accurate predictions, which is often termed the *motion estimation* problem. There are many approaches to motion estimation and good reviews can be found in [131] and [132]. The implicit assumption in performing motion estimation is that there are corresponding features from one frame to the next. This assumption is invalid when there is an abrupt scene change or when features appear or disappear due to occlusion. Most sequence coding schemes provide some means for detecting the lack of correspondence and encode such regions in a different manner than those that can be predicted.

VIII. IMAGE COMPRESSION STANDARDS

An image compression standard provides several benefits: (1) it facilitates the interchange of compressed image data between various devices and applications; (2) it permits common hardware/software to be used for a wide range of products, thus lowering costs and shortening development times; and (3) it provides fixed reference points for the expected quality of compressed images. Ideally, a single standard would satisfy the compression needs of any application but given the wide range of image types and differing image quality requirements, this is unrealizable. As a result, image compression standardization efforts are divided among three image types: bitonal images, continuous-tone still images, and image sequences. Because of the importance of standards, it can safely be assumed that additional standards will be defined and adopted over time, so the techniques described in this section are, at best, a snapshot of current efforts.

A. Standards for Bitonal Images

The only widely used digital image compression standards currently in existence are the international digital facsimile group 3 (1980) and group 4 (1984) standards established by the Consultative Committee of the International Telephone and Telegraph (CCITT) [33]. The ubiquitous fax machine is testimony to the impact of these standards on the ability to easily transfer bitonal images such as text and documents. Although these standards have proven their worth, they do have certain limitations that restrict their utility in some applications. As a result, a committee known as JBIG (for Joint Bilevel Imaging Group) was formed under the joint auspices of the International Standardization Organization (ISO) and CCITT in 1988 to develop an algorithm that would compress more efficiently and have more utility than the existing group 3 and group 4 fax standards. For example, the proposed JBIG standard [133] has the capability for progressive representation of images, an indispensable feature in an environment where the compressed data are made available to a number of display or transmission devices with varying degrees of resolution and/or quality requirements. In addition, it uses adaptive arithmetic coding so that it can efficiently compress bitonal images with widely different characteristics (e.g., both text and halftone images). The JBIG-proposed algorithm became a draft international standard in 1992.

B. Standards for Continuous-Tone, Still Images

A committee known as JPEG (for Joint Photographic Experts Group) was formed under the joint auspices of ISO and CCITT at the end of 1986 for the purpose of developing an international standard for the compression and decompression of continuous-tone, still-frame, monochrome and color images. The goal of this committee was to define a general-purpose standard for such diverse applications as photo-videotex, desktop publishing, graphic arts, color facsimile, photojournalism, medical systems, and many others. To meet the needs of these different applications, the proposed JPEG standard [79,134,135] consists of three main components: (1) a baseline system that provides a simple and efficient algorithm that is adequate for most image coding applications, (2) a set of extended system features such as progressive buildup that allows the baseline system to satisfy a broader range of applications, and (3) an independent lossless method for applications requiring that type of compression. The JPEG-proposed system became a draft international standard in 1992.

The JPEG baseline system has already gained broad acceptance as a lossy compression technique, and several manufacturers have introduced JPEG chips and basic DCT engines. Because of its wide application, it is worthwhile to show an example of a JPEG-compressed image to give the reader some idea of the quality and compression ratios that can be achieved. In Fig. 15, an original 512×512, 8-bit monochrome image is shown along with an JPEG-compressed image at 0.5 bit/pixel (16:1 compression ratio). The HVS-based quantization table given in Section 4.2 was used, multiplied by an overall scaling factor to provide the desired bit rate. The DCT coefficients were encoded using Huffman codes as described in the JPEG specifications [79,135]. Due to limitations of the printing process, it may be difficult to see any differences between the two images (an example of visually lossless compression), so enlarged sections of both images are shown in Fig. 16. In this figure, typical artifacts of block-based DCT compression can be seen (e.g., the obvious block boundaries and edge degradation). Nevertheless, even with these artifacts, the enlarged image provides a very recognizable version of the original while requiring 16 times less information.

C. Standards for Continuous-Tone Image Sequences

CCITT has standardized a coding algorithm (Recommendation H.261 [136]) for real-time applications such as video telephony and video conferencing at bit

Figure 15 Original and JPEG-compressed 512 × 512, 8-bit images.

rates ranging from 64 to 1920 kilobits/s (p × 64 ISDN channels). Since 1988, a standardization group known as MPEG (for Moving Picture Experts Group, under the auspices of ISO) has also been working to develop a standard for storage and retrieval of moving images and sound using digital storage media with a combined bit rate of around 1.5 megabits/s (consumer VCR quality). The algorithm used by MPEG is similar to that used in CCITT H.261, but it is somewhat more general-purpose, since it does not have the real-time constraints imposed in H.261. The MPEG-proposed standard [137,138] became a draft international standard in 1992, and another MPEG committee (known as MPEG-2) has been formed to develop a similar standard for higher data rates.

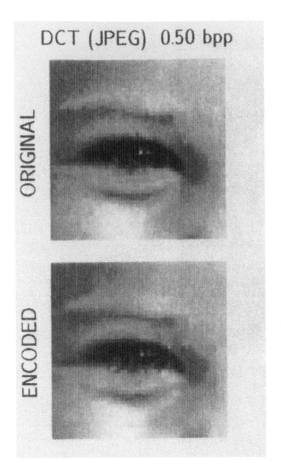

Figure 16   Enlarged sections of images shown in Fig. 15.

IX. SELECTING A COMPRESSION TECHNIQUE

The selection of an appropriate compression technique for a specific application can be a daunting task, due in part to the wide range of basic approaches. Compounding the problem is the existence of many subtle variations on the same theme (such as incorporating adaptivity in different ways), resulting in seemingly similar algorithms that can perform quite differently depending on the application. Similarly, evaluating the claims of a novel technique can be difficult, since the true advantages and disadvantages may not be immediately apparent.

In general, the usefulness of a particular algorithm is heavily dependent on the application requirements. An obvious example is whether lossless or lossy compression is needed. As another example, a storage application might require a constant image quality and a simple decoder structure (at the possible expense of a more complex encoder), while a transmission application might strive for a fixed compression ratio (at the expense of varying image quality) and encoder/ decoder symmetry. The following is a list of factors, by no means exhaustive, that can be consulted as a general guide in the process of evaluating and selecting an appropriate technique.

1. *Sensitivity to input image types.* Input image characteristics such as bit depth, resolution, noise, spatial frequency content, pixel-to-pixel correlation, and other image statistics may all affect the performance and thus the choice of an algorithm. Also, some compression schemes may require parameter tuning to obtain good performance with a given class of images, and performance can degrade significantly if other types of input images are allowed.

2. *Operational bit rate.* In some applications the priority is to achieve a very high degree of compression even at the cost of low image quality. In contrast, other applications may require a high degree of image quality that can only be achieved at modest compression ratios. In general, there is a certain range of output bit rates for which an algorithm is most efficient. Furthermore, some algorithms cannot inherently be operated below a certain bit rate, whereas other algorithms are quite costly to operate at high bit rates. It is also desirable to have the ability to optimally and easily trade off the bit rate for the reconstructed image quality by adjusting a small set of compression parameters (which usually control the degree of quantization). However, some schemes can operate at only a few specific bit rates (or even just one bit rate) and require a redesign of the system to achieve other bit rates.

3. *Constant bit rate versus constant quality.* Algorithms that operate with a constant bit rate are more suitable for transmission applications where a fixed-rate channel with no buffering is used or for storage applications where the storage space is prespecified. Unfortunately, due to the wide variation in the information content of different images, such schemes do not result in constant reconstruction quality. The specified bit rate may be unnecessarily high for some images while resulting in unsatisfactory image quality for others. In contrast, some schemes maintain a constant visual quality or distortion measure (e.g., SNR) at the expense of a variable bit rate. This variable bit rate is often the result of including entropy coding in the scheme. The range over which the bit rate may vary is an important hardware consideration since adequate buffering must generally be provided.

4. *Implementation issues.* This refers to the nature and complexity of the algorithm relative to the particular hardware or software environment in which it is implemented. Three aspects of an algorithm need to be considered: (1) computational complexity (i.e., the number of additions, multiplications, shifts, comparisons, or other operations required per pixel), (2) memory requirements,

and (3) amenability to parallel processing or other efficient processing structures. Typical implementation environments with current technology include PC-based, DSP (digital signal processor) chip-based, and ASIC (application-specific integrated circuit)-based systems. The characteristics of each type of environment determine the suitability of an algorithm and the speed at which it can operate in that environment.

5. *Encoder/decoder asymmetry*. Some approaches to compression result inherently in a complex encoder but a simple decoder; others require an encoder and a decoder of comparable complexity. Although encoders and decoders of equal complexity may be acceptable in many transmission applications, a simple decoder is more desirable in applications where it is used repeatedly, such as image storage and retrieval systems. The inclusion of adaptivity can also alter this balance significantly as well as increase the overall complexity of the system.

6. *Channel error tolerance*. Unfortunately, one of the prices paid for data compression is the increased susceptibility of encoded data to channel errors, and the degree of susceptibility varies widely among the different schemes. In the case of block processing algorithms, the effect of a bit error is often confined to only a small block of the image, while in other schemes it may appear as a streak across the image. If variable-length coding is used, which is often the case in more sophisticated schemes, the effect of channel errors can be catastrophic and can result in the loss of the entire image. Of course, error control coding can be added to any system, but the price to be paid is an increase in the overall complexity and bit rate.

7. *Artifacts*. Different algorithms create different artifacts, depending on their mode of operation. Even a given algorithm may exhibit different artifacts, depending on the bit rate at which it is operated. Some artifacts such as blocking or edge jaggedness may be more visually objectionable than random noise or overall edge smoothing. Also, the visibility of artifacts is highly dependent on the particular image and the conditions under which it is viewed.

8. *Effect of multiple coding*. Some applications may require that an image undergo the compression–decompression cycle many times. For example, an image may be compressed and transmitted to a destination where it is decompressed and viewed. An operator may alter a small portion of the image and then compress the entire image and send it to another destination, where this process is repeated. In such applications it is essential that the repeated compression and decompression of the unaltered portions does not result in any additional degradation beyond the first stage of compression. Some compression schemes do not create further loss of image quality when used successively on the same image, while others cause additional degradations. Fortunately, in most cases, subsequent losses are significantly less than the loss caused by the first stage.

9.   *Progressive transmission capability.* Progressive image transmission allows for an approximate image to be sent at a low bit rate for quick recognition, and then remaining details are transmitted incrementally if desired by the user. Any scheme can artificially be made progressive by using it to encode a low-resolution version of the original image as a first approximation, and then encoding the difference between this stage and either the original image or another intermediate stage to create as many levels of progression as desired. Some schemes are inherently amenable to progressive transmission, due to the embedded structure of the encoder and can operate in that mode without requiring any additional complexity. Other techniques can also operate in a progressive mode but at the expense of additional system complexity.

10.   *System compatibility.* If the system requires compatibility with other manufacturers' products, the choice of a compression scheme may be dictated by the existence of standards that have been proposed and/or adopted (e.g., CCITT facsimile standards or the JPEG standard for continuous-tone images).

REFERENCES

1.   A. Gersho and R. M. Gray, *Vector Quantization and Signal Compression*, Kluwer Academic, Norwell, Mass., 1992.
2.   M. Rabbani and P. W. Jones, *Digital Image Compression Techniques*, Vol. TT7, SPIE Tutorial Text Series, SPIE Press, Bellingham, Wash., 1991.
3.   K. R. Rao and P. Yip, *Discrete Cosine Transform; Algorithms, Advantages, Applications*, Academic Press, San Diego, Calif., 1990.
4.   A. N. Netravali and B. G. Haskell, *Digital Pictures: Representation and Compression*, Plenum Press, New York, 1988.
5.   R. J. Clarke, *Transform Coding of Images*, Academic Press, London, 1985.
6.   N. S. Jayant and P. Noll, *Digital Coding of Waveforms*, Prentice-Hall, Englewood Cliffs, N.J., 1984.
7.   W. K. Pratt, ed., *Image Transmission Techniques, Advances in Electronics and Electron Physics*, Supplement 12, Academic Press, Orlando, Fla., 1979.
8.   R. C. Gonzalez and R. C. Woods, *Digital Image Processing*, Addison-Wesley, Reading, Mass., 1992.
9.   W. K. Pratt, *Digital Image Processing*, 2nd edition, Wiley-Interscience, New York, 1991.
10.   J. S. Lim, *Two-Dimensional Signal and Image Processing*, Prentice-Hall, Englewood Cliffs, N.J., 1990.
11.   A. K. Jain, *Fundamentals of Digital Image Processing*, Prentice-Hall, Englewood Cliffs, N.J., 1989.
12.   A. Rosenfeld and A. C. Kak, *Digital Picture Processing*, Vol. 1, Academic Press, Orlando, Fla., 1982.
13.   M. Rabbani, ed., *Selected Papers on Image Compression*, SPIE Milestone Series, Vol. MS 48, SPIE Press, Bellingham, Wash., 1992.

14. T. R. Hsing and A. G. Tescher, ed., *Selected Papers on Visual Communication*: *Technology and Applications*, SPIE Milestone Series, Vol. MS 13, SPIE Press, Bellingham, Wash., 1990.

15. M. Kunt, A. Ikonomopoulos, and M. Kocher, Second-generation image-coding techniques, *Proc. IEEE*, 73(4):549–574 (1985).

16. A. K. Jain, Image data compression: a review, *Proc. IEEE*, 69(3):349–389 (1981).

17. A. N. Netravali, Picture coding: a review, *Proc. IEEE*, 68(3):366–406 (1980).

18. A. Habibi, Survey of adaptive image coding techniques, *IEEE Trans. Comm.*, COM-25(11):1275–1284 (1977).

19. N. Jayant, ed., Speech and image coding, special issue of *IEEE J. Select. Areas Comm.*, SAC-10(5) (1992).

20. K. -H. Tzou and H. -M. Hang, eds., Visual communications and image processing, special issue of *Opt. Engrg.*, 30(7):863–964 (1991).

21. T. R. Hsing, ed., Visual communications and image processing, special issue of *Opt. Engrg.*, 26(7) (1987) and 28(7) (1989).

22. B. G. Haskell, D. Pearson, and H. Yamamoto, eds., Low-bit-rate coding of moving images, special issue of *IEEE J. Select. Areas Comm.*, SAC-5(7) (1987).

23. A. N. Netravali and B. Prasada, eds., Visual communication systems, special issue of *Proc. IEEE*, 73(4):497–848 (1985).

24. A. N. Netravali and A. Habibi, eds., Picture communication systems, special issue of *IEEE Trans. Comm.*, COM-29(12):1725–2008 (1981).

25. A. N. Netravali, ed., Digital encoding of graphics, special issue of *Proc. IEEE*, 68(7) (1980).

26. A. Habibi, ed., Image bandwidth compression, special issue of *IEEE Trans. Comm.*, COM-25(11 (1977).

27. H. C. Andrews and L. H. Enloe, eds., Digital picture processing, special issue of *Proc. IEEE*, 60(7):763–922 (1972).

28. D. A. Huffman, A method for the construction of minimum redundancy codes, *Proc. IRE*, 40:1098–1101 (1952).

29. M. Hankamer, A modified Huffman procedure with reduced memory requirement, *IEEE Trans. Comm.*, COM-27(6):930–932 (1979).

30. D. E. Knuth, Dynamic Huffman coding, *J. Algorithms*, 6:163–180 (1985).

31. J. S. Vitter, Design and analysis of dynamic Huffman codes, *J. Assoc. Comput. Mach.*, 34(4):825–845 (1987).

32. R. G. Gallager, Variations on a theme by Huffman, *IEEE Trans. Inform. Theory*, IT-24(6):668–674 (1978).

33. R. Hunter and A. H. Robinson, International digital facsimile standards, *Proc. IEEE*, 68(7):854–867 (1980).

34. N. Abramson, *Information Theory and Coding*, McGraw-Hill, New York, 1963.

35. R. E. Blahut, *Principles and Practice of Information Theory*, Addison-Wesley, Reading, Mass., 1987.

36. J. Rissanen and K. M. Mohiuddin, A multiplication-free multialphabet arithmetic code, *IEEE Trans. Comm.*, COM-37(2):93–98 (1989).

37. W. B. Pennebaker, J. L. Mitchell, G. G. Langdon, Jr., and R. B. Arps, An overview of the basic principles of the Q-coder adaptive binary arithmetic coder, *IBM J. Res. Develop.* 32(6):717–726 (1988).

38. I. H. Witten, R. M. Neal, and J. G. Cleary, Arithmetic coding for data compression, *Comm. ACM*, 30(6):520–540 (1987).

39. G. G. Langdon, An introduction to arithmetic coding, *IBM J. Res. Develop.*, 28(2):135–149 (1984).

40. C. B. Jones, An efficient coding system for long source sequences, *IEEE Trans. Inform. Theory*, IT-27(3):280–291 (1981).

41. G. G. Langdon and J. J. Rissanen, Compression of black–white images with arithmetic coding, *IEEE Trans. Comm.*, COM-29(6):858–867 (1981).

42. J. Rissanen and G. G. Langdon, Arithmetic coding, *IBM J. Res. Develop.*, 23(2):149–162 (1979).

43. J. Ziv and A. Lempel, A universal algorithm for sequential data compression, *IEEE Trans. Inform. Theory*, IT-23(3):337–343 (1977).

44. T. A. Welch, A technique for high-performance data compression, *IEEE Comput.*, 17(6): 8–19 (1984).

45. A. Lempel and J. Ziv, Compression of two-dimensional data, *IEEE Trans. Inform. Theory*, IT-32(1):2–8 (1986).

46. S. Bunton and G. Borriello, Practical dictionary management for hardware data compression: development of a theme by Ziv and Lempel, in *Proc. 6th MIT Conference on Advanced Research in VLSI*, pp. 33–50 (1990).

47. L. D. Davisson, Rate-distortion theory and applications, *Proc. IEEE*, 60(7):800–808 (1972).

48. D. J. Sakrison, On the role of the observer and a distortion measure in image transmission, *IEEE Trans. Comm.*, COM-25(11):1251–1267 (1977).

49. D. K. Sharma and A. N. Netravali, Design of quantizers for DPCM coding of picture signals, *IEEE Trans. Comm.*, COM-25(11):1267–1274 (1977).

50. A. N. Netravali and B. Prasada, Adaptive quantization of picture signals using spatial masking, *Proc. IEEE*, 65(4):536–548 (1977).

51. J. O. Limb and C. B. Rubinstein, On the design of quantizers for DPCM coders: a functional relationship between visibility, probability, and masking, *IEEE Trans. Comm.*, COM-26(5):573–578 (1978).

52. S. P. Lloyd, Least squares quantization in PCM, *IEEE Trans. Inform. Theory*, IT-28:129–137 (1982).

53. J. Max, Quantizing for minimum distortion, *IRE Trans. Inform. Theory*, IT-6(1):7–12 (1960).

54. H. Yamaguchi, Optimum quantization of Laplace density signal and its characteristics, *Electron. Comm. Jpn.*, 67-B(5):67–73 (1984).

55. P. Kabal, Quantizers for the gamma distribution and other symmetrical distributions, *IEEE Trans. Acoust. Speech Signal Process.*, ASSP-32(4):836–841 (1984).

56. J. Makhoul, S. Roucos, and H. Gish, Vector quantization in speech coding, *Proc. IEEE*, 73(11):1551–1588 (1985).

57. H. Gish and J. N. Pierce, Asymptotically efficient quantizing, *IEEE Trans. Inform. Theory*, IT-14(5):676–683 (1968).

58. N. Farvardin and J. W. Modestino, Optimum quantizer performance for a class of non-Gaussian memoryless sources, *IEEE Trans. Inform. Theory*, IT-30(3):485–497 (1984).

59. J. -Y. Huang and P. M. Schultheiss, Block quantization of correlated Gaussian random variables, *IEEE Trans. Comm.*, COM-11:289–296 (1963).
60. A. Segall, Bit allocation and encoding for vector sources, *IEEE Trans. Inform. Theory*, IT-22:162–169 (1976).
61. B. Fox, Discrete optimization via marginal analysis, *Management Sci.*, 13:210–216 (1966).
62. Y. Linde, A. Buzo, and R. M. Gray, An algorithm for vector quantizer design, *IEEE Trans. Comm.*, COM-28(1):84–95 (1980).
63. R. M. Gray, Vector quantization, *IEEE Acoust. Speech Signal Process*, 1(2):4–29 (1984).
64. W. H. Equitz, A new vector quantization clustering algorithm, *IEEE Trans. Acoust. Speech Signal Process.*, ASSP-37(10):1568–1575 (1989).
65. R. L. Baker, Vector quantization of digital images, Ph.D. thesis, Department of Electrical Engineering, Stanford University, June 1984.
66. P. Chou, T. Lookabaugh, and R. M. Gray, Entropy-constrained vector quantization, *IEEE Trans. Acoust. Speech Signal Process.*, ASSP-37(1):31–42 (1989).
67. P. A. Chou, T. Lookabaugh, and R. M. Gray, Optimal pruning with applications to tree-structured source coding and modeling, *IEEE Trans. Inform. Theory*, IT-35(2):299–315 (1989).
68. C. C. Cutler, Differential quantization of communication signals, U.S. patent 2,605,361, July 1952.
69. A. Habibi, Comparison of *n*th-order DPCM encoder with linear transformations and block quantization techniques, *IEEE Trans. Comm. Technol.*, COM-19(6):948–956 (1971).
70. S. E. Elnahas, Data compression with applications to digital radiology, D.Sc. dissertation, Washington University, 1984.
71. W. Zschunke, DPCM picture coding with adaptive prediction, *IEEE Trans. Comm.*, COM-25(11):1295–1302 (1977).
72. P. J. Ready and D. J. Spencer, Block adaptive DPCM transmission of images, *NTC Conf. Rec.*, 2:22-10-22–17 (1975).
73. A. Habibi and B. H. Batson, Potential digitization/compression techniques for shuttle video, *IEEE Trans. Comm.*, COM-26(11):1671–1682 (1978).
74. J. R. Sullivan, A new ADPCM image compression algorithm and the effect of fixed-pattern sensor noise, *Proc. SPIE*, Vol. 1075, *Digital Image Processing Applications*, 1989, pp. 129–138.
75. K. Yamada, K. Kinukaba, and H. Sasaki, *Internat. Conf. Comm. Rec.*, 1:76–80 (1977).
76. N. S. Jayant, Adaptive quantization with a one-word memory, *Bell System Tech. J.*, 52:1119–1144 (1973).
77. P. A. Maragos, R. W. Schafer, and R. M. Mersereau, Two-dimensional linear prediction and its application to adaptive predictive coding of images, *IEEE Trans. Acoust. Speech Signal Process.*, ASSP-32(6):1213–1229 (1984).
78. N. Ahmed, T. Natarajan, and K. R. Rao, Discrete cosine transform, *IEEE Trans. Comput.*, C-23(1):90–93 (1974).

79. Digital compression and coding of continuous-tone still images, Part 1: Requirements and guidelines, *ISO/IEC Draft International Standard Draft 10918-1*, Feb. 1992.

80. C. A. Gonzales and J. L. Mitchell, A note on DCT algorithms with low multiplicative complexity, *JPEG*, 150 (1988).

81. W. H. Chen and C. H. Smith, Adaptive coding of monochrome and color images, *IEEE Trans. Comm.*, COM-25(11):1285–1292 (1977).

82. W. H. Chen and W. K. Pratt, Scene adaptive coder, *IEEE Trans. Comm.*, COM-32(3):224–232 (1984).

83. M. Vetterli, Multi-dimensional sub-band coding: some theory and algorithms, *Signal Process.*, 6:97–112 (1984).

84. E. H. Adelson, E. Simoncelli, and R. Hingorani, Orthogonal pyramid transforms for image coding, *Proc. SPIE*, Vol. 845, *Visual Communications and Image Processing II*, 1987, pp. 50–58.

85. B. Mahesh and W. A. Pearlman, Hexagonal sub-band coding for images, *Proc. ICASSP*, 1989, pp. 1953–1956.

86. M. J. T. Smith and T. P. Barnwell, Exact reconstruction techniques for tree-structured subband coders, *IEEE Trans. Acoust. Speech Signal Process*, ASSP-34(3):434–441 (1986).

87. D. Le Gall and A. Tabatabai, Sub-band coding of digital images using symmetric kernel filters and arithmetic coding techniques, *Proc. ICASSP*, 1988, pp. 761–764.

88. J. D. Johnston, A filter family designed for use in quadrature mirror filter banks, *Proc. ICASSP*, 1980, pp. 291–294.

89. P. P. Vaidyanathan, Quadrature mirror filter banks, M-band extensions and perfect-reconstruction techniques, *IEEE ASSP Mag.*, 4(3):4–20 (1987).

90. J. W. Woods and S. D. O'Neil, Subband coding of images, *IEEE Trans. Acoust. Speech Signal Process*, ASSP-34(5):1278–1288 (1986).

91. H. Gharavi and A. Tabatabai, Sub-band coding of monochrome and color images, *IEEE Trans. Circuits Systems*, CS-35(2):207–214 (1988).

92. P. H. Westerink, D. E. Boekee, J. Biemond, and J. W. Woods, Subband coding of images using vector quantization, *IEEE Trans. Comm.*, COM-36(6):713–719 (1988).

93. D. M. Baylon and J. S. Lim, Transform/subband analysis and synthesis of signals, MIT, Cambridge, Mass., Research Laboratory of Electronics, *Technical Report*, June 1990.

94. O. Rioul and M. Vetterli, Wavelets and signal processing, *IEEE Signal Process. Mag.*, Oct. 1991, pp. 14–38.

95. S. G. Mallat, A theory for multiresolution signal decomposition: the wavelet representation, *IEEE Trans. Pattern Anal. Mach. Intelligence*, PAMI-11(7):674–693 (1989).

96. R. L. Baker and R. M. Gray, Differential vector quantization of achromatic imagery, *Proc. International Picture Coding Symposium*, 1983, pp. 105–106.

97. H. -M. Hang and B. Haskell, Interpolative vector quantization of color images, *IEEE Trans. Comm.*, COM-36(4):465–470 (1988).

98.   M. J. Sabin and R. M. Gray, Product code vector quantizers for waveform and voice coding, *IEEE Trans. Acoust. Speech Signal Process.*, ASSP-32(3):474–488 (1984).

99.   B. Ramamurthi and A. Gersho, Classified vector quantization of images, *IEEE Trans. Comm.*, COM-34(11):1105–1115 (1986).

100. J. Foster, R. M. Gray, and M. O. Dunham, Finite-state vector quantization for waveform coding, *IEEE Trans. Inform. Theory*, IT-31(3):348–359 (1985).

101. B. B. Mandelbrot, *The Fractal Geometry of Nature*, W. H. Freeman, San Francisco, 1982.

102. M. F. Barnsley, A. Jacquin, F. Malassenet, L. Reuter, and A. D. Sloan, Harnessing choas for image synthesis, *Comput. Graphics*, 22(4):131–140 (1988).

103. M. F. Barnsley, *Fractals Everywhere*, Academic Press, San Diego, Calif., 1988.

104. M. F. Barnsley and A. D. Sloan, A better way to compress images, *Byte*, Jan. 1988, pp. 215–223.

105. M. F. Barnsley and A. E. Jacquin, Application of recurrent iterated function systems to images, *Proc. SPIE*, Vol. 1001, *Visual Communications and Image Processing*, 1988, pp. 122–131.

106. A. D. Sloan, The fractal image format and JPEG, *Proc. Electronic Imaging International Conference*, 1991, pp. 460–465.

107. A. E. Jacquin, A novel fractal block-coding technique for digital images, *Proc. IEEE International Conference on Acoustics, Speech, and Signal Processing*, 1990, pp. 2225–2228.

108. A. E. Jacquin, Image coding based on a fractal theory of iterated contractive image transformations, *IEEE Trans. Image Process.*, IP-1(1):18–30 (1992).

109. M. Rabbani and P. W. Melnychuck, Conditioning contexts for the arithmetic coding of bit planes, *IEEE Trans. Signal Process.*, SP-40(1):232–236 (1992).

110. M. I. Sezan, M. Rabbani, and P. W. Jones, Progressive transmission of images using a prediction/residual encoding approach, *Opt. Engrg.*, 28(5):556–564 (1989).

111. K. N. Ngan, Image display techniques using the cosine transform, *IEEE Trans. Acoust. Speech Signal Process.*, ASSP-32(1):173–177 (1984).

112. E. Dubois and J. L. Moncet, Encoding and progressive transmission of still pictures in NTSC composite format using transform domain methods, *IEEE Trans. Comm.*, COM-34(3):310–319 (1986).

113. L. Wang and M. Goldberg, Progressive image transmission by transform coefficient residual error quantization, *IEEE Trans. Comm.*, COM-36(1)75–87 (1988).

114. S. E. Elnahas, K. H. Tzou, J. R. Cox, R. L. Hill, and R. G. Gilbert, Progressive coding and transmission of digital diagnostic pictures, *IEEE Trans. Med. Imag.*, MI-5(2):73–83 (1986).

115. S. L. Tanimoto, Image transmission with gross information first, *Comput. Graphics Image Process.*, 9:72–76 (1979).

116. L. Wang and M. Goldberg, Reduced-difference pyramid: a data structure for progressive image transmission, *Opt. Engrg.*, 28(7):708–716 (1989).

117. *ACR/NEMA Standards Publication for Data Compression Standards*, Publication PS-2, NEMA, Washington, D.C., 1989.

118. P. Lux, Redundancy reduction in radiographic pictures, *Optica Acta*, 24(4):349–365 (1977).
119. H. Blume and A. Fand, Reversible and irreversible image data compression using the S-transform and Lempel–Ziv coding, *Proc. SPIE*, Vol. 1091, *Medical Imaging III: Image Capture and Display*, 1989, pp. 2–18.
120. P. J. Burt and E. H. Adelson, The Laplacian pyramid as a compact image code, *IEEE Trans. Comm.*, COM-31(4):532–540 (1983).
121. T. Endoh and T. Yamazaki, Progressive coding scheme for multilevel images, *Proc. Picture Coding Symposium*, 1986, pp. 21–22.
122. P. Roos, A. Viergever, M. C. A. van Dijke, and J. H. Peters, Reversible intraframe compression of medical images, *IEEE Trans. Med. Imag.*, MI-7(4):328–336 (1988).
123. G. Buchsbaum, Color signal encoding: color vision and color television, *Color Res. Appl.*, 12(5):266–269 (1987).
124. R. Schafer, High-definition television production standard: an opportunity for optimal color processing, *SMPTE (Society of Motion Picture and Television Engineers) J.*, July 1985, pp. 749–758.
125. D. H. Kelly, Spatial and temporal interactions in color vision, *J. Imag. Technol.*, 15(2):82–89 (1989).
126. L. M. Hurvich, *Color Vision*, Sinauer Associates, Sunderland, Mass., 1981.
127. H. G. Musmann, P. Pirsch, and H. -J. Grallert, Advances in picture coding, *Proc. IEEE*, 73(4):523–548 (1985).
128. M. Goldberg and H. -F. Sun, Image sequence coding by three-dimensional block vector quantisation, *IEE Proc.*, 133(Pt. F, 5):482–487 (1986).
129. T. R. Natarajan and N. Ahmed, On interframe transform coding, *IEEE Trans. Comm.*, COM-25(11):1323–1329 (1977).
130. G. Karlsson and M. Vetterli, Subband coding of video for packet networks, *Opt. Engrg.*, 27(7):574–586 (1988).
131. J. K. Aggarwal and N. Nandhakumar, On the computation of motion from a sequence of images: a review, *Proc. IEEE*, 76(8):917–935 (1988).
132. J. F. Vega-Riveros and K. Jabbour, Review of motion analysis techniques, *IEE Proc.*, 136(Pt. I, 6):397–404 (1989).
133. Progressive bi-level image compression, *ISO/IEC JTC1 Committee Draft 11544*, Sept. 1991.
134. G. K. Wallace, The JPEG still picture compression standard, *Comm. ACM*, 34(4):30–44 (1991).
135. W. B. Pennebaker and J. L. Mitchell, *JPEG: Still Image Data Compression Standard*, Van Nostrand Reinhold, New York, 1992.
136. Video codec for audiovisual services at $p \times 64$ kbits/s, *CCITT Recommendation H.261, CDM XV-R 37-E*, Aug. 1990.
137. Coding of moving pictures and associated audio for digital storage media at up to about 1.5 Mbits/s, *ISO/IEC JTC1 Committee Draft 11172-2*, Nov. 1991.
138. D. Le Gall, MPEG: a video compression standard for multimedia applications, *Comm. ACM*, 34(4):46–58 (1991).

9

Image-Processing Architectures

Stephen S. Wilson

Applied Intelligent Systems, Inc.
Ann Arbor, Michigan

I. INTRODUCTION

A. Scope

The scope of this chapter is threefold: to survey computer architecture concepts, to analyze these architectures critically, and to disclose some of the methodology that goes into the design or selection process of a computer for machine vision or general imaging. The focus will be on computer architectures that are appropriate to the applications covered in previous chapters and will include technical details, an analysis of efficiencies of the architectures, the relation between hardware types, examples of algorithm approaches, design strategies for hardware, and the difference between research and commercial machines.

There will not be an in-depth survey of the various machines that are available, although a number of processors will be mentioned by way of example. There are simply too many machines to give justice to any of them. Surveys of commercial hardware are soon out of date because the hardware is rapidly changing. For the same reason, some obsolete machines are mentioned in this chapter but are not always indicated as such.

References to the open literature will not always be cited for machines from smaller companies because there may not be any available. However, the status and address of a supplier can most likely be found in the trade journal *Advanced Imaging*, which prints a buyers' guide every year in December. The issue covers all areas of the image-processing and machine vision industry. A recent issue [1] includes 24 vendors of parallel processors, 95 vendors of PC-based systems, and 40 vendors of Unix systems, among many other categories. There are many references in the literature on image-processing and parallel architectures [2–8]. More recent architectures can be found in the proceedings of symposiums [9,10].

If an in-depth understanding of the available products and market segments is required, a comprehensive market report may be the best source, but these are very expensive and not available in libraries. A number of marketing reports exist and are frequently updated. Electronic Trend Publications has reports on image processing [11] and related integrated circuits [12] with vendor surveys and application areas. Frost and Sullivan has an extensive report on image processing with a large section on products [13]. SEAI [14] has a number of publications, including parallel processing and electronic imaging.

B. Different Imaging Hardware Requirements

Aside from the technical nature of the machine, imaging hardware can be classified in a number of different ways. The most important distinctions are:

1. Research and commercial machines
2. Real-time (turnkey) and off-line
3. Imaging and machine vision
4. Process control and inspection

Commercial machines that are used in process control must generally be low in cost and high speed. These are difficult to design and are often application specific. High-performance architectures are also desirable in research machines, but for a different reason. Processor intensive problems must be solved in a reasonable time so that results can be obtained in a few minutes rather than overnight. The machines in research must be more flexible and can be much higher in cost. Often, research machines are different from commercial machines for the simple reason that the research involves a particular architecture, without concern for an immediate practical application.

Off-line machines in machine vision are often used for inspection. In other imaging areas, off-line machines are not constrained to particular dedicated applications, but are used to service a number of applications such as visualization, and animation. These machines should be as fast as economically feasible but do not have to be real time. Analysis of aerospace images is off-line but requires the highest in architecture performance because the amount of data is so large. High cost is overshadowed by the cost of acquiring the images.

One characteristic of a machine is whether it is used in turnkey applications, where a machine is dedicated to one application and is generally tied to parts-handling equipment or some other mechanism that it controls. Commercial machines that are used primarily in research or intensive off-line computations are not used in turnkey applications because they are too expensive.

For a given area there is still really no "best" architecture since an optimal design is algorithm dependent. Real-time machine vision applications such as locating or identifying specific objects often can be done with morphology and can use the local processing of fine-grained SIMD parallelism or pipeline architectures. Image understanding problems are quite different; for example, determining a vehicle path in rough terrain or identifying houses can use local processing at the beginning, but much of the final processing consists of techniques such as subgraph isomorphisms, semantic processing, hypothesis testing, and neural networks. Some techniques are best serviced by machines capable of handling Lisp and perhaps accelerated by MIMD parallelism.

C. Overview

In Section II an informal classification (or taxonomy) of architectures is presented. More formal taxonomies can be found in the literature [8]. Various manufacturers are mentioned as examples of architecture types, but the enumeration is not meant to be exhaustive. Section III is a critical review of architectures and covers their advantages and disadvantages in terms of cost and the efficient use of hardware. The emphasis is on SIMD and MIMD pipelines rather than multiprocessor MIMD, which is better for image understanding. In Section IV we present some sample algorithms to illustrate how thinking must be changed to program fine-grained SIMD machines. The difference between the research and industrial perspectives of image processing is covered in Section V with an emphasis on machine vision. Section VI covers general factors in the design process of new machines, where it is shown how factors such as IC chip area, cost, speed, and pinouts must be understood and carefully orchestrated in order to design an efficient machine. In Section VII several benchmarks designed for image processing are discussed.

II. HARDWARE CLASSIFICATIONS AND EXAMPLES

This section is organized into five basic architecture concepts: (1) serial, sometimes called scalar or von Neumann, (2) MIMD multiprocessors, (3) pipelines, (4) single-bit SIMD, and (5) word-wide SIMD. Representative examples of manufacturers and research systems are given throughout.

A. Von Neumann Serial Processors

The most popular type of image processor uses a low-cost traditional serial processor, probably based on a microprocessor chip with a reduced or complex instruction set (RISC or CISC). An abundance of hardware systems, accelerator

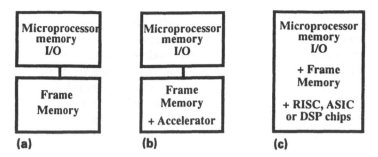

Figure 1 Three types of SISD systems: (a) frame memory in a bus-based system; (b) frame memory and accelerator; (c) single-board system enhanced.

cards, and software for existing systems are being released yearly. Numerous companies have imaging cards for popular bus-based systems. The simplest system, shown in Fig. 1a, is just a plug-in "frame grabber" card. Slightly more complex systems are shown in Fig. 1b, where some form of accelerator is on board the frame grabber. For example, Data Translation (Marlboro, Massachusetts) has a large number of low-cost imaging cards for many popular bus-based systems. EPIX (Northbrook, Illinois) offers an IBM board with a RISC chip accelerator. Coreco (St. Laurent, Quebec) has IBM boards that are based on lookup-table processing.

There are a number of board-level single-processor systems mostly for machine vision. These systems are generally for turnkey applications and interface to a host controller. As shown in Fig. 1c, the microprocessor is often enhanced by a special ASIC (Application Specific IC) coprocessor such as a normalized correlation chip in the machines from Cognex Inc. (Needham, Massachusetts). In the ITRAN (Manchester, NH) systems, an array processor assists a Motorola 68020 to accelerate various techniques, such as subpixel "calipers." Recognition Concepts, Inc., (Incline Village, Nevada) has a single board with a DSP chip that provides image warping, fast Fourier transform (FFT), and convolutions.

B. MIMD Multiprocessors

There are a number of MIMD processors that contain a small array of RISC or CISC elements. These are low enough in cost to have a widespread market. These systems are often characterized by how the separate processors interface with memory and to each other. Figure 2 shows a number of popular interconnection schemes.

The following companies provide multiprocessors on a bus-based board. Sky Computers (Lowell, Massachusetts) has a Q-bus board with two DSP chips. Mercury Computer Systems, Inc. (Lowell, Massachusetts) has a single VME board with four Intel i860s. They also make a number of other bus-based boards,

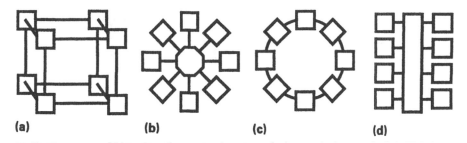

(a) **(b)** **(c)** **(d)**

Figure 2 Multiprocessor interconnections: (a) hypercube; (b) crosspoint switch; (c) ring; (d) shared memory.

some with a pipeline capability. Computer Systems Architects (Provo, Utah) manufacturers PC cards with four Inmos transputers. Visual Information Technologies, Inc. (VIT) (Plano, Texas) sells an image-processing system with a custom VLSI processor. Each system includes four processors, each operating on 8 pixels simultaneously. Ektron (Bedford, Massachusetts) has an expandable SUN/VME board with from 1 to 208 transputers. The ZMOB [15] is a research machine and was completed at the University of Maryland in the early 1980s. It contains 256 Z80 processors in a ring structure. The Butterfly from BBN Advanced Computers, Inc. (Cambridge, Massachusetts) contains up to 256 Motorola 68020 microprocessors that are interconnected by proprietary VLSI switches.

At one time massive parallelism also implied bit-serial processing elements, but that has changed. MIMD parallel processors that consist of a very large number of microprocessor chips are generally too expensive for a broad range of commercial applications, and are considered to be general-purpose supercomputers rather than image-processing engines. Some examples of hypercube massive parallelism in MIMD architectures are the N-Cube Corp. (Beaverton, Oregon) machine that has up to 1024 processors and the recent Thinking Machines Inc. (TMI) (Cambridge, Massachusetts) machine with thousands of SPARC processors.

C. Pipelines

Another MIMD architecture that is widespread in imaging is the pipeline processor with an architecture made popular by the ERIM Cytocomputer (Ann Arbor, Michigan). The Cytocomputer specializes in two- (2D) and three-dimensional (3D) mathematical morphology with a few other support boards. The systems are largely for research. Turnkey cytocomputer systems are used in military imaging applications. Figure 3 shows various forms of the basic structure of the pipeline.

The original pipeline concept used at ERIM is to configure the algorithm so that one long chain of identical processing units are in a sequence as shown in

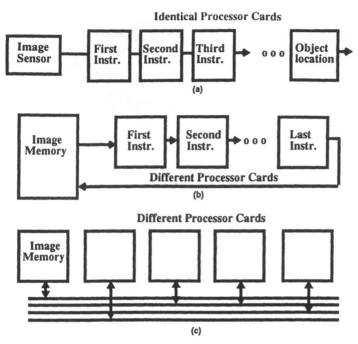

Figure 3 Pipeline processors: (a) long chain of identical modules; (b) shorter chain with image memory and heterogeneous processors; (c) modules with flexible interconnections.

Fig. 3a. Bypass linkages (not shown) allow results of various sequences to be combined at various points in the pipeline. In short, the image goes in at one end of the pipe at the sensor frame rate, passes from one programmable module to the next, and an image with object cues exits the other end at the same rate. VLSI technology is used to allow hundreds of processing elements to be strung together in principle. For this configuration, it is important that all processing sections are identical, so that algorithms are not dependent on a particular order of the processing units.

Datacube (Peabody, Massachusetts) and Imaging Technology, Inc. (ITI, Bedford, Massachusetts) have a large number of different boards for different image-processing functions, such as convolutions, morphology, feature extraction, and histograms. Highly specialized boards are available, such as large-area convolution board with limited kernels, and an image warping board from Datacube. LSI Logic Corp. (Milpitas, California) has chip sets for many specialized functions, such as morphology, convolutions, and histograms. Figure 3b shows a diagram of these systems, which are typically a heterogeneous sequence of processors.

The ASPEX (New York, NY) PIPE [16] is also a system with identical stages. Up to eight modules can be supported, each containing lookup tables, neighborhood, logic, and arithmetic operations. The system differs from the usual pipeline processor in two areas. There are two image buffers in each module, and there are six video buses connecting modules. Reality Imaging Corp. (Solon, Ohio) has a parallel pipeline workstation called the Voxel Flinger, designed for 3D imaging such as volume rendering, oblique reformatting, and multiplanar reconstruction.

There are many more pipeline systems. Matrox (Dorval, Quebec) has a large number of bus-based cards of various types, including pipeline cards for convolutions, morphology, and median filtering. A pipeline processor from Megavision (Santa Barbara, California) provides linear, 2D and 3D processing. VICOM Systems, Inc. (San Jose, California) has a pipeline image processing workstation with a capability for linear and nonlinear processing. Perceptics Corp. (Knoxville, Tennessee) has a pipeline workstation with capabilities for spatial and transform processing. Noesis (St. Laurent, Quebec) sells a VME system based on a SPARC processor with a plug-in board for 2D and 3D morphology.

In practice, commercial development systems have virtual pipelines, where the image is captured in a "frame grabber" memory, sent through a shorter pipe, and then recaptured, as shown in Fig. 3b. The parameters of the module are changed and the process continues, so that longer pipes with arbitrary configurations can be simulated. These systems are more flexible and trade off flexibility for processing time. The various pipeline companies are distinguished largely through the manner in which processing modules are connected. ITI products based on the VME bus have a special backplane that acts as a four-node switch to allow a flexible interconnection of the plug-in modules, as shown in Fig. 3c.

Earlier Datacube systems were interconnected through front-panel cables. Manual cable changing is not necessarily part of the algorithm development, since the modules can be virtually interconnected via the freeze-frame memories, although at a slower throughput. A later system, the Max Video 20, contains a crossbar switch and a number of different function modules on a two-slot VME card.

D. Single-Bit SIMD Parallel Processors

Categories of SIMD architectures are generally given by the interconnection scheme of the processing elements (PEs). Some PEs operate on single bits, and others have a wider word size. Current wisdom is that a more complex linkage, a wider word size, and a larger number of PEs means higher performance, but in practice the issues are more confusing. SIMD systems can be organized into two classes: single-bit and word-wide data flow.

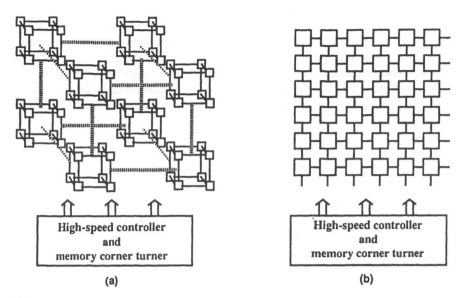

Figure 4   MIMD systems: (a) hypercube; (b) mesh connected.

The earliest form of SIMD machines uses PEs that operate on image data that are represented by one bit. Arithmetic operations are provided by bit-serial addition, where a sum is built up by adding one bit at a time and storing a carry (or borrow) bit in the PE. The most common interconnection schemes are the hypercube, mesh, and linear linkages.

1.   Hypercube

The linkage of the hypercube is shown in Fig.4a and is in a class by itself. There is only one realization of a bit-serial hypercube, and that is built by TMI [17], model number CM-2a. Although this type of architecture has been used in all aspects of imaging, it is too expensive for turnkey applications and is used primarily in research or off-line applications. There are $2^N (N = 12$ or $13)$ PEs on the vertices of a N-dimensional cube, with communication paths along the edges of the cube. In practice, the pathways are virtual since each PE has four communication ports. When data are to be transmitted to other PEs, there is a dynamic and continuous reallocation of the four connections, and the transfer requires many clock cycles. As its name implies, the processor is excellent for problems that require a high connectivity, such as graph matching, complex networks, and finite element analysis. Data I/O in the hypercube and mesh machines are a complex set of "corner-turning" circuits to transform data from a pixel–byte format to a frame–bit format that the PEs can handle directly.

2. Mesh Connected

Historically, the most popular interconnection scheme is the mesh shown in Fig. 4b. The PEs are conceptually arranged in a 2D array and are interconnected to nearest neighbors. The first instance of this architecture is a research system built by Goodyear and called the Massively Parallel Processor (MPP) [18] with a 128 × 128 array of 16K PEs. The BLITZEN chip [19] is an outgrowth of the MPP and has 128 PEs in a 8 × 16 array. Each PE contains a 32-bit cache and a 1024-bit random access memory. Adaptive Memory Technology, Inc. (AMT, Irvine, California) also manufacturers a mesh-connected system with the DAP architecture [20], for which 1K and 4K systems are available. Word-wide co-processors accelerate arithmetic computations. National Cash Register, Inc. (NCR, Fort Collins, Colorado) builds a chip called the GAP (NCR45CG72) which contains 72 bit-serial PEs. In this chip there are two communication layers, horizontal and vertical. There are a number of lesser known realizations of mesh-connected architectures surveyed by Fountain [21]. Many of these are one-of-a-kind research systems.

The Triakis system from Kensal Consulting (Tucson, Arizona) is the only system that provides morphology and other nonlinear transformations directly in a voxel space. Structuring elements are actually connected as a 3D mesh in a 12-neighbor face-centered-cubic tessellation [22]. The types of operations that the system provides covers a wide variety of 3D applications, such as filtering, 3D visualization, and cross sections.

3. Linear Arrays

The most popular example of the bit-serial linear connected SIMD processor, shown in Fig. 5a, is the Applied Intelligent Systems Inc. (AISI, Ann Arbor, Michigan) series of computers using the PIXIE [23] and FIREFLY chips [24]. These systems are 10 times lower in cost than mesh systems and are the only examples of SIMD architectures used in widespread turnkey applications. In some systems (AIS-4000 and AIS-5000) there is one PE for each column in the image. Other systems (AIS-3000 and AIS-3500) divide the processing into smaller ribbons of 64 and 128 PEs. The image memories are on separate chips and are large enough to hold several image frames. Other examples of linear SIMD arrays used in research are the SLAP (Scan Line Array Processor) [25] developed at Carnegie Mellon, and the VIP developed at the University of Linkoping [26]. The Texas Instruments PVP [27] processor is primarily for digital television. These systems are shown in Fig. 5b and differ from the AIS family in that the local tightly coupled memory is on-board the chip and large enough to hold a backlog of a few lines rather than a few frames. The system operates in a mode that is similar to mesh systems, in that a controller broadcasts a new instruction every clock cycle.

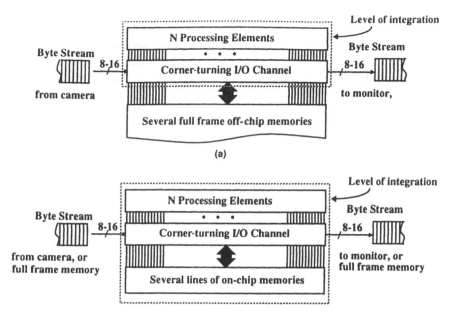

Figure 5   Linear array architectures: (a) with off-chip memories; (b) with on-chip memories.

Linear systems share a common simple data I/O corner-turning technique to transform data from a pixel–byte format to a row–bit format. A corner turner is simply an 8-bit shift register with a length the size of the image width. Pixels are input from the left in a stream of bytes and exit to the memory in the form of single-bit rows. Data output is in the opposite order.

4.   Other Single-Bit SIMD Architectures

The LUCAS (Lund University Content Addressable System) [28] is a 128 processor array that communicates using a "shuffle/exchange" network of switches shown in Fig. 6a. There are many pyramid architectures in the literature, but the only one built is the University of Washington Pyramid Machine [29], which is a research prototype with 341 processors arranged as in Fig. 6b. There are several levels in a pyramid with converging paths of communication. Each level is separately SIMD.

E.   Word-Wide SIMD Processors

In the past, SIMD architectures were synonymous with bit-serial fine-grained parallelism. But with advances in IC gate density and packaging, there are more

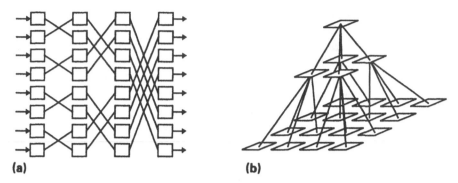

(a) (b)

Figure 6 Other SIMD architectures: (a) shuffle/exchange; (b) pyramid.

and more commercial machines that are massively parallel course-grain processors where each PE operates directly on data words.

1. Hypercube

TMI expanded the earlier architecture into a supercomputer by putting up to 2048 floating-point coprocessor chips in the system. The same interconnection concept is used. There are other systems described above as massively parallel MIMD hypercubes, but the TMI system, the CM-2(G), is SIMD.

2. Mesh Systems

The highest-performance mesh SIMD system is the MASPAR [30] (Sunnyvale, California), which has up to 16,000 floating-point (32-bit) processors and data paths to external memory that are 4 bits wide. As mentioned earlier, the DAP system, which has a bit-serial architecture, also has 32-bit word-wide floating-point processors. The Connection Machine and DAP also have a combined architecture where floating-point chips expand the bit-wide PEs. In this architecture each memory can be interpreted as a single-bit pixel, or adjacent groups of wires can be interpreted as multibit words. Figure 7 illustrates a conceptually simpler hybrid of a 4-bit and a 1-bit machine.

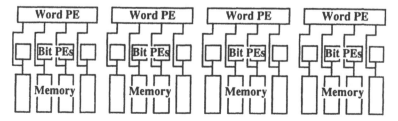

Figure 7 Hybrid bit-wide, word-wide architecture.

3.   Linear Systems

The clip7A [31] is a linear chain of 16-bit PEs with 8-bit external communication paths. There are a number of on-board memories and the capability of larger off-chip memories. The Pixar (Now VICOM) is a well-known system that is capable of many imaging algorithms. The processor board contains four proprietary Chap SIMD processors tightly coupled to an image memory. The Carnegie-Mellon WARP [32] architecture consists of up to 24 identical processing units that receive the same instructions but operate in either a pipeline or data parallel mode. The PEs are complex and contain a floating-point ALU and multiplier. A hybrid combination of bit-wide and byte-wide PEs has been designed at Applied Intelligent Systems, Inc. called the CENTIPEDE [33] chip. A unique interconnection structure in the CENTIPEDE has indirect addressing of the bytes that allows highly parallel operations, such as lookup tables, histograms, and feature extraction.

III. CHARACTERISTICS OF ARCHITECTURES

A. Memory Efficiency Criteria

If the cost/performance ratio is the most important factor, the impression that the best architectures have a wide word size and a versatile interconnection scheme is not always correct. Architecture cost efficiency is closely tied to the specific algorithm being run. There are a number of criteria regarding the system memory that indicate the efficiency of a processor design. The following criteria are a guide to evaluating why or why not a particular algorithm is efficient on a specific architecture:

1.   How many memory wires in the system connect to processors (including cache)? This is a measure of the degree of parallelism.
2.   How fast are memories cycling? This depends on chip technology and includes idle time.
3.   Are all read data values used? This is algorithm dependent.
4.   Is each data value read only once? Processors without enough local cache may reread unchanged data from memory.
5.   How many cycles are needed to move data to the location where they are needed? This is dependent on the interconnection topology of processors.

The foregoing criteria can be quantified as efficiency factors (EFs):

EFw = total number of data memory wires

EFt   = cycle time of memories, including idle time

EFu = percent of useful data bits read

EFr   = average number of times unchanged data is read

EFm = average number of cycles needed to move data

With the criteria above, the memory efficiency of a processor for a given algorithm can be defined as

$$\frac{\text{EFw} \cdot \text{EFu}}{\text{EFt} \cdot \text{EFr} \cdot \text{EFm}}$$

In this definition of efficiency, the word size and complexity of the processor and the number of processors are not factors, and thus all architectures, including SISD, SIMD, and MIMD, can be compared. The memory efficiency equation is only an approximate indication of processor efficiency because the complexity of the processors plays a strong role.

The efficiency equation shows that if there are large word sizes in a computer (factor EFw), but if not all memory pins are being used because of small word operations (factor EFu), the efficiency will drop correspondingly. In a bit-serial processor, the memory usage is efficient because smaller word sizes result in fewer memory cycles. The equation also shows that single-bit-deep memories are not effective as byte-wide memories in a parallel processor design because the bandpass to get the information out of the memory is lower. As silicon gate density increases, the size of memory in megabits is increasing at a rate that is faster than the increase in the number of pixels in image sensors. Figure 8 shows the increase in sensor sizes and memory sizes over a number of years. For parallel processing in the future, there will be a crossover point where larger memory sizes are not useful unless the number of data output wires increases proportionately. Although memory sizes seem to be increasing indefinitely, the need for image resolution above certain limits (depending on the application) will not.

The trade-off in fine- versus coarse-grained processors hinges on the algorithm and application. For a given number of memory wires in a system, how are they best used? Consider two SIMD architectures, where both have the same number of memory wires connected to PEs. Suppose that one architecture comprises 128 PEs with byte-wide addition operations in each PE, and the second

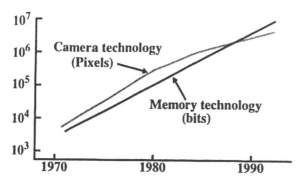

Figure 8   Memory and camera technology.

architecture comprises 1024 bit-serial PEs. EFw is the same for both systems. Suppose that EFt is also the same. Both machines will be able to add two 8-bit numbers at the same speed. The memory efficiency is the same in both cases if we assume that EFu = EFr = EFm = 1. Suppose that the operations on the byte machine require three clock cycles: two to read the operands and one to write the result. Thus 128 additions occur in three cycles. The bit-serial machine requires 24 cycles: 16 to read the two operands and 8 to write the result. Thus 1024 additions occur in 24 cycles. The total number of cycles for 1024 additions is the same for both machines.

However, it may be more difficult to keep the larger number of processors busy if the region of interest is small, as it is, for example, while focusing on a single character in character recognition. If an algorithm requires only 128-byte-wide arithmetic operations on small regions of interest, the first architecture is superior and finishes in only three cycles. The bit architecture still requires 24 cycles and is inefficient because of factor EFu.

Suppose that both machines require N cycles for 2D morphology operations over the full image. The bit-serial processor can handle 1024 pixels in N clock cycles. The byte-wide processor can handle only 128 pixels in the same time and suffers from EFu because 7 of the 8 bits in a pixel are not used in 2D morphology. Thus the bit-wide architecture is superior for large-area morphology. The bottom line is that the efficiency of an architecture is algorithm dependent. For that reason, designs such as the DAP, CM-2(G), and CENTIPEDE have the hybrid combination of the two architectures as shown in Fig. 7.

B. Customization

Custom Application Specific ICs (ASICs) designed for a certain procedure and used in hardware will in principle always be the most efficient use of silicon, compared to a more general programmable processor, but only when the hardware is processing that procedure. Otherwise, the chip will be idle. The idle time of a custom chip would decrease if other procedures in other chips are running at the same time as in a pipeline architecture. Suppose that a convolution is only 10% of the algorithm and running in a convolution module. If there are other modules running the other 90% of the algorithm simultaneously, the idle time of the convolution chip would drop to zero if there is a continuous flow of new images through the pipeline. Then the silicon would be fully utilized.

A major problem with the efficient use of ASICs is that the hardware must be closely mapped to the software. If a custom module is underused, the silicon is idle. If the module is overused, many other ASICs are idle while waiting for the process to complete. EFt is lower in both cases. In pipeline systems EFr and EFm are always the maximum. By definition, data are where they are needed, but the downside is that there is no further flexibility of data movement.

C. Serial Microprocessors

Generic chips such as microprocessors or RISC chips in serial processing are continually cycling instructions and therefore will never be idle. A main source of inefficiency is that they are overpowered for some simple tasks. For example, a 32-bit integer CPU with a floating-point processor will leave many areas of the silicon idle when operating on 8-bit words (EFu). The loss in efficiency can be even greater when doing bit plane morphology. Efficiency can be increased if the word is corner turned into a raster format and consists of, say, 16 adjacent pixels in a row. Then the word-wide logic is more effective. In image processing there are many examples where generic microprocessor chips are grossly under-utilized.

D. Parallel Processing with Generic Chips

Although it can be argued that microprocessor chips do not always use silicon efficiently for the types of operations used in imaging, they are low cost due to the high production volumes. This alone can often make up for the inefficiency. Also, they are well understood, and effective compilers are always available. Thus it makes sense to design MIMD systems around them. A major problem that has been well studied is how to design algorithms and compilers that make full use of a large ensemble of processors, by load balancing to ensure that few are idle.

A further silicon inefficiency arises because of the ensemble. In many types of operations all processors are running the same algorithm on multiple data (SAMD). Then for every processor, there is a redundancy in program memory, address generation, and instruction decoding. Along with the CPU chip, there is a lot of support chips and glue logic.

E. SIMD Fine-Grained Systems

None of these problems exist in fine-grained parallelism. In general, a bit-serial processing unit is so simple that the gates are used almost 100% of the time for local neighborhood imaging algorithms. Because of the single broadcast of an instruction, there is only one program memory, one memory address generator, and one instruction decoder for the entire system. There is a strong efficiency in arithmetic procedures: shorter words use a proportionately shorter cycle time. These systems are very efficient for image-based operations such as edge detection, skeletonizing, morphological processing, order-statistic filtering, character recognition, digital halftoning, and many other nearest-neighbor operations.

1. Efficiencies of SIMD Compared with MIMD

Fine-grained SIMD systems loose efficiency for a number of other reasons. To keep pace with, say, 128 CISC processors doing byte-wide arithmetic, there must be close to 1024 bit-serial processors. There is an absolute dependence on the parallelism since bit-serial operations are slower. In problems that require data-dependent branching, the SIMD processors further lose efficiency. Reading out the locations of a number of cues or features is more difficult due to the distributed nature of the image data memories. Sometimes the parallel processor must be interrupted while a less efficient host serial processor is used to read out data (one word at a time) that are necessary to make decisions before further parallel processing can take place.

There are other algorithm efficiency problems with SIMD fine-grained parallelism that are a subject of continuing research. Since all processors perform the same instruction on an image at any clock cycle, it is at first difficult to imagine how the system can provide operations not related to image space transformations. SIMD processors have a reputation for being good only for "low-level" processing [34]. However, there is much research under way to make SIMD more effective on intermediate- and higher-level procedures such as in neural networks [24], Hough transforms [25], two-dimensional FFT [35], and pattern matching. It is less efficient but still capable of realizing sorting algorithms, histograms, and nonhomogeneous operations such as image rotation and warping. There a continuing development of new SIMD methods that in the past had been reserved only for different architectures. These new methods are generally not obvious, and totally inappropriate for most other types of architectures. There are still many important imaging problems, such as compression and zooming, that are not as effective on a SIMD architecture. Although these procedures are possible, the efficiency is closer to that of a serial processor. Many problems in image understanding such as hypothesis testing, semantic networks, and expert knowledge have not yet been solved on fine-grained SIMD.

2. Linear SIMD

Linear SIMD systems have a number of cost advantages in the controller design over hypercube and mesh systems. The controller for a linear SIMD broadcasts an instruction to all PEs and then cycles through the rows of an image, one at a time, by generating a simple sequence of address patterns. When the processing of all rows in a region of interest are completed, a new instruction is broadcast and the row cycles are repeated. The instruction rate is rather low (tens to hundreds of microseconds), whereas the rate for the simpler task of updating memory addresses corresponding to rows is high (0.1 µs) and is all that the controller needs to handle at a high rate.

A second important advantage of linear SIMD systems is that the linear nature of the PE coupling allows raster scan image data to be entered into the system easily. The image I/O circuits are integrated into the same chip as the PEs. Figure 5 shows how image data in a raster format can be entered into the image memory that is tightly coupled to the PEs. Byte-serial data enter from the left and fills a long shift register that is distributed over the circuit board, but local to each PE. The number of clock cycles needed is equal to the number of pixels in the image width. After a row resides in the shift registers, the bytes are transferred to the local memory in eight clock cycles. This is called corner turning. The data I/O process can occur asynchronously and independently of other processing that is occurring in the array.

A unique characteristic of linear SIMD is that the memory can easily be external to the chip. For a processor chip with 32 PEs, a 512-bit-wide image can be realized with only 16 chips. The use of only 32 PEs per chip results in 32 wires being devoted to tightly coupled external memory. Thus no further staging memory or corner-turning circuits are needed. Mesh-connected system designs rely on memory internal to the chip because the larger number of PEs per chip would result in too many wires for external memory.

3. Mesh and Hypercube

In a mesh, hypercube, or linear SIMD system of the type in Fig. 5b, there is a small number of subimages in the array at any one time. The controller is very complex because it must update both instructions and memory addresses at the maximum speed. Once a subimage is processed, the transformed image must be passed out of the local memory while a new subimage is transferred in. The same batch of high-speed instructions are again applied.

In hypercube or mesh SIMD systems, the image data must be corner turned for each subimage distributed in a multiplicity of PEs. Corner turning circuits are not shown in Fig. 4, but the complexity is obvious. Data cannot enter the array in the form of bytes because of pin limitations in the chips. Thus the I/O cannot be completely integrated into the PEs and complex corner-turning circuitry must be utilized.

Fundamental advantages of hypercube and mesh systems over linear SIMD are the high communication bandwidth between processors (EFr and EFm) and the expendability. The maximum number of PEs that a linear system can have is equal to the width of the image in pixels. A system at this limit is massively parallel in the horizontal dimension and has reached the limit of expansion. An arbitrarily high performance can be achieved by adding a number of linear systems in a pipeline, as shown in Fig. 9. Separate controllers and image memories must be provided for each pipeline segment, and algorithm partitioning must be done as in other MIMD systems.

F. Efficiencies of Pipeline Processors

There are a different set of efficiencies and inefficiencies in pipeline processors. First, as already mentioned, these systems use ASICs that are a very efficient

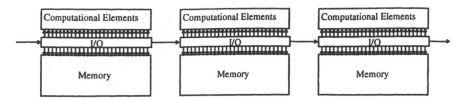

Figure 9   Linear SIMD system in a pipeline to increase performance.

use of silicon in operations for which they were designed. Factor EFw is the number of memory wires reading out data and is a measure of the degree of parallelism. During a clock cycle, a pipeline module input has only one byte of data being read in. However, there is local memory cache on board that generates the data needed for a 3 × 3 neighborhood (at least), so that for each clock cycle, the amount of data read from main memory is effectively multiplied by 9. There is also a 3 × 3 group of numbers in local registers that make up convolution kernels. If these are included as data input, the total bandpass is a respectable 18 bytes per clock cycle for that one module. The use of a multiplicity of modules and larger neighborhoods will further increase the memory bandwidth.

One problem with the pipeline concept arises when there are fundamental serial operations that break the flow of the pipe, such as in adaptive processing. Another problem is that pipeline processors can act only on nearest neighbors, most often in a 3 × 3 window and cannot accommodate other operations, such as the Hough transform or image rotation unless the user has a card in the system specifically designed for that. Arbitrary large neighborhood operations that cannot be obtained by a chaining of smaller neighborhoods can be accommodated by address offsets in the frame memories. That is, one frame memory has the local pixel, and another frame memory that has identical data is given an address offset. When both frame memories are clocked, the distant pixel readout is synchronized with the local pixel so that both are simultaneously available for processing. The EFm cost for this method is very high. Also, local neighborhood memory caches that ordinarily provide a multiplied efficiency cannot be used. Furthermore, the nonneighbor process cannot be put in a pipeline, with the result that one frame time is required for each neighbor point, and all other modules in the system are idle during the process.

The Machine Vision International [36] (MVI) Genesis processor was made specifically to allow nonneighbor operations by using a large onboard cache—a single long programmable delay line—so that a data point from distant neighbor could be associated with a local point in a variety of operations.

IV. DEPENDENCE OF ALGORITHMS ON ARCHITECTURES

An in-depth discussion of algorithm concepts on parallel systems could be considered to be out of the scope of a chapter on architectures. However, SIMD programming is so alien to most people that it is essential to understand the thinking behind some of the algorithms in order to evaluate the efficacy of the architecture. There is a general but inaccurate perception that SIMD systems are restricted to nearest-neighbor operations and low-level processing.

A. Programming Languages

The programming model cannot be disassociated from the architecture. A simple programming environment must be available that implicitly captures the hardware model and hopefully prevents the user from writing code that is not efficient in the hardware. In the simplest case, the processor is a single popular microprocessor, and the programming environment is simply a popular language with function libraries. In MIMD parallelism, the same can be true, but the knowledge of the hardware model is contained in a parallelizing compiler that often is not very effective. Research is being done on the best use of MIMD systems for various imaging problems, such as divide-and-conquer algorithms [37]. Other MIMD systems, such as pipelines, use high-level software routines that control the operation the various modules.

For SIMD systems the processing elements are managed by a controller using microcode. Various high level functions in a popular language call on carefully written microcode. Programming a complex function such as image rotation efficiently in microcode can be a very difficult process. But once done it can be used as a module in a similar manner that in a pipeline system uses a hardware module. Thus a SIMD microcode software designer performs a function equivalent to that of a pipeline hardware designer. In either case the use of function calls for the various operations removes the parallelism to a certain extent from the mind of the programmer. Although new programming paradigms are often portable between classes of similar architectures, the exact code is never portable, in the same way that Fortran or C is on almost all serial machines.

B. Machine-Dependent Thinking

Many algorithms that are perceived to be efficient, are efficient on only one architecture type—generally, serial machines. A totally different algorithm may be best for a parallel machine. The programmer must understand the machine and write the appropriate algorithm. For example, blob analysis using moments has been studied and is possible on a SIMD system. But it is not the best utilization of the parallelism. Maybe morphology, for example, is just as good for that particular problem, but if the programmer has little experience in morphology, the machine is used inefficiently. In fine-grained parallelism, programmers have to understand not "what is a good algorithm" but "what a good SIMD algorithm is on machine X."

For millions of years our creative thinking has been a serial flow of logical deduction, although our brains are highly parallel systems. This ingrained constraint makes the creation of SIMD algorithms difficult. Most parallel algorithms for CISC MIMD break up the two-dimensional data in an image and distribute it to the various processing nodes. An image transformation can then take place as a SAMD process. The breaking of a big task into many small tasks (divide and conquer) can be understood using serial thinking.

C. SIMD Algorithm Examples

SIMD algorithms require a totally different visualization of algorithms. Some examples of an altered visualization follow.

1. Convolution

In a serial computer a convolution might be done using four nested loops. The two outer loops cycle through the x and y coordinates of the pixels; and the two inner loops circulate over the kernel. The computation at the innermost loop provides the weighted sum, where the weights are stored in local CPU registers if possible.

The computation in a mesh-connected SIMD system is shown in Fig. 10. The entire image is translated short distances about a neighborhood defined by the kernel. At each translation, the controller broadcasts a multiplication operation so that each PE multiplies its translated pixel by a constant implicit in the controller. The constant is the kernel weight at that translated position. Finally, the multiplied image is summed to an accumulator memory at each PE. This process is visualized as a whole-image shift and accumulate and is quite different from the use of four nested loops. Convolution efficiencies are high for all SIMD architectures because EFm is close to 1, due to the local communication.

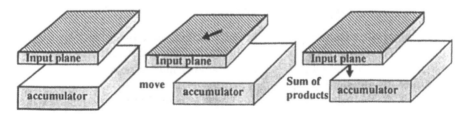

Figure 10 Visualization of a convolution on a SIMD system.

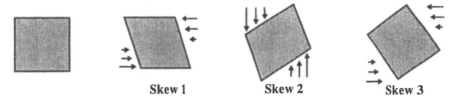

Figure 11 Visualization of a rotation algorithm in a SIMD system, where tree linear skews result in a rotation.

2. Image Rotation

To provide a rotation of an image by angle θ, a serial algorithm might consist of four nested loops, where the outer two loops cycle through the destination pixels of the image. The gray value of each destination coordinate point is given by the gray value of the data at the transformed coordinate point given by

$$X' = X \cos \theta - Y \sin \theta \quad \text{and} \quad Y' = X \sin \theta + Y \cos \theta$$

Since X' and Y' will probably not be an integral value and not refer to an integral memory address, some sort of interpolation must be done about a neighborhood of X' and Y' and is provided by the inner two loops.

This type of algorithm is impossible on a SIMD machine with a simple connection scheme. The solution as it is formulated requires a highly nonhomogeneous movement of data. The image transformations in Fig. 11 will result in a image rotated by $T \Leftarrow 90°$. There are three successive linear skews: a horizontal skew on the input image, a vertical skew on the resultant image, followed by another horizontal skew [38]. This method is totally inappropriate on a serial computer because of the massive data movement. Although this algorithm is still nonhomogeneous, the linear skews are tractable on a SIMD architecture.

A shift along some row of pixels by an integral coordinate value is straightforward. To prevent jagged edges due to noninteger shifts, a weighted average of the nearest integer shifts can be provided. The next problem is how to provide a skew that is a shift by a different amount in each column (row) in the image when each PE is receiving the same instruction. The linear skew is a nonhomogeneous transformation because the behavior of the data manipulation is position dependent. SIMD data-dependent branching is required and is discussed later.

3. Image Transpose

An image transpose on a serial computer is very simple. Two nested loops cycle through the coordinates. The gray value at x, y is stored at the output image coordinate location y, x. The method for a transpose on a SIMD machine is also very inappropriate on a serial machine. It requires $\log_2(N)$ nonhomogeneous op-

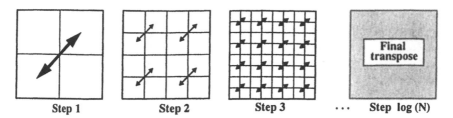

Figure 12   Visualization of a transpose algorithm on a SIMD system.

erations, where N is the linear dimension of a square image. A 512 square image will require nine steps, as shown in Fig. 12. The first step divides the image into quadrants as shown, and off-diagonal quadrants are interchanged. In the second step each quadrant of the previous step is divided into quadrants, and similar off-diagonal subimages are interchanged. The process is repeated until the final quadrants are single pixels.

There may be other more efficient methods for the examples above, and they are most likely equally creative. Other nonhomogeneous operations such as the Hough transform [25] and FFT [35] can be provided on SIMD systems through a variety of other techniques.

D.   Data-Dependent Branching in SIMD

Nonhomogeneous operations require data-dependent branching. In a serial processor, an algorithm flowchart might look like that in Fig. 13a. There are two possible branches of code that the processor might take. These procedures are boxes labeled A and B. One path is chosen and the other is ignored. The length of the box indicates the length of time the procedure requires. The actual processing time is equal to the length of the path taken.

In a SIMD system different branches would be taken in different areas of an image, such as in the rotation, or transpose. There are many kinds of image data that might influence a branch. The SIMD flowchart is shown in Fig. 13b. Both

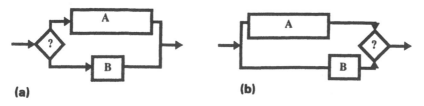

Figure 13   Flowcharts for branching: (a) on a SISD system; (b) on a SIMD system.

procedures are processed, whether they are needed or not. Then the result of the correct procedure is chosen at each pixel site by using Boolean logic. The branch condition C and the results of the two separate computations are combined in a simple expression

output $=$ (C AND A) OR (NOT C AND B)

where A and B are the results of the two paths. The processing time is the sum of the time for the two procedures and the final branch selection process. A major conceptual difference between processing concepts is that in serial processing the branch path is chosen first and then the procedure is run. In SIMD both procedures are run first and then the branch condition is processed.

The linear skew is an example of data-dependent branching and is used in the rotation algorithm. One way of providing a skew is to create a ''skew frame'' of data that has a value at each pixel site equal to the distance of the skew for the particular row (or column). The whole image is then translated one pixel at a time, and the skew frame is decremented for each translation. When the skew frame decrements to zero, the pixels at those points are stored at the destination site. Although this method is very fast on a mesh system, it is not an efficient use of the parallelism because the same data in the moving image and skew frame are continually being read, which results in low EFr and EFm efficiency.

V. COMPUTER VISION IN RESEARCH AND INDUSTRY

There are two perspectives in solving imaging problems that have an influence on the choice or development of the imaging architecture: research in imaging and commercial applications of imaging.

A. Research Versus Commercial

The goal in some research programs is to explore the intricacy of vision, biological models, or the ultimate complexities of various branches of imaging. Understanding the subject of the research problem is of primary importance and the solution to practical problems is secondary. In this case the most advanced (and expensive) architectures are appropriate, and current utility need not be justified.

Some institutions are working on applied research to solve problems that are intended to support near-term applications. This type of research should not use an algorithm strategy that depends on a supercomputer or an inherently expensive architecture. The solution must eventually be mapped to a lower-cost system; otherwise, the commercial perspective is lost. If a problem is solved on a machine that meets cost requirements, but the computation exceeds the cycle-time requirement by a factor of 10, the research is about five years ahead of its time, since computer power increases roughly at a rate of a factor of 10 every five years. It is certainly valid to propel ideas that are not yet practical even on state-of-the-art hardware, but if algorithm ideas are too premature, they will probably be obsolete by the time the hardware becomes realistic.

B. Industrial Perspective in Machine Vision

In this section we focus on examples derived from the machine vision industry rather than the larger and more varied imaging market. The important factors in the development of commercial applications of imaging are cost, speed, precision, and robustness. In new architecture designs, time to market is also important. The commercial imaging market is large and highly segmented. Architectures are also highly segmented, and many vendors have focused on small niche markets, which are best for their architecture. Exceptions occur for high-end systems that have high performance for almost any application, and for low-end systems such as frame grabber boards that interface with a variety of buses. These systems are versatile but of course are lower in cost and slower.

An industrial user deciding on an architecture or algorithm for a low-cost turnkey system must make the following trade-off. For a given cost, inspection cycle time, and development time, what is the optimal algorithm that can be created for that machine? Can it be trained in a few minutes on site? In industrial applications, price is of paramount importance. If price objectives cannot be met, the user will design a special-purpose system in-house. If that is not possible, machine vision will not be used at all.

Cost and inspection cycle time are easily quantified and understood, but what the user wants versus what the architecture and algorithm are capable of providing is much more difficult to quantify. It is often easy to impress someone with a solution that has a 95% success rate using well-chosen image examples in a laboratory demonstration. In fact, almost all architectures and algorithms seem to give the same performance during the course of a demonstration, because 95% is roughly 100%. In this case the choice of architecture is based on low cost and speed. A trade-off to the potential user is not apparent: a machine that requires 100 ms at a 95% success rate may require 150 ms at 99%. If a 99.9% success rate is required, a more expensive machine might be required. In some process control applications, a 6 σ (3.5 parts per million) failure rate is requested. This is unrealistic in the sense that it may take a year to verify that level of failure rate. Informally, 3 σ (99.7%) can be defined as no failure during any demonstration.

The degree of algorithm robustness is often an economic decision by the user who is often aware of exactly how robustness affects the manufacturing process. A 0.3% failure rate seems like a small number to a researcher, but a system that is processing images at a rate of 10 per second will experience a failure (or "operator assist") every 30 s, or 1 million failures per year. If an operator assist costs 10 s, the process speed is 30% slower. If this is the case, it would be better to improve the algorithm at the expense of speed. Processing eight images per second at a 0.1% failure rate is an improved throughput.

VI. DESIGN PROCESS

A. Superscalar Architectures

The propagation times of IC logic gates increases by roughly a factor of 10 every decade. Computer processing power increases at a factor of 100 per decade (see Fig. 14). New high-performance serial "superscalar" architectures have been able to push ahead of the IC curves by taking advantage of higher logic density to promote concurrent operations. Internal pipelines allow the overlap of the various stages of processing instructions. Internal high-speed cache memories hold current data and instructions. Harvard architectures separate the instruction memory from the data memory so that both can be fetched simultaneously. Long instruction words allow further concurrency in processing. Branch prediction allows the instruction for the most probable branch path to be entered into the pipeline so that pipeline efficiency can be maintained.

As a result, the sophistication of scalar architectures allows the instruction rate to come close to that of the clock rate. What happens when serial processors evolve to the point where low-cost processor chips can support architectures that have one instruction per clock cycle? Processing technology will have caught up with IC curves so that annual increases in processing speed will be at the same rate as the slower annual increase in IC speed.

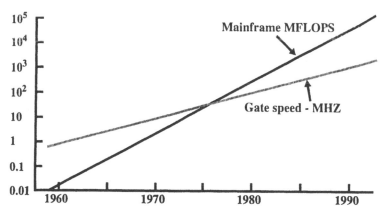

Figure 14 Increase of computer power and chip speed.

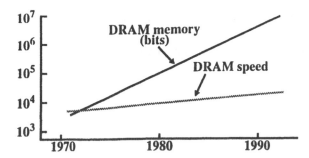

Figure 15   DRAM chip gate area versus chip speed.

B. Trend Toward Parallelism

Parallelism is the only way to maintain the pace of processor technology. The mainframe computer speed increases have been exponential only because of parallelism that was started in the 1970s. It is venturesome to predict the demise of scalar architectures in high-end computing, but 1991 has been heralded as the era of massive MIMD parallelism. The reason for the trend toward parallelism is that logic gate density and chip area is increasing at a much faster rate than the speed. For example, over the last two decades the sizes for the best-selling memories (DRAM) has increased at the rate of a factor of 100 per decade (see Fig. 15), but the speed increase is only a factor of 2 or 3 for the same period. The superiority of gate area over speed dictates parallelism.

Microprocessor systems will have a built-in capability for communication with other microprocessors so that parallel designs will become easier. The Inmos Transputer started with this capability years ago. The Texas Instruments is following suit with the TMS320C40 chip, which has six communication ports.

C. Philosophy of Supercomputers

There is a dilemma in computer architectures: We must use parallelism because of characteristics of chip technology. We should avoid parallelism because of the difficulties in writing efficient algorithms. Traditionally, supercomputers have been designed with less parallelism and more speed. In the efficiency equation, EF_t is maximized by any means possible, while EF_w is more moderate. Less parallelism means that the software of "dusty decks" of many years ago can still be compiled with some efficiency.

The highest-speed technologies are silicon emitter-coupled logic (ECL) and gallium arsenide (GaAs). These technologies have not achieved nearly the same gate densities as those of the slower CMOS technology. Thus the price for speed is the necessity of smaller numbers of gates in ICs and more hot expensive chips in the system. Heat dissipation and thermal management become a difficult problem. Other difficulties involve costly high-speed methods for impedance-controlled circuit traces, carefully timed backplanes, and tricky manufacturing—all in the attempt to minimize parallelism. If supercomputers are to maintain an exponential growth, even they must eventually go to massive parallelism.

D. Application Specific Designs

The definition of a class of algorithm concepts is usually the first phase in the technical design process of a high-performance imaging architecture. The performance of the hardware is optimized for those ideas. After the hardware design is completed, the hardware can no longer evolve quickly, but new algorithms are continuously being developed. From that point, the creative design efforts of a research group often involves working around the constraints of hardware designed for a different set of tasks. Thus initially, the application software drives the hardware design. Later, the hardware model will drive the software design.

For the diversity of image understanding techniques, a single architecture concept is not enough if highest performance is required. A proof-of-concept machine, the Image Understanding Architecture [39] contains an integration of a number of different architectures and is being developed by the University of Maryland. The machine consists of a number of different layers, each optimized for the type of algorithm used in each layer of image understanding, from the lowest image-based functions to the highest symbolic functions of vision.

In the literature and at imaging conferences, there are countless papers of circuit designs for very specific algorithms. Sometimes the designs show an interesting insight into trade-offs that may be more generally relevant, but most often they are too specific to be of general applicability. As mentioned earlier, ASICs may efficiently solve a small part of the entire problem but are seldom cost-effective for an architecture in a general programming environment.

E. Using Technology Efficiently

The very latest in IC technology is not used in the popular affordable machines that are designed for widespread applications. The jellybean silicon processor and memory chips that dominate the market are most cost-effective, and the more expensive ECL and GaAs chips are left to the supercomputers, which have a smaller role in imaging. Speed increases must be tempered by cost-effectiveness.

In designing an architecture with custom chips rather than off-the-shelf VLSI processors, there must be a careful orchestration between the computer architects' idea of a good design and the capabilities and properties of the IC technology. The factors in IC technology that must be kept in mind are chip area, speed, pinouts, and cost. The cost of generic memory chips are much lower than the cost of designing memory on-board an ASIC chip. One advantage of massive parallelism is the purchasing leverage when buying large volumes of the same chip for a system. An example follows of how chip technology influences a SIMD design.

Early SIMD processors were bit-serial because the limitation in gate density on chips restricted the complexity of the processing elements. As technology progresses, the number of gates on a chip is increasing at a faster rate than the number of pin connections for data I/O. The reason is simply that the bonding pads are on the periphery of the chip, so the number of bonding pads increases as the linear dimension of the chip and the number of gates increases as the square of the dimension. This ratio holds roughly even though the areas of gates and bounding pads are shrinking. Furthermore, bonding pads must be limited because they require extra power to drive the pins and circuit board traces. The pins/gates ratio is perfect for mesh-connected systems because interprocessor communication is conceptually also on the periphery of each rectangular sub-block of PEs. Hypercube systems have a higher degree of connectivity and require more silicon for the network connections, more pins, and more circuit board paths and connectors.

In low-cost linear SIMD systems that have external memory, the number of required pins is proportional to the number of PEs. Thus the PEs must increase in complexity as gate density increases if the maximum benefit of the pin/gate ratio is to be realized. These further complexities include on-board multipliers, indirect addressing, look-up tables, and word-wide arithmetic modes. The higher complexity of each PE and the lower number of PEs also favors effective usage of PEs by lowering the dependence on massive parallelism.

In today's technology, the internal speed of gates is approaching 10 ps, the number of pins per chip is several hundred, the number of gates exceeds several tens of millions (in memory technology). However, these extremes in performance cannot be achieved in one chip simultaneously, so a trade-off must be made. Most ASICs used in imaging lag behind the technology a little. GaAs is not used, although it is no longer considered an exotic technology. It is far from the mainstream of silicon, and its use entails long development times. For commercial products, proven memory technology is generally used.

VII. BENCHMARKS AND IMAGE PROCESSING

There is a large amount of research into the efficiencies of various methods for computing various operations on various architectures [40]. Studies often involve expressing the efficiency with formulas that are functions of the image size N, the number of PEs M in a particular architecture, and other factors. The formulas are generally stated in terms of the order $O(\ \ )$ of the highest power of the polynomial or other functions in the efficiency expression: for example, $O(N^2/M)$, or $O((N^3 + \log N^2)/M)$, and so on. The proportionality factor is dropped because it is often hard to determine unless the algorithm is investigated in great detail on the machine.

This formulation is useful in providing an understanding of the asymptotic behavior of the algorithm for very large images; however, architecture assessments may be inaccurate by the missing proportionality factor for smaller images, or in some cases, the formulation is not accurate for image sizes from standard cameras.

Benchmarks give the most accurate evaluation of architecture performance. However, benchmarks also have a number of problems that modify the value of the evaluation. Performance measures such as megaflops or Whetstones are not good enough for the specialized algorithms and processors used in imaging. For supercomputers there are two popular algorithm benchmarks [41]. The LINPAK benchmark tests the capability of the processor in solving large linear equations. The Lawrence Livermore Loops are a suite of code segments that exercise a number of procedures that frequently occur in scientific computing.

To many people the most important factor in evaluating a machine is the cost/performance ratio. However, the cost/performance ratio among various architecture types varies by orders of magnitude, depending on the algorithm that is being exercised. A number of benchmarks have been proposed for imaging that exercise particular tasks.

A. Abingdon Cross

The Abingdon Cross benchmark [42] starts with an image of a cross that is buried in noise schematically, shown in Fig. 16a. The task is to provide a skeleton

(a)

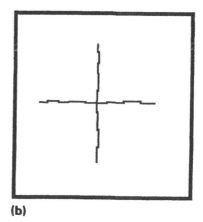

(b)

Figure 16 Abingdon Cross: (a) noisy starting image; (b) final skeletonizing.

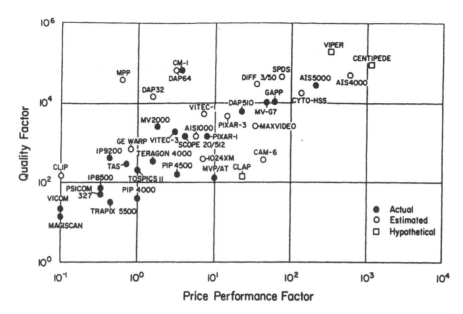

Figure 17   Results of Abingdon Cross benchmark. (From Ref. 42.)

of the cross as shown in Fig. 16b. A number of image-processing steps are needed and are chosen at the discretion of the programmer. These steps involve sophisticated filtering of noise and skeletonizing of the rather wide strokes. One criticism of this benchmark is that different programmers have different premises on how much prior information about the cross, such as SNR, and the width of the strokes, can be used in writing the algorithm.

The response to the benchmark was excellent in that 39 machines from a number of manufacturers and institutions were reported. These results are shown in Fig. 17, where quality is a measure of the performance of the machine relative to image size. The price/performance ratio varies over 10^4. The Abingdon cross benchmark places an emphasis on architectures that are optimized for local neighborhood computations, such as bit-serial massive parallelism and pipelines.

B.   Preston–Seigart Benchmark

A recent benchmark that covers a wider area of image-processing tasks is the Preston–Seigart benchmark [43]. Seventeen machines have been reported. There are 12 tasks, including table lookup, convolutions, nonlinear filters, logic, the Fourier transform, and disk transfer. A difficulty with this benchmark is that with hardware systems that are reconfigurable, results were reported on unrealistic configurations of modules set up especially for each particular benchmark type. In this benchmark, results were quite widespread, as might be expected, where some systems ranked very high in some tasks and very low in others.

C. DARPA Benchmarks

The first DARPA benchmark [44] was developed by the DARPA image understanding community, and contains 10 tasks. Some are low level, such as a convolution. Intermediate-level tasks include connected components and the Hough transform. Half of the tasks involve methods that are related to image understanding, such as minimal spanning tree, minimum cost path, and subgraph isomorphism. The latter techniques do not yet have widespread applications and generally fall under the heading of image understanding rather than image processing. Results of the benchmark are reported for several machines but are difficult to compare because of different underlying assumptions and methods.

For various tasks and machines, there are inconsistent assumptions on how data are initially formatted in memory and how data are left when the task is completed. Data representing the final results of a task may not be in reasonable memory format in the machine so that a subsequent operation on the data may be burdened.

These difficulties of the benchmark have lead to a second DARPA-sponsored benchmark [45]. The tasks are similar to the first DARPA benchmark but are integrated into one image interpretation scenario, so that the benchmark also tests the efficiency in linking the various tasks. The object of the second benchmark is to recognize rectangles floating in space from a range image and an intensity image. Each scene is cluttered so that neither image alone is sufficient to find all the rectangles. A solution to the problem is specified so that there is uniformity in evaluating each machine that participates. The purpose of both DARPA benchmarks is to promote scientific insight into the problems of image understanding.

VIII. CONCLUSIONS

From a user's point of view, a high-performance imaging computer system should have a simple programming model and a low cost that results from a simple architecture. Unfortunately, there is no simple system that is efficient for all algorithms.

The most popular class of affordable image processing architectures are bus-based microprocessors with a plug-in frame grabber card along with a RISC, DSP, or ASIC chip to accelerate some classes of imaging algorithms. These systems are effective on straightforward tasks of imaging and machine vision. On more difficult tasks such as processing large images or noisy or cluttered images, they fail in speed or robust performance. The recourse is one of the hundreds of forms of parallel processing. There are many application-specific architecture designs often using ASICs. These systems are cost-effective if they are used for the purposes for which they were designed. Otherwise, they probably do not make a good general-purpose development environment.

In a sense, supercomputers do not belong in a discussion of imaging architectures because they are designed to be efficient for scientific and engineering applications using floating-point numbers. However, they are covered here because they are often used in imaging applications even though the architectures are too costly for widespread use.

For fundamental physical and economic reasons, LSI advances will favor rapidly increasing gate counts, while speed increases and pin counts will come about more slowly. For this reason the performance increases in imaging architectures will come from massive parallelism and not from raw speed alone. Memory chip capacity is getting large at a rapid pace and will soon outpace its usefulness in high-performance image processing. It is memory data bandpass more than memory capacity that is needed for massive parallelism.

MIMD can handle certain areas of processing, such as symbolic and expert knowledge, but is slower in low-level pixel and bit-level processing. The major difficultly in MIMD is handling the parallelism efficiently in large systems. The load balancing of tasks between processors is often dynamic and therefore not apparent to the person debugging the algorithms. There is a wide diversity in MIMD designs. Pipeline processors are good for most areas of imaging that involve local neighborhood processing. These systems exhibit the lowest flexibility in the choice of algorithms and are largely restricted to those that the available hardware modules can handle. The pipeline architecture is a very efficient use of silicon for those functions as long as the system is configured so that few modules are idle.

SIMD systems can handle local and nonlocal neighborhood processing of large images effectively but are not capable of handling many of the methods used in image understanding. There are no conceptual difficulties in understanding the loading of tasks in SIMD processors. There is only one instruction per clock cycle, and the programs are a serial string of instructions, just as in a RISC chip. The major difficultly is in understanding how to design algorithms that are effective on a large ensemble of processors. Many nonhomogeneous algorithms are possible on SIMD architectures. Their efficiencies are low but often reasonable.

There are three popular types of SIMD systems. Linear SIMD systems are a natural architecture for raster scanned images. Today the cost of fine-grained hypercube SIMD systems is a factor of 10 times that for mesh systems and a factor of 100 times that for linear systems [46]. In the future the cost ratio will probably be similar.

The difficulty of thinking in terms of MIMD algorithms involves how to do hundreds to thousands of different tasks simultaneously. In thinking SIMD, the difficulty is how to do thousands to millions of the same task simultaneously. Although SIMD hardware does not yet have millions of processors, there are millions of virtual processors in the programming model. In the future of imaging, both MIMD and SIMD will thrive. MIMD systems use processors that are well understood individually and very popular. SIMD is so effective in neighborhood operations of all types that it cannot be ignored. It seems that there will be a diversity of parallel imaging architectures well into the future.

REFERENCES

1. Buyers guide, *Adv. Imag.*, 7(12):18–24 (1992).
2. M. J. B. Duff and S. Levialdi, eds., *Languages and Architectures for Image Processing*, Academic Press, London, 1981.
3. M. J. B. Duff, ed., *Computing Structures for Image Processing*, Academic Press, London, 1983.
4. S. Levialdi, ed., *Integrated Technology for Parallel Processing*, Academic Press, London, 1985.
5. L. Uhr, K. Preston, Jr., S. Levialdi, and M. J. B. Duff, eds., *Evaluation of Multicomputers for Image Processing*, Academic Press, New York, 1986.
6. S. Levialdi, ed., *Multicomputer Vision*, Academic Press, London, 1988.
7. J. Kittler and M. J. B. Duff, eds., *Image Processing System Architectures*, Research Studies Press, Taunton, Somerset, England, 1985.
8. R. Duncan, A survey of parallel computer architectures, *Computer*, 23(2):5–16 (1990).
9. *Proc. 18th Annual International Symposium on Computer Architecture*, ACM, New York, May 27–30, 1991.
10. *IEEE International Symposium on Circuits and Systems*, Vol. 1, *Signal, Image, and Video Processing*, 1991.
11. *Image Processing*, Electronic Trend Publications, Saratoga, Calif., 1990.
12. *The Image Processing IC Market*, Electronic Trend Publications, Saratoga, Calif., 1990.
13. The U.S. market for commercial image processing systems, *Report A2310A/D*, Frost and Sullivan, New York, 1991.
14. *Electronic Imaging and Image Processing*, SEAI Technical Publications, Madison, Ga., 1990.
15. T. Kushner, A. Y. Wu, and A. Rosenfeld, Image processing on the ZMOB, *IEEE Trans. Comput.*, C-31:943–951 (1982).
16. R. L. Luck, An overview of the PIPE system, *Proc. 3rd International Conference on Supercomputing*, L. P. Kartashev, and S. I. Kartashev, eds., Vol. III, 1988, pp. 69–78.
17. W. D. Hillis, *The Connection Machine*, MIT Press, Cambridge, Mass., 1985.
18. K. E. Batcher, Design of a massively parallel processor, *IEEE Trans. Comput.*, C-31:377–384 (1982).

19. E. W. Davis and J. H. Reif, Architecture and operation of the BLITZEN processing element, *Proc. 3rd International Conference on Supercomputing*, L. P. Kartashev and S. I. Kartashev, eds., Vol. III, 1988, pp. 128–137.

20. D. Parkinson and J. Litt, Massively parallel computing with the DAP, in *Research Monographs in Parallel and Distributed Computing*, MIT Press, Cambridge, Mass., 1990.

21. T. J. Fountain, A review of SIMD architectures, in *Image Processing System Architectures*, J. Kittler and M. J. B. Duff, eds., Research Studies Press, Taunton, Somerset, England, 1985, pp. 3–22.

22. K. Preston, Jr., Three-dimensional morphology based on Ξ filters, *Proc. SPIE*, Vol. 1658, *Nonlinear Image Processing III*, 1992.

23. L. A. Schmitt and S. S. Wilson, The AIS-5000 parallel processor, *Nonlinear Image Processing III, IEEE Trans. Pattern Anal. Mach. Intelligence*, 10(3):320–330 (1988).

24. S. S. Wilson, Neural network computations on a fine grain array processor, in *Neural Networks for Perception*, Vol. 2, H. Wechsler, ed., Academic Press, New York, 1992, pp. 335–359.

25. A. L. Fisher, Scan line array processors for image computation, *13th Annual International Symposium on Computer Architecture*, Tokyo, June 2–5, 1986, pp. 338–345.

26. K. Chen and C. Svensson, A 512-processor array chip for video/image processing, in *From Pixels to Features*, H. Burkhardt, Y. Neuvo, and J. C. Simon, eds., North-Holland, Amsterdam, 1990, pp. 187–199.

27. H. Miyaguchi, H. Krasawa, S. Watanabe, J. Childers, P. Reinecke, and M. Becker, Digital TV with serial video processor, *IEEE Trans. Consumer Electron.*, CE-36(3):318–326 (1990).

28. C. Fernstrom, I. Kruzela, and B. Svensson, LUCAS associative array processor, *Lecture Notes in Computer Science*, Vol. 216, G. Goos and H. Hartmanis, eds., Springer-Verlag, New York, 1986.

29. S. L. Tanamoto, T. J. Ligocki, and R. Ling, A prototype pyramid machine for hierarchical cellular logic, in *Parallel Computer Vision*, L. Uhr, ed., Academic Press, Orlando, Fla., 1987, pp. 43–83.

30. B. Hogan, High performance image processing on a massively parallel computer, *Adv. Imag.*, 5(10):42–48 (1990).

31. T. J. Fountain, K. N. Mathew, and M. J. B. Duff, The CLIP7A image processor, *IEEE Trans. Pattern Anal. Mach. Intelligence*, PAMI-10(3):310–319 (1988).

32. T. Gross, H. T. Kung, M. Lam, and J. Webb, Warp as a machine for low-level vision, *Proc. 1985 IEEE International Conference on Robotics and Automation*, Mar. 1985, pp. 790–800.

33. S. S. Wilson, One dimensional SIMD architectures—The AIS-5000, in *Multicomputer Vision*, S. Levialdi, ed., Academic Press, London, 1988, pp. 131–149.

34. J. Sanz, Position papers presented at the panel: which parallel architectures are useful/useless for vision algorithms, *Mach. Vision Appl.*, 2:167–173 (1989).

35. L. H. Jamieson, P. T. Mueller, Jr., and H. J. Siegel, FFT algorithms for SIMD parallel processing systems, *J. Parallel Distribut. Comput.* 3:48–71 (1986).

36. S. R. Sternberg, An overview of image algebra and related architectures, in *Integrated Technology for Parallel Image Processing*, S. Levialdi, ed., Academic Press, New York, 1985, pp. 79–100.

37. Q. F. Stout, Supporting divide-and-conquer algorithms for image processing, *J. Parallel Distribut. Comput.*, 4:95–115 (1987).

38. D. H. Ballard and C. M. Brown, *Computer Vision*, Prentice Hall, Englewood Cliffs, N.J., 1982, p. 479.

39. C. C. Weems, S. P. Levitan, A. R. Hanson, E. M. Riseman, D. B. Shu, and J. G. Nash, The image understanding architecture, *Int J. Comput. Vision*, 2:251–282 (1989).

40. V. Cantoni, C. Guerra, and S. Levialdi, Towards an evaluation of an image processing system, in *Computing Structures for Image Processing*, M. J. B. Duff, ed., Academic Press, New York, 1983, pp. 43–56.

41. R. P. Weicker, An overview of common benchmarks, *Computer*, 23(12):65–75 (1990).

42. K. Preston, Jr., The Abingdon Cross benchmark survey, *Computer*, 22(7):9–18 (1989).

43. K. Preston, Jr., Benchmarks for image processing, *Adv. Imag.*, 5(5):30–38 (1990).

44. A. R. Rosenfeld, A Report on the DARPA image understanding architectures workshop, *Proc. 1987 DARPA Image Understanding Workshop*, Los Angeles, Morgan Kaufmann, Los Altos, Calif., 1987, pp. 298–302.

45. C. Weems, E. Riseman, A. Hanson, and A. Rosenfeld, IU parallel processing benchmark, *Proc. Computer Society Conference on Computer Vision and Pattern Recognition*, Ann Arbor, Mich., 1988, pp. 673–688.

46. M. Maresca and T. J. Fountain, Scanning the issue, *Proc. IEEE*, 79(4):395–401 (1991).

10

Digital Halftoning

Paul G. Roetling and Robert P. Loce

Xerox Corporation
Webster, New York

I. INTRODUCTION

In this chapter we present encoding methods, commonly called *halftoning*, that are used to reduce the number of quantization levels per pixel in a digital image while maintaining the gray appearance of the image at normal reading distance. These techniques are widely employed in the printing and display of digital images. The need for halftoning arises either because the physical processes involved are binary in nature or the processes have been restricted to binary operation for reasons of cost, speed, memory, or stability in the presence of process fluctuations. Examples of such processes are most printing presses, ink jet printers, binary cathode ray tube (CRT) displays, and laser xerography. Other technologies, such as computer-generated holography, nonlinear optical image processing, and data compression, also employ halftoning techniques. In this chapter the emphasis is primarily on binary encoding and the major applications of printing and display. In these applications the halftoned image will be composed ideally of two gray levels, black and white. Spatial integration, plus higher-level processing performed by the eye and brain and local area coverage of black and white pixels, provide the appearance of a gray-level or "continuous-tone" image. Halftone techniques are readily extended to color and to quantization using more than two levels, but there is not space to cover these in detail in this chapter.

A brief history of the field is presented, starting with analog methods of halftone image rendering and proceeding to the digital techniques of template dot halftones, noise encoding, ordered dither, and error diffusion. Other, less conventional techniques are also included. Key advances and lessons in the development of the technology are summarized. There is a discussion of visual perception, touching only on the concepts required for the understanding of the halftoning methods. Then, several current methods, such as ordered dither and error diffusion, are described in detail. For each technique there is a discussion of the methodology as well as issues such as tone reproduction, screen visibility, image artifacts, and robustness. Current directions of research are mentioned.

II. HISTORICAL PERSPECTIVE AND OVERVIEW

A. Analog Screening

The technologies for mass reproduction of pictorial imagery include the methods of relief printing, gravure, and lithography. A key step in the evolution of this technology was the invention of the photoengraving halftone process by Fox Talbot [84]. He placed a screen of black gauze between a photosensitive plate and an object whose image he wished to reproduce, and then exposed the resulting sandwich. Upon exposure, the transmittance of the screen and object are multiplied (densities add), producing a screen-encoded latent image (Fig. 1). Chemical development results in a final image that is binarized and screen encoded such that spatial integration of the eye causes the image to have the gray appearance of the original.

In the graphic arts industry, the primitive screen of Talbot has evolved through several stages of design and analysis. For many years, screens were typically ruled glass, employed in an out-of-contact mode, and could be rotated and used in conjunction with other screens in the reproduction of color prints. Streifer et al. [81] describe the theory of operation of such ruled screens. One important improvement in the process was the invention of the contact screen [33]. This invention simplified the process by not requiring careful spacings and simultaneously gave better image quality (diffraction was no longer a degrading factor). Most relevant to the topic of the present chapter is the evolution of the contact screen to digital clustered dot screens. The close relationship allows much of the wisdom of the graphic arts printing industry concerning tone reproduction, screen visibility, moiré, and so on, to be applied directly to digital halftoning methods.

In summary, note that almost one and a half centuries of practical experience with halftone images provide a rich source of information, which was not always fully utilized in optimizing digital methods.

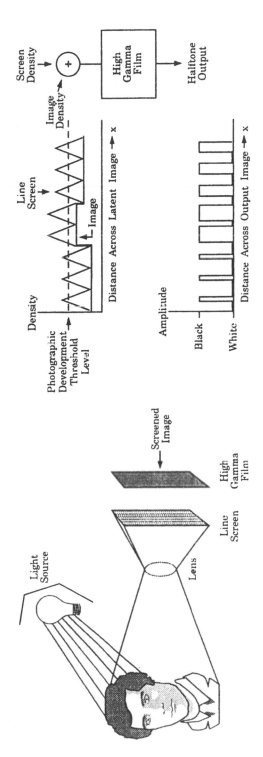

Figure 1 Schematic of analog screening.

B. Template Dots

With the advent of computer-driven line printers, the desire to print images inspired various *template-dot* methods. There was little or no use of graphic arts knowledge in these early attempts. In these methods, each image pixel is represented by a character, or template cell, with its area coverage corresponding to the desired density of that image pixel. To obtain multilevel presentation using a two-level printing apparatus, Perry and Mendelsohn [63] employed alphanumeric characters of different sizes and shapes, combined with one level of overstriking. Figure 2 shows an example of this early template-dot technique. Macleod [55] extended this method, obtaining more gray levels and suppressing contours by allowing a greater number of overstrikes and adding pseudorandom noise to the signal prior to conversion to a basic cell. Schroeder [75] described a technique for producing pictures on a microfilm recorder by adjusting the *f*-stop of the camera and exposing each point on the film several times. Schroeder [76] and Knowlton and Harmon [44] describe other related methods, such as varying dot size, combining spots within a unit cell, and using several output pixels within a cell to represent the brightness of a single input pixel. When printers became more versatile, Arnemann and Tasto [6] proposed a run-length-coding method and Hamill [32] proposed nonalphanumeric cells that can be grown in fine increments with no overstriking. Other names given to this method are *orthographic gray scale*, *surface area modulation* (SAM), and *font dot*. Little use or research has been performed recently with template dots because the resolution of the image is limited to the resolution of a halftone cell, whereas other methods give much better resolution.

In reviewing the history of the template-dot halftone, we recognize that a significant contribution to halftone image quality often comes from spatial detail too fine to be represented by dots that vary only in size. Template-dot methods have not survived because other methods use dot shape (often called *partial dots*) to carry finer detail and thus offer better image quality. To see the relative loss of resolution of the template-dot method, compare the sharpness of Fig. 3a and b, which are halftoned using template dots and ordered dither, respectively (ordered dither is discussed later in the chapter).

C. Noise Encoding

Despite the similarity to graphic arts analog methods, digital screening developed more directly from digital communications and display methods. The link to analog methods was made later. Therefore, we discuss the early digital developments next. *Noise encoding* is a method of bit reduction that was employed early in the fields of digital image display and communications. Goodall [28], working with pulse-code-modulated TV signals, observed that adding a small amount of noise to an input digital image almost eliminated gray-level false con touring in the displayed image, but at the expense of a small increase in granularity. Roberts [66] found that under certain noise and quantization conditions, the bit density for observable contours can be reduced from 6 or 7 bits to 3 or 4 bits per sample. Roberts added pseudorandom noise of amplitude equal to one quantization level to the video signal before* quantizing, then subtracted the same noise at the receiver. Combining this approach with compression and expansion of the intensity scale to match the human eye, Roberts concluded "transmitted data may be reduced to 3 bits per sample for most TV requirements and 4 bits for more demanding applications, and that because the eye tends to average out noise in local areas, distributed quantizing noise is considerably more pleasing to the eye than quantizing contours." A schematic of the noise-encoding process is shown in Fig. 4a, and 4b shows an example of noise encoding a one-dimensional signal using a uniformly distributed random variable. A noise-encoded image is shown in Fig. 5.

Researchers began analyzing the noise, or *dither signal*, in an attempt to optimize its statistical properties. Schuchman [77] has shown that the quantizing noise at the receiver is least dependent on the signal when the dither signal has a uniform amplitude distribution over a quantization level. From an information theory viewpoint, this would be expected to yield the most favorable results, as Widrow [88] has shown that this corresponds to the minimum loss of the statistical data of the picture. It must be noted that the optimal noise amplitude is greater than one quantization level when considering the desire to achieve spatially uniform granularity of a halftoned image. Thompson and Sparkes [85] further optimized and extended this method of quantizing by considering multiple frames in television imaging. Limb [50] considered the visibility of the granulation in the quantizer output. Using a simple model of the visual process, Limb determined the dependence of the visibility of granulation resulting from independent random samples having a uniform probability density function. Limb also showed that visibility can be reduced by introducing negative correlation between samples. Limb [51] then applied his analysis to differential quantization (as opposed to ordinary quantization), *deterministic dither* (ordered), and three-dimensional picturephone applications (2D plane and time). From Limb's analysis of deterministic dither, noise encoding has evolved into various *ordered-dither* methods.

There has been a renewed interest in noise encoding in the way of dithering with *blue noise*, which is related to Limb's negative-correlation dither signal. Steinberg et al. [80] examined the second-order statistics of an image encoded

---

*Note that it is very important that the noise is added before, not after, quantizing.

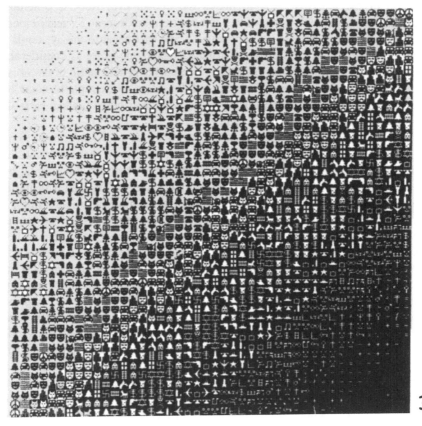

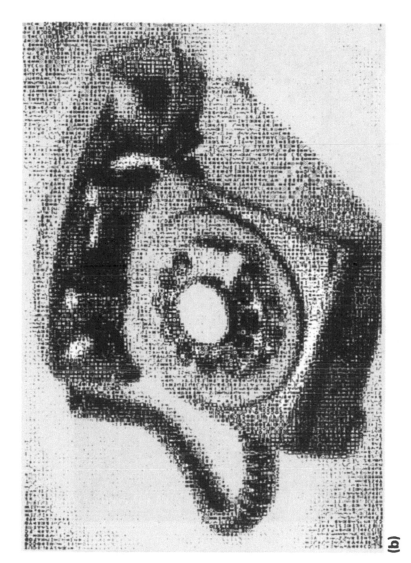

(b)

Figure 2 Example of template-dot halftoning using a set of symbols. (From Ref. 44.)

(a)

Figure 3 Comparison of (a) template-dot and (b) ordered-dither methods. In these examples, both use a 0° 65 level halftone screen at 300 pixels/in. Note the raggedness of the tree trunk and roof in (a) as well as a loss in sharpness on edges of the building.

(b)

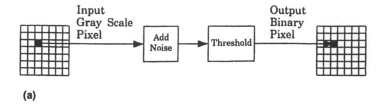

(a)

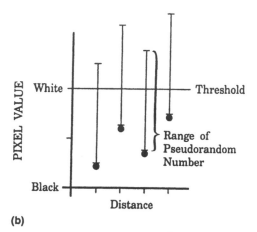

(b)

Figure 4 (a) Schematic of noise-encoding process; (b) one-dimensional noise-encoding example.

with blue noise. To enable use of efficient implementation hardware, Mitsa and Parker [59] developed a blue-noise mask and a comparator method of noise encoding.

Noise encoding with random patterns gave significant new understanding of the binarization process, namely that on a pixel-by-pixel basis, the gray level of each pixel is converted to a probability of that pixel being made white or black by noise addition followed by thresholding. This can readily be seen by noting that for uniformly distributed random numbers, the probability of a pixel value being above or below the threshold is proportional to the length of the interval above or below the threshold, as shown in Fig. 4b. A result of this probabilistic process is that halftone spatial detail may be as fine as one output pixel, not just one halftone cycle. Moreover, the work noise-encoding that tried to improve the process by using nonrandom patterns has led to the development of the half-toning methods that are now called *ordered dither*.

D. Ordered Dither

As just noted, there are advantages to adding patterns other than random noise before quantizing or thresholding. The approach of adding a fixed pattern of

Figure 5 Example of noise-encoded image. Pixels and noise sampling at 75 spots/in. to be similar to typical CRT display resolution.

numbers prior to thresholding to binary has been given the name "ordered dither." Ordered-dither methods may be divided into two categories: *dispersed dot* and *clustered dot*. The dispersed dot was designed for use in binary CRT displays, where isolated pixels are reproduced and the overall response of the system is linear. Although CRTs are capable of gray-scale display, binarization is necessary for compatibility with electronics designed to display text. Clustered-dot patterns were designed for printing devices, which typically have a nonlinear response and have difficulty producing isolated single pixels. Clustering works in conjunction with the nonlinearities to provide fine steps in tone reproduction, and also provide stability in the presence of process fluctuations. The factor of 4 or more in spatial resolution of typical printers of today, compared to CRTs, also changes the optimization of parameters, and allows clustered patterns to be printed in a manner that is visually acceptable. An implementation schematic of ordered-dither halftoning would be similar to the noise-encoding schematic (Fig. 4a), where the "add" and "threshold" are replaced with a step of comparison to a periodically varying threshold. Examples of dispersed-dot and clustered-dot ordered dither are given in Fig. 6 (the methods are described in detail later in the chapter).

A key source of the ordered-dither methodology was the analysis of optimal noise statistics of random noise encoding. When Limb [51] applied visual modeling to minimizing perceived granularity, he examined deterministic noise encoding, which is ordered dither. From Limb's analysis of deterministic dither, noise encoding has evolved into various ordered-dither methods. Other researchers were interested in improving template-dot methods. Klensch et al. [43] proposed dispersed-dot and clustered-dot ordered-dither halftone cells as a method of gaining resolution over template methods. Klensch also proposed angled screens for color applications and made the observation that clustered-dot ordered dither resembled conventional halftone photography. Lippel et al. [53] and Lippel and Kurland [52] verified Limb's prediction of two-dimensional ordered dither producing better visual quality than noise encoding methods. Bayer [8] derived an optimality rule for minimum visibility of halftone patterns in terms of minimizing the low-frequency components. Using Fourier analysis, he showed that the dispersed cells of Limb [51] and Lippel and Kurland [52] followed this rule. Judice et al. [39] further extended the optimally dispersed-dot method by developing a recursion method for designing larger cells.

Widespread use and favorable results of ordered dither led to a great deal of analysis of the method. Especially, firm links were made to the analog methods by several authors and the field began to unify. It was noted that ordered dither is, effectively, a sampled version of the analog contact screen process. Kermisch and Roetling [42], Allebach and Liu [5] and Allebach [2] developed analytic expressions for the Fourier spectrum of a halftone image as a function of the original continuous-tone image and of the halftone process. They showed how *aliasing*, or *moiré*, depends jointly on object contrast, object frequency, and the halftone process. The spectral analyses can be linked to earlier work by Marquet and Tsujiuchi [56], who used coherent optical spatial filtering to examine the spectra of analog halftones. More intuitive explanations of some of the phenomena were provided by Roetling [70]. Spatial resolution and tone reproduction are described by Roetling [71] and Bryngdahl [12]. Methods for characterizing noise and image fidelity are described by Ruckdeschel et al. [74] and Matsumoto and Liu [57], respectively. The effect of "oversized" pixels on tone reproduction was examined by Allebach [3], where "oversized" refers to the fact that printed pixels tend toward circular in shape and spread beyond idealized rectangular lattice sites. Roetling and Holladay [72] employed a model of the printing process to examine tone reproduction, stability of the printing process, and design of clustered-dot screens in raster printers. Rotated halftone screens are used to minimize the screen visual perceptibility and to make the color printing processes less sensitive to multipass registration errors. An algorithm for efficiently generating and storing screens at various angles was published by Holladay [34]. Vision models were employed in determining image coding limits [68] and developing halftone algorithms [4]. Randomly nucleated screens, which are used to minimize moiré, have been described by Allebach [1]. Attempts were made to obtain the advantages of the ordered-dither and error-diffusion methods by combining the processes. This synthesis was described by Billotet-Hoffman and Bryngdahl [9].

Recent research on ordered dither has taken several directions. In an attempt to increase effective halftone frequency, *multiple-nuclei screens* were developed and have been disclosed by Riseman et al. [65]. The dual-dot and quad-dot screens that they described are variants of the traditional clustered-dot algorithm, which uses a single nucleus per cell. In digital printers, geartooth noise, solenoids, and so on, can cause uneven motion and positioning errors as the image signal is being written onto a sensitive medium. Reflectance banding and other image defects caused by pixel placement errors have been analyzed by Melnychuck and Shaw [58], Haas [30], Loce and Lama [54], and for color images, by Bloomberg and Engledrum [10]. Versatility of modern electronics and improved stability of printing processes has prompted research on multiple-bit-per-pixel halftoning algorithms. The image quality of ink jet systems employing trinary pixels was examined by Naing et al. [60]. Tone reproduction capabilities of halftones utilizing trinary and quaternary pixels have been described by Lama et al. [49]. The increased use of office scanners has prompted analysis of moiré caused by digitizing halftoned images [78,79]. Although not all inclusive, these examples should provide a sense of direction of current research.

In several of its embodiments, clustered-dot ordered dither is essentially the sampled equivalent of the analog contact screen halftone process. As such, it is the most popular current technique for digital printing of pictures. Because of

(a)

Figure 6 Examples of (a) dispersed-dot and (b) clustered-dot halftone patterns, where (a) is shown at 75 pixels/in., typical of CRT display operation, and (b) is shown with a 71-dot/in. halftone screen and pixels at 300 per inch, typical of current laser printers.

(b)

the similarity, ordered-dither halftone patterns are often referred to as a "screen." Ordered dither is examined in detail later in the chapter.

E. Error Diffusion

Although the development of digital halftoning techniques did not always follow from the analog methods, the methods described above use processes similar to contact screens. We next consider some methods that use inherently different approaches to binarization. A binary encoding scheme where feedback of quantization error was employed for one-dimensional signals was described by Inose [35,36]. Quantization error of the signal at a given pixel, caused by thresholding, was added onto the signal at the next pixel prior to thresholding that pixel. The Δ-Σ-modulation of Inose, now often referred to as one-dimensional *error diffusion* (ED), propagated accumulated threshold error. (The error that is propagated from a pixel is the difference between its value after thresholding and its modified value, which includes quantization error that was passed from other pixels previously operated upon.) A schematic of the process is shown in Fig. 7. The result is that the encoded image has an average value equal to the average value of the signal prior to thresholding.

An early application of quantization error feedback employed on images was described by Schroeder [75]. Schroeder's method, known as minimum-average-error (MAE) quantization, propagated the difference between the value of a pixel before and after quantization. Not included in the error was the quantization error propagated to that pixel. Floyd and Steinberg [27] are generally credited with the first application of propagating accumulated threshold error in binarizing images, where the accumulated error is calculated from the modified brightness of a pixel. The error was distributed in a weighted manner to four neighboring pixels. An example of an image halftoned by their process is shown in Fig. 8. The number of neighbors and the weights were arrived at empirically. Hale [31] applied the one-dimensional error-diffusion methodology to images in what he referred to as *dot-spacing modulation* (DSM). Billotet-Hoffman and Bryngdahl [9] examined the resolution of the error-diffusion method as well as

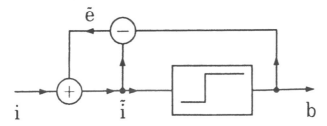

Figure 7 Schematic of error diffusion process for one-dimensional signals.

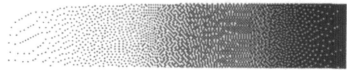

Figure 8 Example of error-diffused image. Pixels at 75/in. to simulate typical CRT display application.

examining the use of a periodic threshold and a larger diffusion matrix (the one employed in MAE studies). To minimize wormlike artifacts, Woo [90] used seven different diffusion masks in dispersing the quantization error. With a similar goal, Dietz [19] described a method to distribute the error randomly to neighboring pixels with certain probability weightings. An analytic description of one-dimensional error diffusion was published by Eschbach and Hauck [23]. It is based on the introduction of a carrier function followed by a signal-dependent threshold. A related quantization technique of *pulse-density modulation* (PDM), where signal is accumulated until a threshold is met, was applied to images by Eschbach and Hauck. They presented an iterative method for achieving pulse-density modulation, then more fully described pulse-density modulation in two dimensions [24,25].

Error diffusion is currently a very active research topic. Recent investigations have studied pulse-density modulation on rasterized media [21] and have employed visual models in error diffusion algorithm design [83]. Optimum intermediate levels in a multilevel pixel scheme have been examined from a minimum granularity perspective [20]. Most notable is that this highly nonlinear process is now being understood in terms of linear processes. The lack of low-frequency screen artifacts was described by Ulichney [86]. Using linear methods, Eschbach and Knox [26] have examined edge enhancement of the error diffusion process, and Weissbach et al. [87] and Knox [46] describe the error diffusion image as a sum of the original image and a high-pass filtered image.

Error diffusion is the most pervasive of binarization methods except for ordered dither. It is examined in detail later in the chapter. Error diffusion offers an attractive alternate way of thinking about binarization. As noted earlier, ordered dither (and noise encoding) converts pixel gray level to a probability of being white, whereas error diffusion directly forces total gray content to be fixed and attempts to localize the distribution of gray content.

F. Unconventional Techniques

We briefly mention several less conventional halftoning methods. A *constrained-average* algorithm was developed by Jarvis and Roberts [37]. Parameters employed in determining the threshold allowed for edge enhancement and contrast control of the halftoned image. The *ARIES* (aliasing reducing image enhancement system) of Roetling [67] forces equality of total gray content of the continuous-tone image and halftone image on a halftone-cell-sized basis to reduce aliasing, then assigns black-and-white pixels in a prioritized way, allowing for reproduction and enhancement of fine detail. A similar approach was described by Pryor et al. [64]. Carnevali et al. [14] described analogies between physical annealing processes and the halftone process. Using their models they were able to apply *simulated annealing* computational techniques to generate halftone images of similar quality to standard techniques. Knuth [48] described a *dot-diffusion* method of halftoning that diffuses quantization error within a halftone cell. Broja et al. [11] describe an iterative Fourier transform procedure for producing halftone images with controlled spectral characteristics. In each iteration a binarization is performed in the spatial domain, and in the frequency domain a low-frequency region of the spectrum is forced to match that of the original gray-scale image. Most recently, Peli [62] described a multiresolution approach to halftoning, where accumulated quantization error is uniformly distributed in all directions. Their symmetric form of error diffusion may be implemented using high-speed parallel architectures. Initial results of these new methods show that images can be obtained that are similar to standard clustered-dot or dispersed-dot ordered dither algorithms.

The freedom offered by digital computers opens the door to innovation. One first needs to examine the tenets of the approach, then the implementation. To a great extent, these methods attempt to mix the tenets of ordered dither and those of error diffusion. A good visual model that defined the optimum binary representation would help to guide such developments.

G. Other Applications

The halftone applications of this chapter are printing and display of digital images. For completeness, we mention briefly several references on the application to optical processing and computer-generated holography. Burch [13], Kato and Goodman [41], and Dashiell and Sawchuck [17,18] describe spatial frequency filtering and nonlinear optical processing with halftones. More recently, Barnard [7] has applied error diffusion to computer-generated holography. The reader is directed to the references in Barnard [7] for a more complete discussion on optical applications.

III. VISUAL PERCEPTION

In the introduction to this chapter, we noted that the primary applications of halftoning to be discussed here are printing and display of digital images. The goal, in particular, is to preserve the visual impression of gray tones despite the fact that pixel by pixel, the image is black or white. Thus before discussing various halftone methods in detail, we consider a few aspects of how the visual system interprets halftone images. It would be inappropriate here to attempt a thorough explanation of the visual system. Owing to space limitations, we outline the problem briefly. Curious readers are directed to texts such as Cornsweet's [16] and the extensive literature on the subject.

To scope the problem, we need to review a few image parameters and their typical values. At normal reading distance, persons with correct (or corrected)

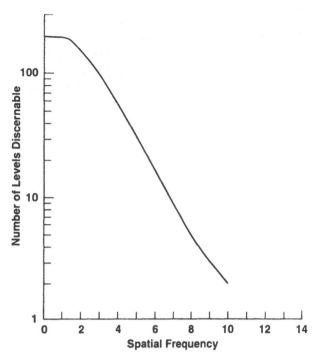

Figure 9   Visual performance limits.

visions can resolve roughly 8 to 10 cycles/mm. The spatial frequency of ordered-dither printed halftones ranges from the low quality of newspapers and many digital printers, using about 65 dots/in. (2.5 dots/mm), to high-performance systems, using 150 to 200 dots/in. (6 to 8 dots/mm). Thus the halftone pattern is usually visually resolvable even for top-quality systems. For displays, Cohen and Gorog [15] have determined that given free choice, people view TV screens at a distance where scan lines are just resolved. If the same holds for computer displays, the halftone patterns are certainly also resolved. From these data it is clear that for both printing and display, the impression of gray does not come simply from the visual system blurring the screen and thereby smoothing the image to a gray appearance. Moreover, dot shape and location, not just dot size, preserve image detail beyond the halftone screen frequency, which is necessary for a sharp appearance of the image.

Fortunately, halftone screens do not have to reproduce every gray level at any spatial frequency. Roetling [68] showed a rough analysis of how many gray levels can be perceived as a function of spatial frequency. His result, shown in Fig. 9, is an overestimate of performance, as a number of simplifying assumptions were biased toward showing an upper limit to viewer capability. Especially, res-

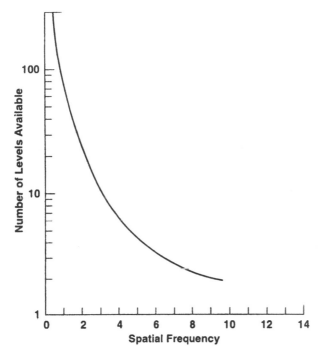

Figure 10 Theoretical limit for number of gray levels that can be represented by a binary image sampled at 20/mm.

olution of the eye is roughly a factor of 2 lower for frequencies oriented at 45° to the horizontal as compared to resolution at 0 or 90°, a fact usually employed in practice by placing the screen (or the black screen, in color images) at 45°. Allebach [4] has used this as a guide in designing some ordered dither patterns as well.

Roetling also showed a theoretical limit of how many gray levels could be represented by an ideal binary digital image as a function of spatial frequency, shown in Fig. 10. For a periodic digital screen, only the area inside the rectangle (terminated at one-half of the screen frequency) is reproduced by simple dot-size differences (template dots). Higher-frequency detail requires "partial dots"; that is, the pattern must change within a halftone period (for ordered dither). More gray levels are achieved by employing adjacent halftone cycles, where the cycles are not identical (i.e., by introducing lower spatial frequencies into the screen pattern). Both of these constraints are discussed in the algorithm explanations later in this chapter.

Two other points need mention in discussing perceived gray-scale rendition. The discussion above treated perceived gray level as though the eye averaged the

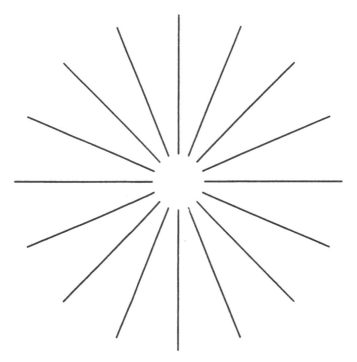

Figure 11   Example of induced contour and ''whiter-than-white'' appearance in center area.

amount of area covered, whereas we already noted that the screen is usually visible to some degree. One effect of being able to resolve the screen is a tendency to treat it like a window screen, ignoring it to the extent it is observed. This creates the perception of a lighter image for the same dot area coverage in coarser screens, one of several reasons why the TRC needs to be adjusted as a function of screen frequency. The second point is that in highlight areas, one can always see a contour between areas with the smallest printable dots and areas without dots. As shown in Fig. 11, where lines are used instead of dots, one distinctly sees a circular shape with an induced contour where none exists physically [40]. Not only is the contour seen, but if one compares the perceived lightness inside the circle with the surrounding white paper, the circle appears brighter. This illusion can be used to give the impression of specular highlights within an image, but otherwise is normally avoided in halftone images by never completely dropping out highlight dots.

Having briefly discussed gray scale and detail perception, we turn to noise and artifact visibility. In general, these are low spatial frequency effects. In noise encoding or random screening, the low spatial frequencies of the screen create a perceptible mottle in the image. In attempts to minimize screen visibility of ordered dither, Bayer [8] and Allebach [4] have tried to maximize the frequency of the lowest-frequency harmonic or the strongest harmonics in the screens that they have designed (see ordered-dither descriptions). Note that built-in shape recognition of the visual process can upset expected results. A low spatial frequency possessing low amplitude can often yield disproportionate visibility when patterns are recognized (e.g., in a checkerboard of pixels, changing one pixel creates a very obvious plus sign that stands out). For the error-diffusion process, Ulichney [86] has examined the spatial frequency spectra of binarized images in terms of the amount of power occurring at low spatial frequencies. Several workers (see Section II, especially Section II.E) have examined methods of reducing wormlike patterns, which are relatively low spatial frequency structures.

A common problem in digital imaging is the appearance of false gray-level contours, caused by an inadequate number of levels attainable at a particular spatial frequency. A related but separable problem is that of texture contouring. Especially in many ordered-dither techniques, the fundamental frequency of the screen pattern can rotate 45° between adjacent gray levels. A texture change will show as a contour even if the gray levels themselves would not cause a perceived contour. Allebach [4] has included a condition in some screen designs to minimize such artifacts. Experimentally, one can distinguish between texture and gray-level contours by backing away from the image. Gray-level contours occur at very low spatial frequencies and are visible at a considerable distance. Texture contours, on the other hand, disappear when one cannot resolve the textures and therefore are not visible at larger viewing distances.

Unfortunately, no one has yet determined a single or small set of optimization criteria for binary image perception. We know, for example, that the popular idea of minimizing mean-square error is incorrect. Note that at each pixel, simple midrange thresholding minimizes the expected error between the gray pixel value and the binary result, thus minimizing the total mean-square error over the image. However, this result does not maintain the "impression of gray" referred to as desirable in the introduction to the chapter. First filtering both the continuous-tone and the resulting halftone through a spatial filter that models the visual system before determining mean-square error improves the estimate of quality, but this clearly fails to take into account the observer's ability to ignore some visual artifacts, while others are very disturbing. It also ignores effects such as human perception not matching physical averaging, such as the effect described above.

The noise-encoding and ordered-dither tenet of converting gray to a probability of being white offers a means to give the "impression of gray" but does not provide a good means to determine the relative merits of systems that converge to the correct gray over different-sized areas, although convergence over smaller areas is clearly better. Error diffusion achieves the total gray content, but like the above, does not give a quantitative measure, depending on the distance over which the gray is spread. None of the methods give clear criteria in terms of the visibility of undesirable image noise patterns, nor do they give methods to trade-off between image content and appearance of noise.

Despite the state of the art in the understanding of visual interpretation, much halftone design is still performed empirically. The fact that a halftone image is usually affected by the physical printing process in ways that are also not well modeled simply increases the uncertainty of mathematical prediction of output image quality.

IV. METHODS OF HALFTONING

In the present section we describe details of several key halftoning methods. We discuss the algorithms as well as describing when the algorithm is applicable, effects on lines and edges, spatial spectral effects, tone reproduction, and other notable effects. Algorithm schematics and examples of the methods are given. The algorithms discussed are variations of ordered-dither and error-diffusion techniques. Before ordered dither, we discuss noise encoding for its simple embodiment of the dither method as well as for historical significance. In describing resolution (frequency response) capabilities of ordered dither, we also discuss template-dot methods for comparison.

A. Noise Encoding

Recall that the problem at hand is to reduce the number of quantization levels in a digital image while maintaining the appearance of grays. Consider simple midrange thresholding of an 8-bit/pixel image that contains a uniform gray region producing a binary [0,1] representation of that image. Although as noted previously, this thresholding method results in minimum mean-square-quantization error, the uniform region suffers a severe shift in gray appearance, becoming either completely black or white. To obtain a truer-appearing image we may employ the probabilistic method of thresholding known as *noise encoding*. Noise encoding is a point-process halftoning method intended to maintain gray appearance and edge sharpness.

To understand the gray appearance effect, consider a uniform gray image $f(i,j)$, where prior to thresholding, a pseudorandom number in the range $[-a, a]$ is added to each pixel, to produce the intermediate random image $f'(i,j)$,

$$f'(i, j) = f(i, j) + \text{ran}[-a, a] \tag{1}$$

Thresholding each pixel at level t is then applied to produce

$$b(i, j) = \begin{cases} 0, & f'(i, j) < t \\ 1, & f'(i, j) \geq t \end{cases} \tag{2}$$

where b is the binary halftoned image.

The value of a pixel at (i, j) will be dependent on the original value of the pixel, $f(i, j)$, the threshold level, t, and the range and particular value of the pseudorandom number, a. For the simple case where the random number is uniformly distributed, t is midway in the gray-scale range $[0, M]$, and $a = M/2$, we can write probability expressions for the gray level of a binarized pixel:

$$P[b(i, j) = 1] = \frac{f(i, j)}{M} \tag{3}$$

$$P[b(i, j) = 0] = 1 - \frac{f(i, j)}{M} \tag{4}$$

Assuming a uniform continuous-tone image region and stationarity of the noise process, the fractional area coverage of 1's and 0's in the halftone image region will be equal to the probabilities given in Eqs. (3) and (4). When viewed on a print or display, spatial integration performed by the eye would produce a perceived gray level that is proportional to the fractional area coverage, which Eq. (3) and (4) show to be proportional to the original gray level of the continuous-tone image.

The edge preservation capabilities of noise encoding can be also be seen in Eqs. (3) and (4). At high-contrast edges, one side will have high brightness values and tend to be encoded to 1, and conversely for the low-brightness edge. Although all edge transitions tend to occur at the correct pixel locations, higher-contrast edges will be better rendered because the greater difference in the probabilities will make the halftoned edge transition more defined about the true edge location.

Although noise encoding has the virtues described above, it is not ideal in the sense that granularity is introduced into the image. In early applications of noise encoding to bit reduction, Goodall [28] and Roberts [66] noted the trade-off between false contouring and graininess. The statistical method described above is a binary implementation of their method, where they employed a pseudorandom dither signal with a white (uniform) power spectrum and flat histogram. This method applied to an 8-bit/pixel image was shown in Fig. 5. Although grayness and edge definition are preserved, low-frequency components of the noise are not filtered out by the eye and a very grainy image results. A goal of early noise encoding research was the design of optimal dither signals. Limb [50], employing a human visual model, designed minimum visibility dither signals that evolved into ordered dither.

There has been a renewed interest in noise encoding in the way of screening with a blue noise mask [59]. *Blue noise* is a term derived from optics denoting that only high spatial frequency components are present. Thus, in a blue-noise encoded image, low-frequency granularity is not present and the underlying structure is less distracting to the viewer [86]. As opposed to pseudorandom-number generation and addition, efficient encoding is performed by comparison to a mask consisting of carefully designed thresholds. The design requirement of the mask is that frequency components introduced by the encoding must be primarily high frequency for all gray levels of the original image. To achieve the proper statistics, Mitsa and Parker have employed relatively large masks (compared to typical ordered dither masks) and an iterative Fourier transform method for choosing the thresholds.

B. Ordered Dither

As the statistical understanding of noise encoding evolved, deterministic dither signals were developed. In currently employed techniques, a two-dimensional *dither mask* contains threshold values that are used in a comparison algorithm to determine the value of an output halftone pixel. The dither mask, *threshold array*, or *halftone cell* is replicated to tile the image plane, thereby producing periodically varying thresholds throughout the image. Key factors in designing an ordered-dither halftone are the arrangement of thresholds within the cell, cell size and shape, and offset of successive rows of cells. We first describe the basic algorithm of implementation. Then we discuss two threshold arrangement methodologies. One, the *dispersed-dot* method, minimizes objectionable low-frequency components induced by the halftone screen. It is ideally suited for display devices where isolated pixels tend to be reproduced faithfully and the overall response of the system is linear. The *clustered-dot* method groups pixel types within a cell to form, typically, a single black dot per cell. In an attempt to suppress low-frequency components induced by clustering, some variations of the method form several small clustered dots within the cell (multinuclei cell). Clustered dots are employed in printing devices, which tend to be nonlinear in response and are subject to process fluctuations. This method of halftoning yields repeatable, low noise images with acceptable tone reproduction characteristics. The clustered-dot method, in its basic form, is essentially a sampled version of analog halftone techniques. Note that low-resolution printers (<300 pixels per inch) sometimes use dispered-dot dither because the printing processes tend to be more stable relative to the larger spots, thereby eliminating the necessity for clustering. For the clustered-dot method, we include a discussion on angled screens for both monochrome and four-color images. We end the ordered-dither section with discussions on topics common to either threshold arrangement: the trade-off between screen visibility and the number of achievable gray levels, and spatial resolution effects such as frequency response and moiré.

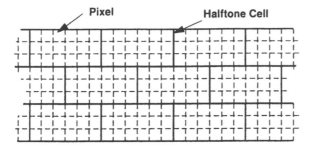

Figure 12   Schematic of tiling a halftone cell on a page.

1.   Implementation

When halftoning a digital image, the threshold array is replicated to tile the image plane (Fig. 12) and then a comparison is performed (Fig. 13). The comparison is performed on a pixel-by-pixel basis to retain spatial resolution in the output image. For a threshold array T_{mn} having dimensions m by n, the image $f(i,j)$ is halftoned through the operation

$$b(i, j) = \begin{cases} 0, & f(i, j) < T_{mn} (k, l) \\ 1, & f(i, j) \geq T_{mn} (k, l) \end{cases} \tag{5}$$

where $k = \text{mod}(i, m)$ and $l = \text{mod}(j, n)$ for screens at $0°$ [the notation $\text{mod}(i,m)$ denotes i modulo m]. When screens are desired at some other rational tangent angle, successive cells of the tiling are offset by an amount s (see the discussion of the Holladay [34] algorithm in Section IV.B.3), filling the page much like bricks in a wall. Using this local comparison method results in *partial dots*, where the internal shape of a printed cell varies depending on the image detail within the cell boundary. Sharp edges and discontinuities, to which the eye is very sensitive, are preserved in a manner similar to the noise-encoding method. This high resolution (or high-frequency response) is in contrast to older *template-dot* methods, which used many output pixels in fixed patterns to represent a single input pixel. An example of the frequency-response difference of the methods was given in Fig. 3. The relationship to noise encoding may be seen by allowing T to be the size of the image and employing uniformly randomly distributed numbers for the threshold values.

2.   Dispersed-Dot Ordered Dither

Dispersed-dot ordered dither is used primarily in display devices where isolated pixels can be reproduced faithfully, or the system response is approximately linear. The threshold arrays are designed to minimize low-frequency texture in-

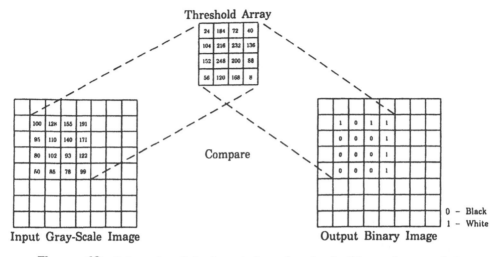

Figure 13 Schematic of implementation of ordered dither using sample-by-sample comparison.

duced by the screen. Threshold arrays are typically designed to meet optimality criteria defined by Bayer [8]. We first state Bayer's criteria and then show how cells may be designed using the recursion relationship of Judice et al. [39].

As an index of merit for texture induced by a halftone cell, Bayer employed the longest finite wavelength of the nonzero-amplitude sinusoidal components of a halftoned uniform area. This quantity, denoted by Λ, can be written

$$\Lambda = \max[\lambda_{uv} : A_{uv} \neq 0, \lambda_{uv} < \infty] \tag{6}$$

where λ_{uv} and A_{uv} are the wavelength and amplitude of the u, v frequency component, respectively. The preferred halftone cell is defined to be one such that Λ is as small as possible for all possible gray levels. This choice of definition is based on the premise that if we view an entire sequence of gray levels at one time, we base our first impression of overall texture on those patterns having the largest value of Λ.

To express these conditions, let k equal the number of black dots in a printed cell halftoned by threshold array T having dimensions $2^m \times 2^m$ ($0 \leq k \leq 2^{2m}$). For each k, let 2^n by the largest power of 2 that divides k. That is, $k = p2^n$, where p is an odd integer. With $b(i, j)$ denoting the halftoned image, the optimality conditions can now be stated in two parts:

1. For even n, b must obey

$$b(i + 2^{m-n/2}, j) = b(i, j + 2^{m-n/2}) = b(i,j) \tag{7}$$

2. For odd n, b must obey

$$b(i + 2^{m-(n+1)/2}, j + 2^{m-(n+1)/2}) =$$
$$b(i + 2^{m-(n+1)/2}, j - 2^{m-(n+1)/2}) = b(i, j) \tag{8}$$

The halftone cell will yield the maximum index of merit if and only if these conditions are satisfied for all k. Furthermore, under such conditions, the maximum wavelength is given by

$$\Lambda = 2^{m-n/2} \tag{9}$$

for each k, $0 < k < 2^m$, $k = p2^n$, p odd.

Dither arrays developed by Limb [51] and Lippel and Kurland [52] were found to satisfy Bayer's condition. Judice et al. [39] developed a recursion method of designing larger dither arrays. Given Limb's 2×2 array

$$T^2 = \begin{bmatrix} 0 & 2 \\ 3 & 1 \end{bmatrix} \tag{10}$$

(normalized to four levels) and defining U^n to be dimension $n \times n$,

$$U^n = \begin{bmatrix} 1 & 1 & \cdots & 1 \\ & & & \\ 1 & 1 & \cdots & 1 \\ & & & \\ 1 & 1 & \cdots & 1 \end{bmatrix} \tag{11}$$

allows us to state the recursion relationship,

$$T^n = \begin{bmatrix} 4T^{n/2} + T^2[0, 0]U^{n/2} & 4T^{n/2} + T^2[0, 1]U^{n/2} \\ 4T^{n/2} + T^2[1, 0]U^{n/2} & 4T^{n/2} + T^2[1, 1]U^{n/2} \end{bmatrix} \tag{12}$$

Where $T^2[i, j]$ denotes the i,jth element of Limb's 2×2 array. Using the recursion relationship, the 4×4 dither array is found to be

$$T^4 = \begin{bmatrix} 0 & 8 & 2 & 10 \\ 12 & 4 & 14 & 6 \\ 3 & 11 & 1 & 9 \\ 15 & 7 & 13 & 5 \end{bmatrix} \tag{13}$$

An example of dispersed-dot dither was shown in Fig. 6a. We see that spatial resolution is high and gray tones are well rendered. An artifact of the process can be seen in the gray wedge, where texture contouring can be observed between the gray levels.

3. Clustered-Dot Ordered Dither

Clustered-dot ordered dither is the most widely employed method of halftoning in the digital printing industry. In one of its basic forms, it is a sampled version of analog contact screen printing. In many digital printing devices, the reflectance response is nonlinear: due to the size, shape, and density-saturated nature of printed pixels, the reflectance is not equal to the ratio of number of white pixels to the total number of pixels in a cell. In such a device, tone reproduction is optimized by clustering pixels, thereby forcing maximum overlap. Other considerations are pixel position accuracy, single-pixel reproduction fidelity, and dot overlap, which are engineering issues. Most often, due to cost or physical limitations, a more stable image, free of noise artifacts and nonuniformities is obtained by clustering black pixels within a cell. The cost of correct tone reproduction and robustness in the presence of process fluctuations is a screen with lower-frequency content, hence more visible than a dispersed-dot halftone. We first discuss the 45° dot screen along with fundamental tone reproduction and spatial resolution issues. Then Holladay's method of describing angled screens for color applications is presented. Finally, there is a discussion on various multidot clustering schemes.

Dot Growth Pattern and Tone Reproduction   To understand the effect of the dot growth pattern on print reproducibility, consider the 4×8 clustered-dot threshold array example of Eq. 14, which simulates an analog screen and grows approximately in a spiral pattern. As image darkness varies from white to black, the corresponding cell turns black one pixel at a time in a spiral fill pattern. At low density, a small cluster of black pixels on the left side is grown to a larger cluster by spiraling outward. At 50% area coverage, a checkerboard pattern is formed. At progressively higher densities, black pixels are grown inward on the right side to fill the remaining white space. For the given threshold array, successive rows of cells are offset by four pixels so that the lowest nonzero screen frequency component is at 45° and thus is less visible to the viewer. A halftone dot grown in this manner tends to have a minimal perimeter. Since most deviations from ideal printing occur around the perimeter of the dot, this growth scheme stabilizes the gray levels produced.

$$T = \begin{bmatrix} 116 & 108 & 100 & 124 & 132 & 172 & 164 & 156 \\ 60 & 52 & 68 & 84 & 220 & 228 & 236 & 188 \\ 28 & 20 & 44 & 76 & 212 & 252 & 244 & 180 \\ 4 & 12 & 36 & 92 & 140 & 204 & 196 & 148 \end{bmatrix} \tag{14}$$

When designing a threshold array, a key concern is tone reproduction. Tone reproduction is affected by the threshold-value sequence (fill order) as well as the threshold values themselves. The goal is to produce a halftone image free of gray-level false contours and have the same apparent tone rendition as the gray-

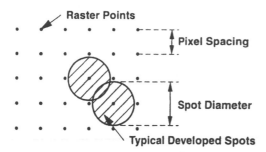

Figure 14 Illustration of raster spacing and pixel spacing, showing how individual spots are turned black and overlap.

scale image. Roetling and Holladay [72] and others have discussed these issues in detail. First, let us discuss contouring with respect to fill order.

The eye is more sensitive to gray steps in reflectance in highlight regions of an image than in the midtones or shadows. For that reason we must attempt to achieve the maximum number of gray levels per unit of reflectance in the high-light region of the tone reproduction curve. To analyze tone reproduction, printer models have been employed. For example, for an electrographic printer, Roetling and Holladay assumed that black pixels are circular, infinitely dense, and centered on addressable grid locations. For such printers, spot diameter is generally greater than pixel spacing to allow full coverage for the all-black case, so diagonal lines are printed without discontinuities (see Fig. 14). Since neighboring spots overlap, area coverage (and optical density) increases less rapidly in the highlights when the spots are clustered as in the growth pattern described above. This printing system is in contrast to a linear device, where with each additional pixel turned black, reflectance would decrement by a fixed amount.

Accurate tone reproduction also depends on gray-level spacing between threshold levels. The printer model just described indicates that spot overlap causes cell reflectance to vary nonlinearly with increasing fill number. Thus the threshold levels should not be equispaced in reflectance. A common method of selecting threshold levels is first to measure the set of possible reflectances for the chosen fill order, $\{R_i : i \in [0, p]\}$, where p is the number of pixels in the cell and reflectance R_i increases with i. Using this notation, in an idealized system, $R_0 = 0$ and $R_p = 1$. To minimize printed gray error, thresholds are picked approximately midway between reflectance levels:

$$t_n = \frac{M(R_{p-n+1} + R_{p-n})}{2} \tag{15}$$

where n denotes fill order and M is the largest digital value of image reflectance (e.g., 255 for 8 bits/pixel). The threshold of the first pixel blackened, t_1, is often chosen in a slightly different manner, being set very close to M. This is done to maintain at least one black pixel per cell in the highlight region, thereby avoiding an abrupt texture change. The all-white cell case is reserved to represent only very bright portions of a digital image. Note that Eq. (15) presents an idealized scenario. In practice, t_2, as well as the other threshold values, are adjusted to compensate for the chosen value of t_1.

We end this section by mentioning two variations of ordered dither. The variants, *multidot screens* and *line screens*, involve different cell growth patterns than those described above. The multidot screen is an attempt to increase the apparent frequency of a clustered-dot screen. Spiral-type growth is employed, but alternating from, typically, two or four nuclei in the cell. The virtues are less low-frequency content and almost equal stability compared to a single clustered-dot cell. The primary drawback is slight texture contouring that results from the growth alternating from nucleus to nucleus. Also, highlight tone reproduction will have slight larger gray steps than the single dot cell and may be somewhat more sensitive to process fluctuations. Examples of *dual-dot T_2* and *quad-dot T_4* threshold arrays are as follows:

$$
T_2 = \begin{bmatrix}
64 & 78 & 134 & 192 & 148 & 120 \\
6 & 20 & 234 & 248 & 206 & 48 \\
34 & 92 & 176 & 220 & 162 & 106 \\
199 & 155 & 127 & 71 & 85 & 141 \\
254 & 213 & 56 & 13 & 27 & 241 \\
227 & 169 & 113 & 41 & 99 & 184
\end{bmatrix} \tag{16}
$$

$$
T_4 = \begin{bmatrix}
158 & 215 & 144 & 44 & 30 & 72 & 165 & 222 & 151 & 51 & 37 & 79 \\
126 & 97 & 111 & 175 & 190 & 133 & 119 & 90 & 104 & 183 & 197 & 140 \\
12 & 26 & 69 & 232 & 247 & 204 & 5 & 19 & 62 & 239 & 254 & 211 \\
55 & 40 & 83 & 161 & 218 & 147 & 48 & 33 & 76 & 168 & 225 & 154 \\
179 & 193 & 136 & 122 & 94 & 108 & 172 & 186 & 129 & 115 & 87 & 101 \\
236 & 250 & 207 & 8 & 23 & 65 & 229 & 243 & 200 & 2 & 16 & 58
\end{bmatrix} \tag{17}
$$

Line screen digital halftones are sometimes employed in reducing motion quality and registration requirement in printers. A line screen that runs in the process direction (e.g., direction of photoreceptor motion in a laser printer) is less susceptible to mechanical-vibration-induced reflectance banding than is a dot-screen halftone. Also, misregistration of color separations will produce less color error than would occur in a dot screen when misregistration is restricted to the process direction. (Line screens are very susceptible to color shifts when the misregistration occurs in the cross-process direction, but little error usually occurs in this direction in real systems.) Although line screens may be more visible than dot screens, the trade-off may be worthwhile when considering the robustness against printer defects and cost savings in motion quality engineering.

Clustered-dot halftones are often confused with template-dot screens. For a given density the template-dot method always prints the same internal halftone cell pattern. The technique arose out of the early days of computer halftones, where one image pixel generated one cell. Although the constant dot shape allows for well-controlled tone reproduction, it causes a loss in spatial resolution in comparison to clustered-dot techniques. The partial dots of the modern techniques produce images of the same high contrast resolution as the gray-scale image. For comparison, template-dot and clustered-dot halftoned images were shown in Fig. 3.

Screen Angle and Color Separations To minimize visibility, single color images (e.g., black and white) use screens that are typically designed to have the fundamental frequency at 45° from horizontal. Color prints require an overlaying of three or four color separation images, each halftoned. Typical colorants have overlapping spectral absorptions, thus interfering with each other. When combining periodic signals, such as two colored halftone screens of vector frequency f_1 and f_2, interference produces a beat at the vector difference frequency, $f_b = f_1 - f_2$. If the individual color screens were made at the same angle and frequency, any slight frequency modulation due to vibration in the printing process forms a low-frequency visually objectionable beat known as color banding. If misregistration error between color separations occurred ($f_1 = f_2$, but the relative phase of the frequencies deviates from design conditions) a color error will result across the entire page ($f_b = 0$). To avoid these types of color errors, the screens are typically oriented at different angles, usually about 30° apart. At 30° separation, the moiré is at about half the screen frequency, thereby producing a high-frequency "rosette"-shaped beat pattern. Misregistration errors cause color errors only at high frequency and therefore are not visible. Thus rotated screens are much less sensitive to misregistration and vibration effects but at a cost of moiré at half the screen frequency. To achieve the highest frequency and least visible rosette beat pattern it is typical to orient cyan and magenta at ±15°, yellow at 0°, and black at 45°. Because yellow and black have the least and most impact on visual density, respectively, they are oriented at angles where the eye is most and least sensitive, respectively, to observing the screen pattern. Note that although yellow is at 15° to the other screens its intercolor moiré is not objectionable because of low contrast. Deviating from these ideal angles can introduce undesirable beats at frequencies lower than the rosette frequency.

Another approach to color screening has also been taken. Because the rosettes are roughly half the screen frequency, it is tempting to eliminate these lower frequencies by using aligned screens (also called dot-on-dot screens) for lower-resolution digital printers. If registration is good enough, this is a viable trade-off (and necessary at low addressability), but at higher resolutions, printers usually use rotated screens.

The particular choice of screen angle and the desired angular accuracy have algorithmic implications. Angled screens are defined as *rational* or *irrational*, where rational indicates that the tangent angle of a given screen is a ratio of integers. In this case, halftone dot centers fall neatly on the pixel grid. While rational screens are unable to achieve $\pm 15°$ separations accurately, and therefore offer poorer moiré control, they offer algorithmic advantages over irrational screens. Larger halftone cells allow screens to be chosen at finer angular increments (this approach is used in a method termed superscreening, where very large multidot cells are used to achieve angles of arbitrary precision). Irrational screens may be produced by randomly dithering between rational screens, or by other methods [61]. In this approach, many different cells occur and screen storage problem is encountered. Holladay [34] described an efficient method of specifying a rationally angled screen where only one rectangular cell of area equal to one halftone period is stored and used in processing. Not considering fill order, only three parameters are needed to describe the screen: height, width, and shift of the rectangle. A general angled parallelogram-shaped cell may be described by two vectors corresponding to its side lengths and directions: $Z = z_1 x + z_2 y, W = w_1 x + w_2 y$. Given Z and W, the height p, width l, and offset s may be calculated from

$$p = \text{GCD}(z_2, w_2) \tag{18}$$

$$l = \frac{A}{p} \tag{19}$$

and

$$s = l - \frac{tA - pw_1}{w_2} \tag{20}$$

respectively, where A is the area of the parallelogram, $A = z_1 w_2 - z_2 w_1$, GCD refers to greatest common divisor, and t is an integer chosen so that $0 < (tA - pw_1)/w_2 \le l$. An example of an equivalent rectangular cell tiling the image plane is shown in Fig. 15.

In the graphic arts industry it is common to use what is known as an *orthogonal screen*. This widely used special case of a rotated screen consists of square cells oriented at an arbitrary angle. The orthogonal screens may also be described by a minimal rectangular cell and a shift. The screens are typically specified by the components x and y of the vector describing the base of the square cell. Rational screens are obtained for integer x and y. Table 1 gives the number of pixels in the cell and the screen angle for x and y in the range 0 to 7, where N/θ signifies N gray levels and a rotation of $\theta°$ rounded to the nearest integer. Screens are given for the angle range of 0 to 45°, and other angles may be obtained by interchanging x and y. These same screens are described by Holladay's

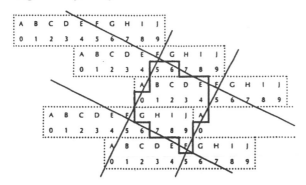

Figure 15 Example showing how an angled halftone screen may be represented by rectangular cells aligned with a raster. Method described by Holladay (1980). (From Ref. 45.)

parameters in Table 2. The entry $p/l/s$ gives the height, width, and offset of the cell. The parameters are calculated from Eqs. (18) to (20), where $Z = xx - yy$, $W = yx + xy$.

4. Screen Frequency Versus Number of Gray Levels

Two conflicting design requirements exist for ordered-dither halftones. It is desirable to be able to print many gray levels (e.g., roughly 100 levels for high-quality pictorials). It is also desirable to maximize the screen frequency, thus minimizing its visibility. High-quality printers may employ screens at 100 to 180 cells per inch. Consider that as the number of pixels in a cell increases, a halftone dot can represent more gray levels, thereby decreasing the likelihood of false contours, but as the cell size increases for single clustered dots, the screen becomes coarser and more visible. The coarser screen is also less able to rep-

Table 1 Number of Pixels in the Cell and Screen Angles for the Orthogonal Screens

	N/θ					
	2	3	4	5	6	7
0	4/0°	9/0°	16/0°	25/0°	36/0°	49/0°
1	5/27°	10/18°	17/14°	26/11°	37/9°	50/8°
2	8/45°	13/34°	20/27°	29/22°	40/18°	53/16°
3		18/45°	25/37°	34/31°	45/27°	58/23°
4			32/45°	41/39°	52/34°	65/30°
5				50/45°	61/40°	74/36°
6					72/45°	85/41°
7						98/45°

Table 2   Minimal Rectangle Parameters for the Orthogonal Screens

y/x	p/l/s					
	2	3	4	5	6	7
0	2/2/0	3/3/0	4/4/0	5/5/0	6/6/0	7/7/0
1	1/5/2	1/10/3	1/17/4	1/26/5	1/37/6	1/50/7
2	2/4/2	1/13/8	2/10/4	1/29/17	2/20/6	1/53/30
3		3/6/3	1/25/18	1/34/17	3/15/6	1/58/41
4			4/8/4	1/41/32	2/26/16	1/65/18
5				5/10/5	1/61/50	1/74/31
6					6/12/6	1/85/72
7						7/14/7

resent low-contrast fine detail (for low-contrast periodic patterns the spatial resolution limit is half the screen frequency, as stated in Section IV.B.5). Conversely, as the number of pixels decreases, the screen frequency becomes higher and less visible, and finer detail may be represented. This gain is at the cost of a potential for false contours. Examples of low- and high-frequency screens are shown in Fig. 16. The trade-off between screen frequency and number of gray levels is ever present.

A direction of recent research in ordered-dither halftoning concerns alleviating the compromise between screen frequency and number of gray levels. Utilization of multilevel (>2) pixels has been examined [49] and it has been shown that where printers can support limited gray-level printing, the number of cell gray levels produced by a halftoning system may be increased significantly through the use of gray pixels. Hybrid halftone systems with trinary and quaternary pixels (having one and two intermediate gray levels in addition to black and white) produce many more unique reflectance cells than a binary system with the same cell size. For a cell with p number of pixels and r reflectance levels per pixel, the number of cell gray levels is given by

$$N = \prod_{k=1}^{r-1} \frac{p + k}{k} = \frac{(p + r - 1)!}{p! \, (r - 1)!} \tag{21}$$

For binary, trinary, and quaternary pixels ($r = 2, 3,$ and 4) the number of cell gray levels becomes

$$\text{Binary:} \quad N_B = \sum_{k=1}^{p+1} 1 = \frac{(p + 1)!}{p!} = p + 1 \tag{22}$$

Table 3 Number of Cell Gray Levels for $r = 2$, 3, and 4

Dimensions $m \times n$	Number of pixels p	N_B (binary) $r = 2$	N_T (trinary) $r = 3$	N_Q (quaternary) $r = 4$
2×4	8	9	45	165
3×6	18	19	190	1330
4×8	32	33	561	6545

Trinary:
$$N_T = \sum_{k=1}^{p+1} k = \frac{(p+2)!}{2p!} = \frac{(p+2)(p+1)}{2} \qquad (23)$$

Quarternary:
$$N_Q = \sum_{k=1}^{p+1} (p+2-k)k$$
$$= \frac{(p+3)!}{6p!} = \frac{(p+3)(p+2)(p+1)}{6} \qquad (24)$$

Several numerical examples of the number of gray levels produced for a given cell size are given in Table 3. We see that increasing the number of pixel levels greatly increases the number of cell reflectance levels. At the current stage of this research, there is an attempt to develop algorithms that can utilize many of these hybrid cells while retaining the image resolution of partial dot methods.

5. Aliasing and Moiré Effects

Spatial resolution and aliasing are two key considerations of digital image processes. People often refer to spatial resolution as the highest frequency that may be reproduced. To be more precise, we should separate resolution and addressability. Resolution is the highest spatial frequency that can be seen, and half the addressability frequency is the highest the system may represent. Aliasing, or moiré, refers to spurious low-frequency components introduced by the digital imaging process. These fundamental sampled system issues are often best understood from a frequency-domain perspective. Such is the case in ordered-dither halftoning. The Fourier transform of the halftone image can be expressed in series form, where one term is the Fourier transform of the original image. Resolution, information density, and aliasing can be understood by examining the other terms of the series. These terms depend on the halftone-dot fill order. Due to nonlinearity of the process, the analysis is quite lengthy and complicated. Here we present only a qualitative description of the results. The reader is directed to the references for a detailed mathematical analysis [2,5,42,71].

Spatial resolution is dependent on original image contrast and filling order of the halftone dot. At a high contrast limit, the halftone process resolution is

(a)

Figure 16 Examples of screen frequency and gray-level trade-off: (a) lower screen frequency with minimal false contours; (b) higher screen frequency, but very visible false contours.

(b)

equal to the resolution of the original digital image, while at low contrast the resolution is comparable to sampling at the halftone frequency. To understand these limits, consider a light-dark alternating bar pattern. For a high contrast input, the black-and-white bars of the original digital image will be thresholded to black and white, respectively, regardless of the spatial frequency. At low contrast, the gray-scale steps of an entire cell are needed to represent the gray difference between a light bar and dark bar. There would be only one pixel changed in the cell representing the light bar when compared to the cell representing the dark bar. The resolution for intermediate contrast images is dependent on the filling order of the cell, where dispersed cells exhibit higher spatial resolution and information density than clustered-dot cells for the same cell size.

Aliasing occurs at the (vector) difference frequency between frequencies in the signal and the halftone pattern. Unlike the sampled-signal case, the aliasing for halftones may be absent or at different amplitudes, depending on the dot-filling pattern. The maximum contrast of the aliasing is the same as the contrast of the signal creating it for common halftoning algorithms. However, if dots are used in which bits are moved rather than simply added as gray levels change, the amplitude of aliasing patterns is unlimited.

C. Error Diffusion

Thus far, the halftone methods considered in detail were point processes. Here, we describe the neighborhood process of error diffusion, which uses the concept of fixing the total gray content of the image by calculating the brightness error incurred upon binarizing a pixel and incorporating this error in the processing of subsequent pixels. Due to resulting isolated white and isolated black pixels produced by the basic algorithm, the application of error diffusion has been primarily in display technologies (certain variations of the method do provide some clustering of like pixels, thereby rendering a printable image). The error diffusion method mitigates the trade-off of screen visibility versus gray-level contouring that is inherent in clustered-dot ordered-dither methods. Smoothly varying gray-scale images as well as sharp discontinuities are well rendered. A pleasant "blue noise" granular structure is observed with the exception of some undesirable worm-shaped artifacts. An example of error diffusion was shown in Fig. 8, where the pixel resolution is one quarter of that used in the clustered-dot examples so as to render the images printable and show the method at roughly the resolution of typical use. We discuss the basic algorithm introduced by Floyd and Steinberg [27] and some of its modifications, such as the use of other masks, perturbed thresholds, randomly distributed error, edge enhancement, and so on. To develop a deeper understanding of the fundamental technique, a one-dimensional frequency modulation model and a spectral analysis of the general technique are presented.

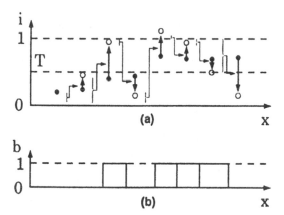

```
  ⋮
···X X X X X X X X X      Diffusion Mask
   X X X X X X X X X              ┌─┐
   X X X X X X X X X              │A│
   X X X X P┤A│O O O        ┌─────┴─┤
   O O O│D⁄CB│O O O        │D C B│
   O O O O O O O O O
   O O O O O O O O O
```

A: 7/16	X: Processed Pixel
B: 1/16	O: Unprocessed Pixel
C: 5/16	P: Pixel Being
D: 3/16	Processed

Figure 17 Schematic of error-diffusion process operating on an image.

Figure 18 Example of error diffusion applied to a one-dimensional signal: (a) original signal and how the error is propagated forward; (b) signal after thresholding.

1. Fundamental Algorithm

In error diffusion, as a pixel is thresholded at level T to a binary state, the resulting brightness quantization error is distributed in a weighted manner to neighboring pixels that have not yet been processed. Using the original Floyd–Steinberg diffusion mask, Fig. 17 shows schematically an image plane, where pixel P is being processed, X denotes a pixel already processed, $\{A, B, C, D\}$ denotes the set of unprocessed pixels that will receive the quantization error, and O denotes unprocessed pixels not relevant to the current calculation. The processing of pixels proceeds in raster fashion, diffusing the error forward and downward. In the original algorithm, error is distributed to $\{A, B, C, D\}$ weighted by $\{\frac{7}{16}, \frac{1}{16}, \frac{5}{16}, \frac{3}{16}\}$, respectively. A schematic of the algorithm was shown as a feedback loop in Fig. 7 and is demonstrated graphically for one-dimensional error diffusion in Fig. 18a and b. The threshold and diffusion processes are shown in Fig. 18a and b shows the output binary signal.

2. Modifications of Error Diffusion

Many modifications of the original algorithm have been proposed, primarily to eliminate the structured artifacts. First, consider the diffusion mask. The original diffusion weights were derived by trial and error and were found to give reasonably good results. Other diffusion masks commonly employed are

$$D = \frac{1}{48}\begin{bmatrix} & & P & 7 & 5 \\ 3 & 5 & 7 & 5 & 3 \\ 1 & 3 & 5 & 3 & 1 \end{bmatrix} \tag{25}$$

$$D = \frac{1}{42}\begin{bmatrix} & & P & 8 & 4 \\ 2 & 4 & 8 & 4 & 2 \\ 1 & 2 & 4 & 2 & 1 \end{bmatrix} \tag{26}$$

which were proposed by Jarvis et al. [38] and Stucki [82], respectively. It has been shown that the artifacts decrease somewhat with the larger masks, but the directional orientation becomes more pronounced [86]. The mask of Jarvis was also shown to enhance edges strongly, a characteristic of these larger masks [47].

One key method used to reduce the structured artifacts involves processing the image in other than a common raster fashion. Artifacts are reduced simply by applying error diffusion in a "serpentine raster," where the successive lines are processed in alternating directions. Note that the diffusion mask is flipped as the direction alternates so that the error is always propagated forward onto unprocessed pixels. Other space-filling curves can be used to define the order of processing. For example, Witten and Neal [89] process and diffuse error along a Peano curve, which greatly reduces structured artifacts but requires large memory buffers.

Other methods of minimizing artifacts vary the threshold either deterministically or randomly, or employ randomization in distributing error and setting thresholds. Billotet-Hoffman and Bryngdahl [9] demonstrated several advantages of utilizing an ordered-dither threshold array in an error diffusion algorithm, the advantages include reduction of worm artifacts. Also, more clustered pixel arrangements can be formed, allowing error diffusion to be utilized in electronic printing process [22]. This scheme provides more gray levels and thus less chance of false gray contours than ordered dither alone. Woo [90] demonstrated that randomly alternating between seven diffusion masks reduced artifacts. At the processing of each pixel, a random number chosen from the set $\{-3, -2, -1, 0, 1, 2, 3\}$, scaled by $\frac{1}{42}$, was added to each coefficient in Stucki's diffusion mask, Eq. (26). Dietz [19] reduced artifacts by treating the mask weights as a probability density function. Upon binarization of a given pixel, all the error is distributed to one neighboring pixel. The neighbor is chosen randomly according to the probability density function. Positions of the weights and the threshold could also be randomly varied.

3. Mathematical Description of Error Diffusion and Edge-Sharpening Effects

A one-dimensional model developed by Eschbach and Hauck [23] shows that the error-diffusion process can be described as converting the image to a frequency-modulated carrier signal with a threshold that is dependent on the input signal. This would be analogous to an ordered-dither algorithm where the screen frequency (carrier signal) adapts to the signal. A more recently developed model also shows how controlled edge enhancement may be incorporated into the algorithm [26]. Finally, a spectral analysis shows the inherent high-pass filter effect on the texture patterns of error diffusion [46].

To understand the frequency-modulation nature of error diffusion, first consider that error diffusion is a spatially quantized form of *pulse-density modulation*. The binary output of pulse density modulation can be written

$$b_{PDM}(x) = \text{step}[c(x) - t(x)] \tag{27}$$

where $c(x)$ and $t(x)$ are the carrier and threshold functions, respectively, and $\text{step}(x) = 0$ for $x < 0$ and $\text{step}(x) = 1$ for $x \geq 0$. For pulse-density modulation, we may write $c(x) = \cos[\Phi(x)]$, where the derivative of the phase of the cosinusoidal carrier is proportional to the gray-scale image $f(x)$. Solving the differential equation $\partial c/\partial x \propto f(x)$, leads to

$$c(x) = \cos\left[\frac{2\pi}{\Delta f_0} \int_0^x f(x')\, dx'\right] \tag{28}$$

where f_0 is the maximum intensity of the image and Δ is the pulse width in the binarized output image. To maintain constant pulse width, the threshold function must be

$$t(x) = \cos\left[\frac{f(x)\pi}{f_0}\right] \tag{29}$$

which upon substitution yields

$$b_{PDM}(x) = \text{step}\left[\cos\left[\left(\frac{2\pi}{\Delta f_0}\right)\int_0^x f(x')\, dx'\right] - \cos\left[\frac{f(x)\pi}{f_0}\right]\right] \tag{30}$$

Error diffusion requires spatial quantization within a fixed pixel grid. The effect of the grid can be described by a multiplication by an array of Dirac delta functions and a convolution by a rect function:

$$b_{ED} = b_{PDM}\left(x + \frac{\Delta}{2}\right)\frac{1}{\Delta}\sum \text{rect}\left(\frac{x - n\Delta}{\Delta}\right) \tag{31}$$

Figure 19 shows the frequency modulation nature of error diffusion, where (a) shows a linear ramp as the input gray signal, (b) and (c) show the carrier frequency and threshold functions, (d) is the thresholded carrier (PDM output), and (e) is the spatially quantized version of (d) (error diffusion).

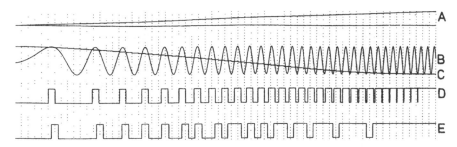

Figure 19 Generation of a binary pulse sequence from a one-dimensional gray-wedge input (a); (b) and (c) carrier and threshold functions, respectively; (d) pulse-density modulation output signal; and (e) spatially quantized binary sequence (error diffusion output). (From Ref. 23.)

The error diffusion algorithm may be slightly modified to allow for a controlled degree of edge enhancement. A formalism related to the frequency-modulation description given above may be employed to examine the edge-enhancement capabilities. The binary output of one-dimensional error diffusion, where all quantization error is passed onto the next pixel only, can be written

$$b_{ED}(n) = step[f(n) - e(n - 1) - t_0] \tag{32}$$

where $f(n)$ is the gray-scale value of the nth pixel with the appropriate modified brightness values, $e(n - 1)$ is the error incurred upon binarizing the $(n - 1)$th pixel, and t_0 is a constant threshold. The cumulative error satisfies the following recursive relationship:

$$e(n) = b(n) - f(n) + e(n - 1) \tag{33}$$

and also may be written

$$e(n) = \sum_{l=0}^{n} b(l) - f(l) \tag{34}$$

which upon substitution into Eq. (32) yields

$$b_{ED}(n) = step\left[f(n) - \sum_{l=0}^{n-1} [b(l) - f(l)] - t_0 \right] \tag{35}$$

In Eq. (35) we can see that the sum of the difference between the binary output and the original serves to preserve the overall gray content of the image. We may also compare Eq. (35) to Eq. (30) to better understand the relationship between error diffusion and frequency modulation. The error sum may be associated with the frequency-determining integral, and the slowly varying component $f(n) - t_0$ may be considered an additional phase modulation. Modifying the error diffusion algorithm by generalizing the phase modulation term allows for control over the image microstructure [26]. In generalized form, the error diffusion process may be written as

$$b_{ED}(n) = \text{step}\left[\sum_{l=0}^{n-1} [f(l) - b(l)] - t[f(n)] \right] \tag{36}$$

where the threshold is dependent on both $f(n)$ and n. The generalized threshold function allows the process to adapt the local pulse distribution to certain requirements, such as edge enhancement. A simple form for t that allows for controlled edge-enhancement capability is

$$t[f(n), n] = t_0 - kf(n) \tag{37}$$

where the amount of edge enhancement increases linearly with the constant k.

We now provide a spatial frequency analysis of the two-dimensional error diffusion process and describe an inherent high-pass filtering effect [47,87]. In two dimensions the process equation may be written

$$e(m, n) = b(m, n) - \left[f(m, n) - \sum_{a_{jk}} e(m - j, n - k) \right] \tag{38}$$

where $a_{jk} e(m - j, n - k)$ are the weighted errors from the previous pixels. Fourier transforming yields

$$E(u, v) = B(u, v) - F(u, v) + \left\{ \sum_{a_{jk}} \exp[-i (uj + vk)] \right\} E(u, v) \tag{39}$$

where the capital letters denote Fourier transforms of the lowercase functions, u and v are the frequency-domain variables, and a linear-phase term has arisen from shifted errors that were passed onto the neighboring pixels. Equation (39) may be rearranged to yield

$$B(u, v) = F(u, v) + H(u, v)E(u, v) \tag{40}$$

where the filter function $H(u,v)$ is defined as

$$H(u, v) = 1 - \sum_{a_{jk}} \exp[-i (uj + vk)] \tag{41}$$

We see that the spectrum of the output binary image equals the input image spectrum plus the filtered error term. Figure 20 shows a contour plot of $H(u, v)$ for the traditional Floyd–Steinberg four-element mask where the high-amplitude spectral values are denoted by lighter shades. Note that the filter is zero at the

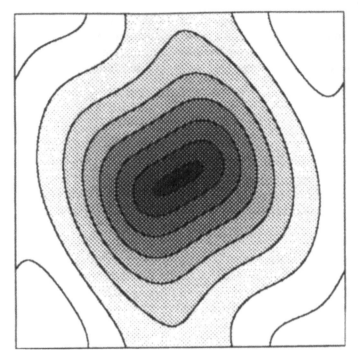

Figure 20   Modulus of high-pass filter built into the error-diffusion algorithm for the standard four-element error weights. Zero frequency is located in the center and has zero value. The corners are half the scan frequency. where the maximum value of 1.5 occurs (From Ref. 47.)

origin and 1 at the corners and therefore has a high-pass filter effect on the error image. This filtering effect may be used to explain inherent high-pass effects of the algorithm and also the blue-noise characteristics of the output image. Because quantization artifacts are described by the filtered error term $H(u, v) E(u, v)$, the binarized image tends to consist of high-frequency patterns, hence a blue-noise appearance.

Current research in error diffusion is aimed primarily at understanding and controlling the spatial filtering properties of the process from a linear system perspective. Other research has been directed toward minimizing wormlike image artifacts.

V. CONCLUSIONS

In this chapter we have presented encoding methods that are used to reduce the number of quantization levels per pixel in a digital image while maintaining the gray appearance of the image. These techniques are widely employed in the printing and display of digital images. The need for halftoning arises either because the physical processes involved are binary in nature or the processes have been restricted to binary operation for reasons of cost, speed, memory, or stability in the presence of process fluctuations. In these applications, the halftoned image is composed ideally of two gray levels, black and white. Spatial integration, plus higher-level processing performed by the eye and brain, and local area coverage of black-and-white pixels, provide the appearance of a gray-level or "continuous-tone" image.

A brief history of the field was presented, starting with analog methods of halftone image rendering and proceeding to digital techniques. There was a discussion of visual perception, touching only upon the concepts required for the understanding of the halftoning methods. The current methods, ordered dither and error diffusion, were described in detail. Other, less conventional techniques were also described briefly. For each technique there was a discussion of the methodology as well as issues that include tone reproduction, screen visibility, screen angle, image artifacts, and robustness. Current directions of research were discussed.

REFERENCES

1. J. P. Allebach, Random nucleated halftone screening, *Photogr. Sci. Engrg.*, 22(2): 89–91 (1978).
2. J. P. Allebach, Aliasing and quantization in the efficient display of images, *J. Opt. Soc. Am.*, 696:869–877 (1979).
3. J. P. Allebach, Binary display of images when spot size exceeds step size, *Appl. Opt.*, 119 No. 15:2513–2519 (1980).
4. J. P. Allebach, Visual model-based algorithms for halftoning images, *Proc. SPIE*, 310:151–158 (1981).
5. J. P. Allebach, and B. Liu, Analysis of halftone dot profile and aliasing in the discrete binary representation on images, *J. Opt. Soc. Am.*, 67 No. 9:1147–1154 (1977).
6. S. Arnemann and M. Tasto, Generating halftone pictures on graphic terminals using run length coding, *Comput. Graphics Image Process.*, 2:1–11 (1973).
7. E. Barnard, Optimal error diffusion for computer-generated holograms, *J. Opt. Soc. Am.*, 5(11):1803–1817 (1988).
8. B. E. Bayer, An optimum method for two-level rendition of continuous-tone pictures, *IEEE International Conference on Communications*, Vol. 1, 1973, pp. 26–11, 26–15.
9. C. Billotet-Hoffman and O. Bryngdahl, On the error diffusion technique for electronic halftoning, *Proc. SID*, 24(3):253–258 (1983).
10. S. J. Bloomberg, and P. G. Engledrum, Estimation of color errors due to random pixel placement errors, *J. Imag. Technol.*, 16(2):75–79 (1990).

11. M. Broja, F. Wyrowski, and O. Bryngdahl, Digital halftoning by iterative procedure, *Optics Communications*, 69, No. 3,4, pp. 205–210 (1989).

12. O. Bryngdahl, Halftone images: spatial resolution and tone reproduction, *J. Opt. Soc. Am.*, 68 No. 3, 416–422 (1978).

13. J. J. Burch, A computer algorithm for synthesis of spatial frequency filters, *Proc. IEEE*, 55:599–601 (1967).

14. C. Carnevali, L. Coletti, and S. Patarnello, Image processing by simulated annealing, *IBM J. Res. Develop.*, 29 No. 6:569 (1985).

15. R. W. Cohen and I. Gorog, Visual capacity: an image quality descriptor for display evaluation, *RCA Eng.*, 20(3):72–74 (1974).

16. T. N. Cornsweet, *Visual Perception*, Academic Press, New York, 1970.

17. S. R. Dashiell and A. A. Sawchuck, Nonlinear optical processing: analysis and synthesis, *Appl. Opt.*, 16 No. 4:1009–1025 (1977).

18. S. R. Dashiell and A. A. Sawchuck, Nonlinear optical processing: nonmonotonic halftone cells and phase halftones, *Appl. Opt.*, 16 No. 7:1936–1943 (1977).

19. H. Dietz, *Randomized Error Distribution*, Industrial Associates at Brooklyn Polytechnic Institute, New York, Apr. 1985.

20. P. G. Engledrum, Optimum density levels for multilevel halftone printing, *Jrnl. of Imaging Sci.*, Vol. 30, No. 5, pp. 220–222 (1987).

21. R. Eschbach, Pulse-density modulation on rastered media: combining pulse-density modulation and error diffusion, *J. Opt. Soc. Am. A*, 7(4):708–716 (1990).

22. R. Eschbach, Active binarization techniques in printing applications, presented at *Annual Meeting of DGaO*, Germany, June 8–13, 1992.

23. R. Eschbach and R. Hauck, Analytic description of the 1-D error diffusion technique for halftoning, *Opt. Comm.*, 52(3):165–168 (1984).

24. R. Eschbach and R. Hauck A 2-D pulse density modulation by iteration for halftoning, *Opt. Comm.*, 62(5):300–304 (1987).

25. R. Eschbach and R. Hauck, Binarization using a two-dimensional pulse-density modulation, *J. Opt. Soc. Am. A*, 4(10):1873 (1987).

26. R. Eschbach and K. T. Knox, Error-diffusion algorithm with edge enhancement, *J. Opt. Soc. Am.*, 8(12):1844 (1991).

27. R. Floyd and L. Steinberg, An adaptive algorithm for spatial grey scale, presented at *SID*, Washington, D.C., Apr. 1975; *Proc. Soc. Inform. Display*, 17(2): 75–77 (1976); *SID International Symposium Digest of Technical Papers*, 1975, pp. 36–37.

28. W. W. Goodall, Television by pulse code modulation, *Bell System Tech. J.*, 30 No. 1:33–49 (1951).

29. R. N. Goren, High quality pictorial xerographic reproduction by halftone screening, *J. Appl. Photogr. Engrg.*, 8:6 (1982).

30. D. Haas, Contrast modulation in halftone images produced by variation in scan line spacing, *J. Imag. Technol.*, 5(1):46–55 (1989).

31. J. A. G. Hale, Dot spacing modulation for the production of pseudo gray pictures, *Proc. SID*, 17:63–74 (1976).

32. P. Hamill, Line printer modification for better grey level pictures, *Comput. Graphics Image Process.*, 6:485–491 (1977).

33. M. Hepher, A comparison of ruled and vignetted screens, *Penrose Annual*, 47:166–177 (1953).

34. T. M. Holladay, An optimum algorithm for halftone generation for displays and hard copies, *Proc. SID*, 21(2)185–192 (1980).
35. H. Inose, and Y. Yasuda, A unity bit coding method by negative feedback, *Proc. IEEE*, Nov. 1963, pp. 1524–1535.
36. H. Inose, Y. Yasuda, and J. Murakami, A telemetering system by code modulation–Δ-Σ modulation, *IRE Trans. Space Electron. Telemetry*, SET-8:204–209 (1962).
37. J. F. Jarvis and C. S. Roberts, A new technique for displaying continuous-tone images on a bilevel display, *IEEE Trans. Comm.*, COM-24:891–898 (1976).
38. J. F. Jarvis, C. N. Judice, and W. H. Ninke, A survey of techniques for the display of continuous-tone pictures on bilevel displays, *Comput. Graphics Image Process.*, 5:13–40- (1976).
39. C. N. Judice, J. F. Jarvis, and W. H. Ninke, Using ordered dither to display continuous tone pictures on an ac plasma panel, *Proc. SID*, 15(fourth quarter):161–169 (1974).
40. G. Kaniza, Subjective contours, *Sci. Am.*, Apr. 1976.
41. H. Kato and L. Goodman, Non-linear transformations and logarithmic filtering in coherent optical systems, *Opt. Comm.*, 8(4) (1973).
42. D. Kermisch and P. G. Roetling, Fourier spectrum of halftone images, *J. Am. Opt. Soc.*, 65 No. 6:716–723 (June 1975).
43. R. J. Klensch, D. Meyerhofer, and J. J. Walsh, Electronically generated halftone pictures, *RCA Rev.*, pp. 512–533, September (1970).
44. K. Knowlton, and L. Harmon, Computer-produced greyscales, *Comput. Graphics Image Process.*, 1:1–20 (1972).
45. K. T. Knox, Digital haltoning algorithms and parameters, Lasers '90, December San Diego (1990).
46. K. T. Knox, Spectral analysis of error diffusion, *Proc. IS&T Annual Meeting*, St. Paul, Minn., May 16, 1991, p. 448.
47. K. T. Knox, Error image in error diffusion, *Proc. SPIE/IS&T's Symposium on Electronic Imaging Science and Technology*, Vol. 1657, San Jose, Calif., Feb. 9–14, 1992.
48. D. E. Knuth, Digital halftones by dot diffusion, *ACM Trans. Graphics*, 6(4):245–273 (1987).
49. W. Lama, S. Feth, and R. Loce, Hybrid (gray pixel) halftone printing, *J. Imag. Technol.*, 15(3):130–135 (1989).
50. J. O. Limb, Coarse quantization of visual signals, *Austral. Telecomm. Res.*, 1(1/2):32–42. (1967).
51. J. O. Limb, Design of dither waveforms for quantized visual signals, *Bell System Tech. J.*, 48:2555–2582 (Sept. 1969).
52. B. Lippel, and M. Kurland, The effect of dither luminance quantization of pictures, *IEEE Trans. Comm.*, COM-19 (6):879–888 (1971).
53. B. Lippel, M. Kurland, and A. H. Marsh, Ordered dither patterns for coarse quantization of pictures, *Proc. IEEE*, 59(3):429–431 (1971).
54. R. Loce, and W. Lama, Halftone banding in a xerographic image bar printer, *J. Imag. Technol.*, 16(1):6–11 (1990).
55. I. D. G. Macleod, Pictorial output with a line printer, *IEEE Trans. Comput. (Short Notes)*, Feb. 1970, pp. 160–162.

56. M. Marquet and J. Tsujuichi, Interpretation de aspects particuliers des images obtenues dan une experience de detremage, *Optica Acta*, *8*, No. 3 pp. 267–277 (1961).

57. S. Matsumoto, and B. Liu, Analytical fidelity measures in the characterization of the halftone processes, *J. Opt. Soc. Am.*, 70(10):1248–1254 (1980).

58. P. Melnychuck, and R. Shaw, Fourier spectra of digital halftone images containing dot position errors, *J. Opt. Soc. Am. A.*, 5 No. 8:1328–1338 (1988).

59. T. Mitsa, and K. J. Parker, Digital halftoning with a blue noise mask, *SPIE/IS&T Symposium on Electronic Imaging Science and Technology*, San Jose, Calif., Feb. 24–Mar. 1, 1991.

60. W. Naing, Y. Miyake, T. Taniguchi, and S. Kubo, An evaluation of image quality of tri-level images obtained by a new algorithm, *J. Imag. Technol.*, 15(1):23–28 (1989).

61. M. Nishikawa, Method of forming oblique dot pattern, U.S. patent 4,805,003, Feb. 14, 1989.

62. E. Peli, Multiresolution, error-convergence halftone algorithm, *J. Opt. Soc. Am. A.*, Vol. 8, No. 4, pp. 625–636 (1991).

63. B. Perry, and M. L. Mendelsohn, Picture generation with a standard line printer, *Comm. ACM*, 7(5):311–313 (1964).

64. R. W. Pryor, G. M. Cinque, and A. Rubinstein, Bilevel displays: a new approach,'' *Proc. SID*, 19 No. 2:127–131 (1978).

65. J. Riseman, J. Smith, A. d'Entremont, and C. Goldman, An apparatus for generating an image from a video signal, U.S. patent 4,800,442, Jan. 24, 1989.

66. L. G. Roberts, Picture coding using pseudo-random noise, *IRE Trans. Inform. Theory*, IT-8:145–154 (Feb. 1962).

67. P. G. Roetling, Halftone method for edge enhancement and moire suppression, *J. Opt. Soc. Am.*, 66:985–989 (1976).

68. P. G. Roetling, Visual performance and image coding, *Proc. SID*, 17(2):111–114 (1976).

69. P. G. Roetling, Visual effects in binary display of continuous-tone images, *SID International Symposium Digest of Technical Papers*, Vol. 8, 1977.

70. P. G. Roetling, Binary approximation of continuous-tone images, *Photogr. Sci. Engrg.*, 21(2):60–65 (1977).

71. P. G. Roetling, Analysis of detail and spurious signals in halftone images, *J. Appl. Photogr. Engrg.*, Vol. 3, No. 1 pp. 12–17 (1977).

72. P. G. Roetling, and T. M. Holladay, Tone reproduction and screen design for pictorial eletrographic printing, *J. Appl. Photogr. Engrg.*, 5(4):179–182 (1979).

73. P. G. Roetling, and T. M. Holladay, Tone reproduction and screen design for pictorial eletrographic printing, *J. Appl. Photogr. Engrg.*, 5(4):179–182 (1979).

74. F. R. Ruckdeschel, A. M. Walsh, O. G. Hauser, and C. Stephan, Characterizing halftone noise: a technique, *Appl. Opt.*, Vol. 17 No. 24 pp. 3999–4002 (Dec. 1978).

75. M. R. Schroeder, Images from computers and microfilm plotters, *Comm. ACM*, 12:95–110 (Feb. 1969).

76. M. R. Schroeder, Images from computers, *IEEE Spectrum*, 6:66–78 (Mar. 1969).

77. L. Schuchman, Dither signals and their effect on quantization noise, *IEEE Trans. Comm. Technol.*, Vol. COM-12:162–164 (Dec. 1964).

78. S. J. Shu, R. Springer, and C. L. Yeh, Moire factors and visibility in scanned and printed halftone images, *Opt. Engrg.*, 28(7):805–812 (1989).
79. A. Steinbach, and K. Y. Wong, Moire patterns in scanned halftone pictures, *J. Opt. Soc. Am.*, 72(9):1190–1198 (1982).
80. E. Steinberg, R. Easton, and R. Rolleston, Analysis of random dither patterns using second-order statistics, *IS&T 44th Annual Conference*, St. Paul, Minn., May 1991.
81. W. Streifer, R. N. Goren, and L. M. Marks, Analysis and experimental study of ruled halftone screens, *Appl. Opt.*, June 1974, vol. 13 no. 6 pp. 1299–1317.
82. P. Stucki, MECCA, A multiple-error correcting computation algorithm for bilevel image hardcopy reproduction, Research Report RZ 1060, IBM Research Laboratory Zurich, Switzerland (1981).
83. J. Sullivan, and R. Miller, New algorithm for image halftoning using a human visual model, *Proc. SPSE's 43rd Annual Conference*, Rochester, N.Y., May 20–25, 1990, pp. 145–148.
84. W. Fox Talbot, Improvements in the art of engraving, British patent 565, 1852.
85. J. E. Thompson, and J. J. Sparkes, A pseudo-random quantizer for television signals, *Proc. IEEE*, 55:353–355 (Mar. 1967).
86. R. Ulichney, *Digital Halftoning*, MIT Press, Cambridge, Mass., 1987.
87. S. Weissbach, F. Wyrowski, and O. Bryngdahl, Error diffusion as a filter in digital optics, *Conference of the German Society for Applied Optics*, June 1990.
88. B. Widrow, Statistical analysis of amplitude-quantized sampled-data systems, *Trans. AIEE (Appl. Ind.), Pt. II.*, 79:555–568 (1960) (Jan. 1961 Section).
89. I. H. Witten and M. Neal, Using Peano curves for bilevel display of continuous-tone images, *IEEE CG&E*, (1982) pp. 47–52.
90. B. Woo, A survey of halftoning algorithms and investigation of the error diffusion technique, S.B. thesis, MIT, 1984.

11

Glossary of Computer Vision Terms

Robert M. Haralick and Linda G. Shapiro

University of Washington
Seattle, Washington

1. THE IMAGE

1. An *image* is a spatial representation of an object, a two-dimensional or three-dimensional scene, or another image. It can be real or virtual as in optics. In computer vision, "image" usually means recorded image such as a video image, digital image, or picture. It may be abstractly thought of as a continuous function I of two variables defined on some bounded and usually rectangular region of a plane. The value of the image located at spatial coordinates (r, c) is denoted by $I(r,c)$. For optic or photographic sensors, $I(r, c)$ is typically proportional to the radiant energy received in the electromagnetic band to which the sensor or detector is sensitive in a small area around (r, c). For range finder sensors, $I(r, c)$ is a function of the line of sight distance from (r, c) to an object in the three-dimensional world. For a tactile sensor, $I(r, c)$ is proportional to the amount that the surface at and around (r, c) deforms the sensor. When the image is a map, $I(r, c)$ is an index or symbol associated with some category such as a color, a thematic land use category, a soil type, or a rock type. A recorded image may be in photographic format, video signal format, or digital format.

2. A *video image* is an image in electronic signal format capable of being displayed on a cathode ray tube screen or monitor. The video signal can be generated from devices like a CCD camera, a vidicon, a flying spot scanner, a tactile sensor, a range sensor, or a frame buffer driving a digital to analog converter. Video images have two common formats. In the frame format, the video signal

itself is a sequence of signals, the ith signal representing the ith line of the image. The ith signal is separated from the $(i + 1)$st signal by a horizontal sync or pulse. Each video frame is separated from the next video frame by a vertical sync pulse. In the interlaced format, the video signal is divided into two fields. The first field contains all the odd numbered lines and the second field contains all the even numbered lines. As in the frame format, the ith line of the field is its ith signal, and it is separated from next line of the field by a horizontal sync pulse. Successive fields are separated by vertical sync pulses.

3.   The *gray level, gray shade, gray tone, gray tone intensity, image intensity, image density, brightness*, or *image value* is a number or value assigned to a position on an image. For optic or photographic sensors, the image intensity at (r, c) is proportional to the integrated output, reflectance, or transmittance of a small area, usually called a resolution cell or pixel, centered on the position (r, c). Its value can be related to transmittance, reflectance, a coordinate of the tristimulus, ICI, YIQ, or RGB color coordinate system, brightness, radiance, luminance, density, voltage, or current.

4.   *Resolution* is a generic term which describes how well a system, process, component, material, or image can reproduce an isolated object consisting of separate closely spaced objects or lines. The *limiting resolution, resolution limit* or *spatial resolution* is described in terms of the smallest dimension of the target or object that can just be discriminated or observed. Resolution may be a function of object contrast and spatial position as well as element shape (single point, number of points in a cluster, continuum, or line etc.).

5.   A *resolution cell* is the smallest most elementary areal constituent having an associated image intensity in a digital image. A resolution cell is referenced by its spatial coordinates which are the center coordinates of its area. The resolution cell or spatial formations of resolution cell constitute the basic unit for low level processing of digital image data. Resolution cells usually have areas which are square, rectangular, or hexagonal.

6.   *Acutance* is a measure of the sharpness of edges in a photograph or image. It is defined for any edge by the average squared rate of change of the image intensity across the edge divided by the total image intensity difference from one side of the edge to the other side of the edge.

7.   The *contrast* of an object against its background can be measured by: (1) its *contrast ratio*, which is the ratio between the higher of object transmittance or background transmittance to the lower of object transmittance or background transmittance; (2) its *contrast difference*, which is the difference between the higher density of object or background to the lower density of object or background; (3) its *contrast modulation*, which is the difference between the darker of object or background image intensity and the lighter of the two divided by the sum of object image intensity and background image intensity.

8. A *pixel, picture element,* or *pel* is a pair whose first member is a resolution cell or (row, column) spatial position and whose second member is the image intensity value or vector of image values associated with the spatial position.

9. A *voxel,* short for volume element, is an ordered pair whose first component is a (row, column, slice) location of a volume rectangular parallelepiped and whose second component is the vector of properties in the rectangular parallelepiped volume.

10. An *edgel,* short for edge element, is a triplett whose first component is the (row, column) location of a pixel, whose second component is the position and orientation of an edge running through the pixel, and whose third component is the strength of the edge.

11. *Raster scan order* refers to the sequence of pixel locations obtained by scanning the spatial domain of an image in a left to right scan of each image row with the rows taken in a top to bottom ordering. Frame format video images are images scanned in raster scan order.

12. A *range image* is an image in which each pixel value is a function of the distance between the pixel and the object surface patch imaged on the pixel. Depending on the sensor and preprocessing used to create the range image, the distance can be the distance between the image plane and the ranged surface patch, the line of sight distance between the pixel and its corresponding ranged surface patch or some function of these distances and the pixel's position.

13. A *digital image, digitized image,* or *digital picture function* is an image in digital format and is obtained by partitioning the area of the image into a finite two-dimensional array of small uniformly shaped mutually exclusive regions called resolution cells and assigning a representative image value to each such spatial region. A digital image may be abstractly thought of as a function whose domain is the finite two-dimensional set of resolution cells and whose range is the set of possible image intensities.

14. *Rangel* is the range data element produced by a range sensor. It is a pair whose first member is a row column spatial position and whose second member is the range value or a vector whose first component is the range value and whose second component is the image intensity value.

15. A *depth map* or *range map* is a digital range image in which the range value in each pixel's position is the distance between the image plane and the ranged surface patch corresponding to the pixel.

16. An *orientation map* is a digital image in which each pixel contains the 3D orientation vector of the normal to the 3D surface patch corresponding to the pixel position.

17. A *multi-image set* or *multi-band image* is a set of registered images each related to the same subject but taken at different times, from different positions,

with different lighting, with different sensors, at different electromagnetic frequencies, with different polarizations, or from different sections of the subject. Although there is a high degree of information redundancy between images in a multi-image set, each image usually has some information not available in any one of or combination of the other images in the set. If the multi-image set has *N* images, then each resolution cell is associated with an *N*-tuple of image values.

18.   A *multi-spectral image* is a multi-band image in which each band is an image taken at the same time, but sensitive in a different part of the electro-magnetic spectrum.

19.   A *time varying image, multi-temporal image, dynamic imagery*, or *image time sequence* is a multi-image set in which each successive image in the set is taken of the same scene at a successive time. Between successive snapshots, the objects in the scene may move or change and the sensor may move.

20.   A *binary image* is an image in which each pixel takes either the value zero or the value one.

21.   A *gray scale image* or a *gray level image* is an image in which each pixel has a value in a range larger than just 0 or 1. Gray scale images typically have values in the range 0 to 63, 0 to 255, or 0 to 1023 corresponding to 6 bit, 8 bit, or 10 bit digitizations.

22.   A *symbolic image* is an image in which the value of each pixel is an index or symbol.

23.   A *histogram* or *image histogram* is a function *h* defined on the set of image intensity values to the non-negative integers. The value $h(k)$ is given by the number of pixels in the image having image intensity *k*. For images having a large gray tone range, the image will often be quantized before being histogrammed or will be quantized on the fly during the histogramming process.

2.   PHOTOMETRY AND ILLUMINATION

24.   *Luminous flux* is radiant power evaluated according to its capacity to produce visual sensation. *Luminous intensity* in a given direction is measured in terms of luminous flux per steradian. The unit of luminous intensity is the *candela*. The luminance of a black body radiator at the temperature of solidification of platinum is 60 candelas per square centimeter. The unit of luminous flux is the *lumen*. The luminous flux emitted by a uniform point light source of luminous intensity of one candela in one steradian solid angle is one *lumen*.

25.   The *illumination* at a point on a surface is the luminous flux incident on an infinitesimal element of the surface centred at the given point divided by area of the surface element. The unit of illumination is the *lux* or meter-candle. The lux is equal to one lumen per square meter. Another unit of illumination is the *foot candle* and it is equal to one lumen per square foot. The illumination at a

point on a surface due to a point source of light is proportional to the luminous intensity of the source in the direction of the surface point and to the cosine of the angle between this direction and the surface normal direction. It is inversely proportional to the square of the distance between the surface point and the source.

26. The *illuminance* in a given direction at a surface point is the luminous intensity in that direction of an infinitesimal surface element containing the given point divided by the area of the orthogonal projection of the element on a plane perpendicular to the given direction.

27. The *radiance* of an object is a measure of the power per unit foreshortened surface area per unit solid angle radiated or reflected by the object about a specified direction. Radiance can be a function of the viewing angle and the spectral wavelength and bandwidth.

28. The *radiant intensity* of a point object is a measure of the radiant power per steradian radiated or reflected by the object. Radiant intensity can be a function of the viewing angle and the spectral wavelength and bandwidth.

29. *Irradiance* is the power per unit area of radiant energy incident on a surface.

30. The *reflectance*, the *reflection coefficient*, or the *bidirectional reflectance distribution function* of a surface is the ratio of the radiant power per unit area reflected by the surface to the radiant power per unit area incident on the surface. The reflectance can be a function of the incident angle of the radiance, the viewing angle of the sensor, and the spectral wavelength and bandwidth.

31. A *reflectance image* or *reflectance map* is a digital image in which the value in each pixel's position is proportional to the reflectance of the surface patch imaged at the pixel's position for a given illumination and viewing direction.

32. A *Lambertian surface* is a uniformly diffusing surface. It appears as a matt surface and it has a reflectance function which is a constant. The reflectance function of a Lambertian surface does not depend on the viewing angle and, therefore, a planar surface having a Lambertian reflectance appears equally bright from all viewing angles. For a Lambertian surface, the luminous intensity per unit area in a given direction varies as the cosine of the angle between the direction and the surface normal direction.

33. *Backlighting* refers to an illumination arrangement in which the light source is on the opposite side of the object from the camera. Backlighting tends to produce images which are black and white silhouettes.

34. *Frontlighting* refers to an illumination arrangement in which the light source is on the same side of the object as the camera.

35. *Ambient light* refers to the light which is present in the environment around a machine vision system and which is generated from sources outside of the system. From the point of view of the machine vision system, ambient light

is unplanned light which might adversely affect the image processing. Care is usually taken to minimize its effect.

36.   The *r, g, b chromaticity* coordinates of a multispectral pixel, where red brightness is *R*, green brightness is *G* and blue brightness is *B*, is given by

$$r = \frac{R}{R + G + B}$$

$$g = \frac{G}{R + G + B}$$

$$b = \frac{B}{R + G + B}$$

37.   The *H, S hue saturation* coordinates of a multispectral pixel whose chromaticity coordinates are *r, g, b* is given by

$$H = \begin{cases} \theta \\ 360 - \theta \end{cases} \quad \text{if} \quad \begin{matrix} b < g \\ b > g \end{matrix}$$

where

$$\theta = \arccos \frac{2r - g - b}{\sqrt{6}\left[\left(r - \frac{1}{3}\right)^2 + \left(g - \frac{1}{3}\right)^2 + \left(b - \frac{1}{3}\right)^2\right]^{1/2}}$$

$$S = 1 - 3 \min\{r, g, b\}$$

38.   The *YIQ coordinate* used in NTSC color TV transmissions is related to the RGB coordinates by the following linear transformation:

$$\begin{pmatrix} Y \\ I \\ Q \end{pmatrix} = \begin{pmatrix} 0.299 & 0.587 & 0.144 \\ 0.596 & -0.274 & -0.322 \\ 0.211 & -0.523 & 0.312 \end{pmatrix} \begin{pmatrix} R \\ G \\ B \end{pmatrix}$$

3.   PHOTOGRAMMETRY

39.   *Analytic photogrammetry* refers to the analytic mathematical techniques which permit the inference to geometric relations between points and lines in the two-dimensional perspective projection image space and the three-dimensional object space.

40.   *Digital photogrammetry* refers to the computer processing of perspective projection digital images with analytic photogrammetry techniques as well as other computer techniques for the automatic interpretation of scene or image content.

41.   *Relative orientation* in analytic photogrammetry is the relative position and orientation of one common reference frame with respect to another. When two common reference frames are in known relative orientation, the rays emanating from the same object point located on each camera's image will intersect exactly at a point in 3D space.

42. The *exterior orientation* or *outer orientation* refers to the position and orientation of a camera reference frame with respect to a world reference frame.

43. *Absolute orientation* in analytic photogrammetry is the rotation and translation transformation(s) by which one or more camera reference frame(s) can be made to correspond to a world reference frame.

44. The *optic axis* or *principal axis* is the straight line which passes through the centers of curvatures of the lens surfaces.

45. The *principal point* is that point on the image which is the intersection of the image plane with the optic axis.

46. The *center of perspectivity* of a perspective projection is the common point where all rays meet.

47. The *principal distance* is the distance between the center of perspectivity and the image projection plane. For a fixed image, it is sometimes called the *camera constant* or *Gaussian focal length*.

48. The *inner orientation* or *internal orientation* is given by the triple (u_0, v_0, f) whose (u_0, v_0) is the position of the principal point in the measurement image plane coordinate system and f is the principal distance. Inner orientation may also include the values of the free parameters which describe the lens distortion.

49. A *perspective projection* is defined in terms of a projection plane and a center of perspectivity. Imaging sensors whose ideal model is the pinhole camera generate perspective projection images. There are three commonly used frameworks for defining the relationship between a 3D point and its 2D perspective projection. All take the projection plane to be perpendicular to the z-axis. In the first framework, the center of perspectivity is taken to be the origin and the projection plane is a distance f from the origin on the positive z-axis. In this case the projection (P_x, P_y) of the point (x, y, z) is given by

$$P_x = \frac{fx}{z}, \qquad P_y = \frac{fy}{z}$$

In the second framework, the center of perspectivity is taken to be $(0, 0, -f)$ and the projection plane passes through the origin. In this case the projection (P_x, P_y) of the point (x, y, z), $z > 0$, is given by

$$P_x = \frac{fx}{f + z}, \qquad P_y = \frac{fy}{f + z}$$

In the third framework, the center of perspectivity is taken to be $(0, 0, f)$ and the projection plane passes through the origin. In this case the projection (P_x, P_y) of the point (x, y, z), $z < 0$, is given by

$$P_x = \frac{fx}{f - z}, \qquad P_y = \frac{fy}{f - z}$$

50.   A *parallel* or *orthographic projection* onto a plane perpendicular to the z-axis of a point (x, y, z) produces the projected (P_x, P_y) defined by

$$P_x = sx, \qquad P_y = sy$$

where s is the scale factor of the projection.

51.   The *parallax* is the observed positional difference of a projected 3D point on a pair of 2D perspective images. The difference in position is caused by a shift in the position of the perspective centers and optical axis orientation. That portion of the parallax in the direction of the x-axis is called the *x-parallax*. That portion of the parallax in the direction of the y-axis is called the *y-parallax*. For a pair of stereo images where the line joining the centers of perspectivity is parallel to the x-axis, the parallax will be entirely in the direction of the x-axis when the two image planes are in the same orientation. Should one image plane be tilted with respect to the other, the y-parallax will not be zero.

52.   A *vanishing point* is the point in the 2D perspective projection image plane where a system of 3D parallel lines converge. The vanishing points of all systems of 3D parallel lines parallel to a given plane will lie along a corresponding line in the 2D perspective projection image plane called the *vanishing line* for the given plane.

4.   IMAGE OPERATORS

53.   An *image operator*, *image transform*, or *image transform operator* is a function which takes an image for its input and produces an image for its output. The domain of a transform operator is often called the spatial or space domain. The range of the transform operator is often called the transform domain. Some image transform operators have spatial and transform domains of entirely different geometry or character; the image in the spatial domain may appear entirely different from and have a different interpretation from the image in the transform domain. Specific examples of these kinds of image transforms include Fourier, Sine, Cosine, Slant, Haar, Hadamard, Mellin, Karhunen-Loeve, and Hough transforms. Image operators which have spatial and transform domains of similar geometry or character include point operators, neighborhood operators, and spatial filters.

5.   POINT OPERATORS

54.   A *point operator* is an image operator in which the output image value at each pixel position depends only on the input image value at the corresponding pixel position.

55.   *Thresholding* is an image point operation which produces a binary image from a gray scale image. A binary one is produced on the output image

whenever a pixel value on the input image is above a specified minimum threshold level. A binary zero is produced otherwise. Alternatively thresholding can produce a binary one on the output image whenever a pixel value on the input image is below a specified maximum threshold level. A binary zero is produced otherwise.

56.   *Level slicing* or *density slicing* is a point operation which employs two thresholds and produces a binary image. A binary one is produced on the output image wherever a pixel value on the input image lies between the specified minimum and maximum threshold levels. A binary zero is produced otherwise.

57.   *Multi-level thresholding* is a point operator employing two or more thresholds. Pixel values which are in the interval between two successive threshold values are assigned an index associated with the interval.

58.   *Contrast stretching* refers to any monotonically increasing point operator whose effect is to increase or enhance the visibility of an image's detail.

59.   *Quantizing* is a monotonically increasing point operator by which each image intensity value in a digital image is assigned a new value from a given finite set of quantized values. The quantized image has fewer distinct gray levels but may make better use of the dynamic range. Thus quantizing often enhances the image's appearance. There are four often used methods of quantizing: equal interval quantizing, equal probability quantizing, minimum variance quantizing, and histogram hyperbolization. In each method, the range of image values from the maximum to the minimum value is divided into contiguous intervals and each image value is assigned either the mean value of the quantized class to which it belongs or the index of the quantized class to which it belongs.

(a)   60.   In *equal interval quantizing* or *linear quantizing*, the range of image values from maximum value to minimum value is divided into contiguous intervals each of equal length, and each image value is assigned to the quantized class which corresponds to the interval within which it lies;

(b)   61.   In *equal probability quantizing*, the range of image values is divided into contiguous intervals such that after the image values are assigned to their quantized class there is an equal frequency of occurrence for each quantized value in the quantized digital image; equal probability quantizing is sometimes referred to as or *histogram equalization*;

(c)   62.   In *minimum variance quantizing*, the range of image values is divided into contiguous intervals such that the weighted sum of the variances of the quantized intervals is minimized. The weights are usually chosen to be quantized class probabilities which are computed as the proportional areas on the image which have values in the quantizing intervals.

(d)   63.   In *histogram hyperbolization quantizing*, the range of image values is divided into contiguous intervals and each image value is assigned to the mean of its quantized class. The division is done in such a way that the quantized image has a uniform perceived brightness. Histogram hyperbolization takes into account the nonlinearity of the human eye brain combination.

64. *Masking* is a point operator applied to a two-band image. One image band is a binary image B and is called the mask image band. The second image band I is called the image to be masked. Masking produces a resulting image J whose pixels take the value zero wherever the mask image has value zero and whose pixels take the value of the image I wherever the mask image has value one. That is,

$$J(r, c) = \begin{cases} 0 & \text{if} \quad B(r, c) = 0 \\ I(r, c) & \text{if} \quad B(r, c) = 1 \end{cases}$$

65. *Change detection* is the process by which two registered images may be compared, pixel by pixel, and a binary one value given to the output pixel whenever corresponding pixels on the input images have significantly different enough gray levels. Corresponding pixels on the input images which do not have significantly different enough gray levels generate a binary zero value on the output image. A change detection operator is a point operator.

6. SPATIAL OPERATORS

66. *Two-dimensional signal processing* refers to that area of image processing in which the one-dimensional signal processing techniques of noise filtering, restoration, data compression, and detection have been generalized to two dimensions and thereby made applicable to image data.

67. A *neighborhood operator* is an image operator in which the output image value at each pixel position depends only on the input image values in a neighborhood containing or surrounding the corresponding input pixel position.

68. A *spatial filter* is an image operator in which the spatial and transform domain have similar geometries and in which the image output value at each pixel depends on more than one pixel value in the input image. Usually, but not always, the image output value has its highest dependence on the image input values in some neighborhood centered in the corresponding pixel in the input image.

69. A *linear spatial filter* is a spatial filter for which the image intensity at coordinates (r, c) in the output image is some weighted average or linear combination of the image intensities located in a particular spatial pattern around coordinates (r, c) of the input image. A linear spatial filter is often used to change the spatial frequency characteristics of the image. For example, a linear spatial filter which emphasizes high spatial frequencies will tend to sharpen the edges in an image. A linear spatial filter which emphasizes low spatial frequen-

cies will tend to blur the image and reduce salt and pepper noise. When the purpose of the filter is to enhance neighborhoods having certain shapes, the operation is sometimes called *mask matching*.

70.   The *kernel* of a linear spatial filter is a function defined on the domain of the spatial pattern of the filter and whose value at each pixel of the domain is the weight or coefficient of the linear combination which defines the spatial linear filter.

71.   A *box filter* is a linear spatial smoothing filter in which each pixel in the filtered image is the equally weighted average of the pixels in a rectangular window centered at its spatial position in the input image.

72.   The *Gaussian filter* is a linear spatial smoothing filter whose kernel is given by the two-dimensional Gaussian

$$k(r, c) = \frac{1}{2\pi}e^{-\frac{1}{2}\left(\frac{r^2}{\sigma_r^2} + \frac{c^2}{\sigma_c^2}\right)}$$

Filtering an image with a Gaussian filter will smooth the image.

73.   *Convolving* an image I with a kernel k having support or domain K produces a convolved image, denoted by $I * k$, which is defined by

$$(I * k)(r, c) = \sum_{(i,j) \in K} I(r - i, c - j)k(i, j)$$

Convolution is a linear operator.

74.   A two-dimensional filtering operation is called *separable* if the convolution can be decomposed into two successive one-dimensional convolutions, one convolution operating on the image row by row and the second convolution operating on the image column by column.

75.   *Correlating* an image I with a kernel k having support or domain K produces a correlated image J defined by

$$J(r, c) = \sum_{(i,j) \in K} I(r + i, c + j)k(i, j)$$

Correlation is a linear operator.

76.   The *discrete Fourier transform* $\hat{I}$ of a digital image I represents the image in terms of a linear combination of complex exponentials. The Fourier transform $\hat{I}$ is defined by

$$\hat{I}(w_r, w_c) = \frac{1}{RC} \sum_{r=0}^{R-1} \sum_{c=0}^{C-1} I(r, c)e^{-2j\pi\left(\frac{rw_r}{R} + \frac{cw_c}{C}\right)}$$

$\hat{I}(w_r, w_c)$ is the coefficient of the complex exponential $e^{2j\pi}(rw_r/R + cw_c/C)$ in the linear combination representing I as can immediately be seen from the corresponding relation

$$I(r, c) = \sum_{w_r=0}^{R} \sum_{w_c=0}^{C} \hat{I}(w_r, w_c) e^{2j\pi\left(\frac{rw_r}{R} + \frac{cw_c}{C}\right)}$$

which is called the *inverse discrete Fourier transform*. The variables w_r and w_c have the interpretation of being row and column *spatial frequencies*.

77. A *high pass filter* is a linear spatial filter which attenuates the low spatial frequencies of an image and accentuates the high spatial frequencies of an image. It is typically used to enhance small details, edges, and lines.

78. A *low pass filter* is a linear spatial filter which attenuates the high spatial frequencies of an image and accentuates the low spatial frequencies of an image. It is typically used to supress small undesired details, eliminate noise, enhance coarse image features, or smooth the image.

79. A *band pass filter* is a linear spatial filter which attenuates those spatial frequencies outside the band and accentuates those spatial frequencies within the band. It is typically used to enhance details of the image whose spatial size characteristics are related to the spatial frequencies within the band.

80. A *pyramid* or *image pyramid* is a sequence of copies of an image in which both sample density and resolution are decreased in regular steps. The bottom level of the pyramid is the original image. Each successive level is obtained from the previous level by a filtering operator followed by a sampling operator.

In the *Gaussian image pyramid*, the resolution is decreased by successive convolutions of the image at the previous level of the pyramid with a Gaussian-like kernel. After the low pass Gaussian convolutions, the sample density is typically decreased by sampling every other pixel.

In the *morphological image pyramid*, each successive level is obtained by an opening and closing operation on the previous level followed by sampling.

In the *Laplacian image pyramid*, each successive layer is obtained by taking the Laplacian of the corresponding level on the Gaussian pyramid. The Laplacian convolution kernel here is typically defined as the kernel obtained by taking the Laplacian of a Gaussian having an appropriately chosen value for its standard deviation.

81. An *edge operator* or *step edge operator* is a neighborhood operation which determines the extent to which each pixel's neighborhood can be partitioned by a simple arc passing through the pixel where pixels in the neighborhood on one side of the arc have one predominant value and pixels in the neighborhood on the other side of the arc have a different predominant value. Some edge operators can also produce a direction which is the predominant tangent direction of the arc as it passes through the pixel.

There are four classes of edge operators: *gradient operators*, *Laplacian operators*, *zero-crossing operators*, and *morphologic edge operators*. The gradient operators compute some quantity related to the magnitude of the slope of the underlying image gray tone intensity surface of which the observed image pixel values are noisy discretized sample. The Laplacian operators compute some quantity related to the Laplacian of the underlying image gray tone intensity surface. The zero-crossing operators determine whether or not the digital Laplacian or the estimated second direction derivative has a zero-crossing within the pixel. The morphologic edge operators compute a quantity related to the residues of an erosion and/or dilation operation.

82.   An *edge image* is an image in which each pixel is labelled as "edge" or "non-edge". In addition to this basic labeling, pixels in an edge image may carry additional information such as edge direction, edge contrast, or edge strength.

83.   *Edge linking* refers to the process by which neighboring edge labeled pixels can be aggregated to constitute a chain or sequence of edge pixels.

84.   *Boundary detection* or *boundary delineation* refers to any process which determines a chain of pixels separating one image region from a neighboring image region.

85.   An *occluding edge* is an image edge which arises from a range or depth discontinuity. This typically happens where one object surface projects to a pixel on one side of the edge and another object surface which is some distance behind the first object surface projects to a pixel on the other side of the edge. Step edges in depth maps are always occluding edges.

86.   A pair of straight edges are said to be *antiparallel* if there are no edges between them and the edges have opposite contrast.

87.   *Homomorphic filtering* is a filtering process in which the filter is applied to the logarithm of the image and the output image is obtained by exponentiating the filtered logarithm image.

88.   A *median filter* is a non-linear neighborhood image smoothing spatial filter in which the value of an output pixel is the median value of all the input pixels in the supporting neighborhood of the filter about the given pixel's position. Median filters are used to smooth and remove noise from images.

89.   *Image smoothing* refers to any spatial filtering producing an output image which spatially simplifies and approximates the input image. Image smoothing suppresses small image details and enhances large or coarse image structures.

90.   A *scale space image* is an image in which each pixel's value is a function indicating for each standard deviation σ the value at the pixel's position of the convolution of the image with a Gaussian kernel having standard deviation σ.

91.   *Scale space structure* refers to that analysis of a scale space image in which each pixel's value is a function specifying for each possible standard deviation σ whether the pixel contains a zero crossing of some combination of a fixed order of spatial partial derivatives evaluated at the pixel's position.

7. MORPHOLOGIC OPERATORS

92.   *Mathematical morphology* refers to an area of image processing concerned with the analysis of shape. The basic morphologic operations consist of dilating, eroding, opening, and closing an image with a structuring element.

93.   The *structuring element* of a morphologic operator is a function defined on the domain of the spatial pattern of the morphologic operator and whose value at each pixel of the domain is the weight or coefficient employed by the morphologic operator at that pixel position. The structuring element of a morphologic operator has a role in morphology exactly analogous to the role of the kernel in a convolution operation.

94.   *Dilating* an image I by a structuring element s having support or domain S produces a dilated image denoted by $I \oplus s$ which is defined by

$$(I \oplus s)(r, c) = \max_{(i, j) \in S} \{I(r - i, c - j) + s(i, j)\}$$

Dilating is a commutative, associative, translation invariant, and increasing operation. Dilating is the dual operation to eroding.

95.   *Eroding* an image I by a structuring element s having support or domain S produces an eroded image denoted by $I \ominus s$ which is defined by

$$(I \ominus s)(r, c) = \min_{(i, j) \in S} \{I(r + i, c + j) - s(i, j)\}$$

Eroding is a translation invariant and increasing operation. It is the dual operation to dilating.

96.   *Opening* an image I with a structuring element s produces an opened image denoted by $I \bigcirc s$ which is defined by

$$I \bigcirc s = (I \ominus s) \oplus s$$

Opening is an increasing, anti-extensive, and idempotent operation. It is the dual operation to closing. Opening an image with a disk shaped structuring element smooths the contour, breaks narrow isthmuses, and eliminates islands and capes smaller in size or width than the disk structuring element.

97.   *Closing* an image I with a structuring element s produces a closed image denoted by $I \bullet s$ which is defined by

$$I \bullet s = (I \oplus s) \ominus s$$

Closing is an increasing, extensive, and idempotent operation. It is the dual operation to opening. Closing an image with a disk shaped structuring element smooths the contours, fuses narrow breaks and long thin gulfs, eliminates holes smaller in size than the disk structuring element and fills gaps on the contour.

98.   A *thinning operator* is a symbolic image neighborhood operator which deletes, in some symmetric way, all the interior border pixels of a region which

do not disconnect the region. Successive applications of a thinning operator reduces a region to a set of arcs which constitute a *skeleton* of the region.

99. A *thickening operator* is a symbolic image neighborhood operator which in some symmetric way aggregates all background pixels near enough to a region into the region.

8. HOUGH TRANSFORM

100. The *discrete Radon transform* $R : Q \rightarrow [0, \infty)$ of a function $I : X \rightarrow [0, \infty)$ relative to a functional form $F : X \times Q [0, \infty)$ is defined by

$$R(q) = \sum_{\{x \in X | F(x,q) = 0\}} I(x)$$

101. The *Hough transform* $H : \mathcal{B} \rightarrow [0, \infty)$ of a function $I : X \rightarrow [0, \infty)$ relative to a functional form $F : X \times Q [0, \infty)$, where $\mathcal{B}$ is a partition of the parameter space Q, is defined by

$$H(B) = \sum_{\{x \in X | \text{for some } q \in B, F(x, q) = 0\}} I(x), \; B \in \mathcal{B}$$

When I has the form $I(x) = 1$, for every $x \in X_0$, where $X_0 \subseteq X$, and $I(x) = 0$ elsewhere, the Hough transform of X_0, relative to a functional form $F : X \times Q \rightarrow [0, \infty)$, has the simple form defined by

$$H(B) = \#\{x \in X_0 \mid \text{for some } q \in B, F(x, q) = 0\}$$

The Hough transform is a transform which can aid in the detection of image arcs of a given shape or form or 3D object shapes. Each shape or form has some free parameters which when specified precisely define the arc, shape or form. The shape having free parameter q corresponds to the set $\{x \in X | F(x, q) = 0\}$. The free parameters constitute the transform domain or the parameter space of the Hough transform. Depending on the information available to the Hough transform, each neighborhood of the image or object surface being transformed will map to a point or a set of points in the Hough parameter space. The Hough transform discretizes the Hough parameter space into bins and counts for each bin how many neighborhoods on the image or object surface has one of its transformed points lie in the volume assigned to the bin.

102. The *Gaussian sphere* refers to a unit sphere and its associated spherical coordinate system. The quantities usually represented on the Gaussian sphere are orientation vectors.

103. *Gradient space* is a two-dimensional space whose axes represent the first order partial derivatives of a surface of the form $z = f(x, y)$. Each point in gradient space corresponds to the orientation of a possible surface normal.

104. The *extended Gaussian image* or *orientation histogram* of a 3D object is a two-dimensional histogram or Hough transform of the surface normal orientations of the object. It is computed by tessellating the surface of a sphere into cells and assigning to each cell a value which estimates the total area of the object's surface having surface normal orientation which falls within the cell.

9. DIGITAL GEOMETRY

105. The *Euclidean distance* between two points $p = p_1, \ldots, p_N)$ and $q = (q_1, \ldots, q_N)$ is defined by

$$d(p, q) = \sqrt{\sum_{n=1}^{N} (p_n - q_n)^2}$$

106. The *block* or *city distance* between two points $p = (p_1, \ldots, p_N)$ and $q = (q_1, \ldots, q_N)$ is defined by

$$d(p, q) = \sum_{n=1}^{N} |p_n - q_n|$$

107. The *square* or *max distance* between two points $p = (p_1, \ldots, p_N)$ and $q = (q_1, \ldots, q_N)$ is defined by

$$d(p, q) = \max_{n=1, \ldots, N} |p_n - q_n|$$

108. The *distance transform* of a binary image is an image having in each pixel's position its distance from the nearest binary zero pixel of the input image. Distance can be city block distance, Euclidean distance, or square distances.

109. A *figure F* or a *subimage F* in a continuous or digital image *I* is any function *F* whose domain is some subset *A* of the set of spatial coordinates or resolution cells, whose range is the set *G* of image intensities and which is defined by $F(r, c) = I(r, c)$ for any (r, c) belonging to *A*.

110. A *region R* of an image is any subset of resolution cells in the spatial domain of the image.

111. A neighboring pair of pixels are said to be 4-*connected* if they share a common side. A neighboring pair of pixels are said to be 8-*connected* if they share a common side or a common corner.

112. A region *R* is *connected* if there is a path between any two resolution cells contained in *R*. More precisely, *R* is 4-connected (8-connected) if for each pair of resolution cells (r, c) and (u, v) belonging to *R*, there exists some sequence $\langle (a_1, b_1), (a_2, b_2), \ldots, (a_m, b_m) \rangle$ of resolution cells belonging to *R* such

that $(r, c) = (a_1, b_1)$, $(u, v) = (a_m, b_m)$, and (a_i, b_i), is 4-connected (8-connected) to (a_{i+1}, b_{i+1}), $i = 1, 2, \ldots, m - 1$.

113.   A *blob* or *connected component* is a maximal sized connected region.

114.   A *digital straight line segment* between resolution cells (r_1, c_1) and (r_2, c_2) is that subset of all pixels such that some part of the line segment joining (r_1, c_1) and (r_2, c_2) has a non-empty intersection with the pixel's area.

115.   A region R is *convex* if for every pair of resolution cells in R, R contains the digital straight line segment which joins the pair of resolution cells.

116.   A pixel is an *interior border pixel* of a region R if the pixel belongs to R and neighbors a pixel outside of R.

117.   A pixel is an *exterior border pixel* of a region R if the pixel does not belong to R and neighbors a pixel belonging to R.

118.   A pixel is an *interior pixel* of a region R if every pixel it neighbors belongs to R.

119.   A *simple boundary* is an oriented closed curve which does not touch or cross itself. Pixels which are on the inside of a simple boundary constitute a connected region having no holes.

120.   The *bounding contour* of a region R consists of the simple boundary which surrounds the pixels of R.

121.   A set of pixels H constitutes a *hole* of a region R if H is a maximal connected set of pixels which do not belong to R but which are surrounded by R.

122.   The *border* or *boundary* of a connected region R consists of its bounding contour and the (possibly empty) set of simple boundaries each of which surrounds the pixels belonging to some hole of R.

123.   *Boundary following* refers to the sequential procedure by which the chain of the boundary pixels of a region can be determined.

124.   A *concurve* is a continuous curve, usually representing a blob boundary, consisting of a connected chain of simply described arcs.

125.   The *minimum perimeter polygon* of a digital curve C is the polygon of shortest length whose digitization is C. The shape of the minimum perimeter polygon is often similar to the general perceived shape of the digital curve.

126.   *Contour tracing* is a searching or traversing process by which the bounding contour of a blob can be identified.

10.   2D SHAPE DESCRIPTION

127.   The *perimeter* of a connected region R is the length of the bounding contour of R.

128.   The *area A* of a region R is defined by

$$A = [\#R] \cdot s$$

where s is the scale factor which specifies the area of a pixel.

129.   The *centroid* $(\bar{r}, \bar{c})$ of a region R is the center of mass of the region. It is the mean (row, column) position for all pixels in the region and is given by

$$\bar{r} = \frac{1}{\#R} \sum_{(r,c)\in R} r, \; \bar{c} = \frac{1}{\#R} \sum_{(r,c)\in R} c$$

130.   The (j, k)th *moment* M_{jk} of a digital shape S is given by

$$M_{jk} = \sum_{(r,c)\in S} r^j c^k$$

The center of gravity $(\bar{r}, \bar{c})$ of S can be expressed in terms of the moments of S:

$$\bar{r} = \frac{M_{10}}{M_{00}} \qquad \bar{c} = \frac{M_{01}}{M_{00}}$$

131.   The (j, k)th *central moment* μ_{jk} of a digital shape S is given by

$$\mu_{jk} = \sum_{(r,c)\in S} (r - \bar{r})^j (c - \bar{c})^k$$

132. (j, k)th *normalized central moment* of S is given by

$$\eta_{jk} = \frac{\mu_{jk}}{\mu_{00}^\gamma}, \qquad \text{where} \quad \gamma = \frac{j + k}{2} + 1$$

133.   *Rotation invariant moments* of S are given by

$$\phi(1) = \eta_{20} + \eta_{02}$$

$$\phi(2) = (\eta_{20} - \eta_{02})^2 + 4\eta_{11}^2$$

$$\phi(3) = (\eta_{30} - 3\eta_{12})^2 + (3\eta_{21} - \eta_{03})^2$$

$$\phi(4) = (\eta_{30} + \eta_{12})^2 + (\eta_{21} + \eta_{03})^2$$

$$\phi(5) = (\eta_{30} - 3\eta_{12})(\eta_{30} + \eta_{12})[(\eta_{30} + \eta_{12})^2 \\ - 3(\eta_{21} + \eta_{03})^2] \\ + (3\eta_{21} - \eta_{03})(\eta_{21} + \eta_{03})[3(\eta_{30} + \eta_{12})^2 \\ - (\eta_{21} + \eta_{03})^2]$$

$$\phi(6) = (\eta_{20} - \eta_{02})[(\eta_{30} + \eta_{12})^2 - (\eta_{21} + \eta_{03})^2] \\ + 4\eta_{11}(\eta_{30} + \eta_{12})(\eta_{21} + \eta_{03})$$

$$\phi(7) = (3\eta_{21} - \eta_{03})(\eta_{30} + \eta_{12})[(\eta_{30} + \eta_{12})^2 \\ - 3(\eta_{21} + \eta_{03})^2] \\ - (\eta_{30} - 3\eta_{12})(\eta_{21} + \eta_{03})[3(\eta_{30} + \eta_{12})^2 \\ - (\eta_{21} + \eta_{03})^2]$$

134. The *Euler number* of a region is the number of its connected components minus the number of its holes.

135. The *compactness* of a blob can be measured by the length of its perimeter squared divided by its area or alternatively measured by the standard deviation of the radii from the centroid to the boundary divided by the mean radius. The clasical measure of perimeter squared divided by area has the disadvantage that in the digital domain it takes its smallest value not for a digital circle but for a digital octagon or diamond depending on whether 8-connectivity or 4-connectivity is used in calculating the perimeter.

136. The *bounding rectangle* of a region R is a rectangle which circumscribes R. It has its sides aligned with the row and column directions, its leftmost side aligning with the lowest numbered column of R, its rightmost side aligning with the highest numbered column of R, its topmost side aligning with the lowest numbered row of R, and its bottommost side aligning with the highest numbered row of R.

137. An *extremal pixel* of R is a pixel of R having from among all pixels in R either

(a) An extremal row coordinate value r and an extremal column coordinate value taken from among all the column positions c such that $(r, c) \in R$, or
(b) An extremal column coordinate value c and an extremal row coordinate value taken from among all the row positions r such that $(r, c) \in R$.

A region may have as many as eight distinct extremal points, each of which must be lying on the bounding rectangle of the region. Extremal pixels can be used to represent the areal extent of a region and to infer the dominant axis length and orientation of the region.

138. The *second moment matrix*

$$\Sigma = \begin{pmatrix} \mu_{rr} & \mu_{rc} \\ \mu_{rc} & \mu_{cc} \end{pmatrix}$$

of a region R is defined by

$$\mu_{rc} = \frac{s_r^2}{A} \sum_{(r,c) \in R} (r - \bar{r})^2$$

$$\mu_{rc} = \frac{s_r s_c}{A} \sum_{(r,c) \in R} (r - \bar{r})(c - \bar{c})$$

$$\mu_{cc} = \frac{s_c^2}{A} \sum_{(r,c) \in R} (c - \bar{c})^2$$

where A is the area of the region, s_r is the row scale factor, and s_c is the column scale factor, $\bar{r}$ is the row centroid, and $\bar{c}$ is the column centroid.

139.   The *elongation* or *elongatedness* of a blob or connected region can be measured in a variety of ways. One technique is to use the ratio of the length of the maximum length chord in the blob to the length of its maximum length perpendicular chord. A second technique is to use the square root of the ratio of the largest to smallest eigenvalue of the second central moment matrix of the blob. A third technique is to use the ratio of the largest distance between an opposing pair of extremal points to the distance between that opposing pair of extremal points having next to largest distance.

140.   The *symmetric axis* or *medial axis* of a blob is a subset of blob pixels which are the centers of maximal lines, squares or disks which are contained in the blob. Associated with each pixel which is part of a symmetric axis may also be additional information such as the size of the maximal line or square or the radius of the maximal disk of which it is the center.

141.   The *connected component* operator has as input a binary image and produces as output an image in which each binary one pixel is given a unique label of the maximally connected component of pixels having a binary one value to which it belongs.

142.   In *connected component analysis* or *blob analysis*, the position and shape properties of each connected component are measured. Typical shape properties include area, perimeter, number of holes, bounding rectangle, extremal points, centroid, second moments, and orientation derived from second moments or extremal points. The connected components are then identified or classified by a decision rule on the basis of its measured properties.

143.   *Signature analysis* of a binary image analyses the binary image in terms of its projections. Projections can be vertical, horizontal, diagonal, circular, radial, spiral, or general projections. The analysis consists of computing the projections, segmenting each projection, and taking property measurements of each projection segment. Signature analysis may also use the projection segmentation to induce a segmentation of the image.

144.   A *binary image projection* is the histogram of the gray scale image produced by masking a projection index image with the given binary image. Each pixel of the *projection index image* contains a number which is the index of the projection bin to which the pixel belongs. A histogram of the masked projection index image then contains in projection bin i the number of pixels which on the binary image have binary value one and which on the projection index image have index value i.

11.   CURVE AND IMAGE DATA STRUCTURES

145.   The *Fourier descriptors* of a closed planar curve are the coefficients of the Fourier series of the spatial positions of the curve as a function of arc length. Typically the low frequency coefficients are the ones of greatest interest.

146.   *Iterative end point curve fitting* refers to an iterative process of segmenting a curve into a set of piecewise linear segments which approximate the curve. The process begins by constructing a straight line between the end points of the curve. If the furthest distance between the curve and the straight line is less than a specified tolerance then the approximation is considered to be suitable and the curve segment is divided no further. If the furthest distance between the curve and the straight line is greater than a specified tolerance, then the approximation is considered to be not suitable. The curve is then divided into two segments at this furthest distance point and the straight line fitting process independently continues on each segment.

147.   The *chain code* representation of a digital arc or blob boundary is a sequence in which each element is a symbol representing the vector joining two neighboring pixels of the digital arc or blob boundary. The most common chain code uses the symbols 0 to 7 to represent the vectors $(0, 1)$, $(-1, 1)$, $(-1, 0)$, $(-1, -1)$, $(0, -1)$, $(1, -1)$, $(-1, -1)$, and $(-1, 0)$ of row column coordinates, which can join two neighboring pixels. More complex chain codes have more symbols and can represent the vector joining two more distant pixels which define the beginning and ending of a digital straight line segment which is part of a digital arc or blob boundary.

148.   A *quadtree* is a tree data structure which represents an image. Each node of the quadtree represents a square subset of the image's spatial domain. The root node of the quadtree represents the spatial domain of the entire image. If all the pixels of the spatial domain subset represented by a node have the same value, then the value of the node is the value of the pixels in the subset. Such a node is called a pure node. If the node is a mixed or impure node, then the square represented by the node is partitioned into four quadrants and the node has four children nodes, one child node for each quadrant. If the image being represented is a binary image, then the corresponding quadtree is called a *binary quadtree*. If the image being represented is a gray scale image, then the corresponding quadtree is called a *gray scale quadtree*.

149.   A *curve pyramid* consists of a sequence of symbolic images representing curves at multiple resolutions. The main operation in a bottom-up construction consists of locally connecting the short segments of the curve into longer ones. If the segments are described by binary "curve relations" of the labeled sides of a square cell, the concatenation of the short segments can be achieved formally by taking the transitive closure of the curve relations. Overlapping pyramids are necessary if the resulting pyramid is to have the "length reduction property": long curves with many segments survive to high levels, whereas short curves disappear after a few reduction steps.

150.   An *octree* is a data in a tree data structure which represents a function defined as a three-dimensional space volume. Each node of the octree represents a cube subset of the volume. The root node of the quadtree represents the entire

volume. If all the voxels represented by a node have the same function values, then the value of the node is their function value. Such a node is called a pure node. If the node is a mixed or impure node, then the cube represented by the node is partitioned into eight volume octants and the node has eight children, one child for each octant. If the function is binary, then the corresponding octree is called a *binary octree*. The binary octree is useful for representing three-dimensional volumes. If the function being represented is a non-binary, such as a real or integer valued function, then the corresponding octree is called a *gray scale octree*.

151. *Run length encoding* is a way to compactly represent binary images. There are a variety of run length encoding formats. Each format has a way of representing the starting column position of a maximally long horizontal string of binary one valued pixels as well as the number of pixels in a run. Many vision systems which recognize objects from their binary images use run length encoding to reduce the volume of data to be processed.

152. A *generalized cone* or *generalized cylinder* is a data structure for volumetric representation of a 3D object. The volume is generated by sweeping an arbitrarily shaped cross section along a 3D curve called the *generalized cone axis* or *generalized cylinder spine*. The axis passes through the centroids of the cross sections and is at a fixed angle (usually orthogonal) to them. The cross sectional shape is permitted to have some free parameters such as size or elongation. These values are specified for each axis point by the *cross section function* or the *sweeping rule*.

153. *Constructive solid geometry* is a mechanism for representing three-dimensional volumes by a constructive process which begins with simple shaped volumes and which are combined and subtracted from each other by the set of operations consisting of union, intersection and set difference.

154. A *boundary surface description* or *boundary representation* of a three-dimensional object or volume is a representation which contains each of the surface boundaries of the volume. Each surface boundary is represented in terms of simply described pieces, each of which has its own arc boundary. Each arc itself is represented in terms of simply described pieces which begin and terminate at end points or vertices.

155. A *superquadric* is a closed surface spanned by a vector whose x, y, and z components are specified as functions of the angles η and ω via the spherical product of two two-dimensional parametrized curves

$$h = \begin{pmatrix} h_1(\eta) \\ h_2(\eta) \end{pmatrix}$$

and

$$m = \begin{pmatrix} m_1(\omega) \\ m_2(\omega) \end{pmatrix}$$

which come from one of three basic trigonometric forms. The *spherical product* of h with m scaled by the vector $\begin{pmatrix} a_1 \\ a_2 \\ a_3 \end{pmatrix}$ is defined by

$$h \otimes m = \begin{pmatrix} a_1 h_1(\eta) m_1(\omega) \\ a_2 h_1(\eta) m_2(\omega) \\ a_3 h_2(\omega) \end{pmatrix}$$

156. A *strip tree* is a binary tree data structure for hierarchically representing a planar arc segment. The root node of the tree represents the minimal sized rectangle which bounds the arc. Each non-terminal node of the tree splits the arc which its bounding rectangle approximates into two continuous pieces. Its two children nodes then each contain the minimal sized rectangle which bounds the arc segment piece belonging to it. Terminal nodes of the tree have bounding rectangles which are sufficiently close to the arc segment they contain.

157. The *primal sketch* is a data structure for representing gray level intensity changes, their geometrical distribution, and organization in each image neighborhood. Primitives of the primal sketch include zero-crossings, blobs, terminations and discontinuities, edge segments, virtual lines, groups, curvilinear organizations, and boundaries.

158. In *facet model* image processing, the digital image's pixel values are regarded as noisy discretized sampled observations of its underlying and unknown gray tone intensity surface. Any operation to be performed on the image is defined in terms of this underlying gray tone intensity surface. Thus, in order to do any processing, the underlying intensity surface must be estimated. This requires a model which describes what the general form of the surface would be in any image neighborhood if there were no noise. To estimate the surface from the neighborhood around a pixel then amounts to estimating the free parameters of the general form. Useful image processing operations which then can be performed using the facet model processing approach include gradient edge detection, zero-crossing edge detection, image segmentation, line detection, corner detection, 3D shape estimation from shading, and determination of optic flow.

12. TEXTURE

159. A *discrete tonal feature* on a digital image is a connected set of resolution cells all of which have the same or almost the same image intensity.

160. *Texture* is concerned with the spatial distribution of the image intensities and discrete tonal features. When a small area of the image has little variation of discrete tonal features, the dominant property of that area is gray tone. When a small area has wide variation of discrete tonal features, the dominant property of that area is texture. There are three things crucial in this distinction:

(1) the size of the small areas, (2) the relative sizes of the discrete tonal features, and (3) the number of distinguishable discrete tonal features. Texture can be described along dimensions of uniformity, density, coarseness, roughness regularity, intensity and directionality.

161.   A *texel*, short for texture element, is a triplet whose first component is the (row, column) location of a small neighborhood, whose second component is the size of the neighborhood, and whose third component is the vector of texture properties.

162.   The *gray level dependence matrix* or *gray level co-occurrence matrix* characterizes the micro-texture of an image region by measuring the dependence between pairs of gray levels arising from pixels in a specified spatial relation. For gray level pair (i, j) the gray level dependence matrix P for region R of image I has value $P(i, j)$ where $P(i, j)$ is the number of pairs of pixels in the region having the desired spatial relation where the first pixel has gray level i and the second pixel has gray level j. If S designates the set of all pairs of pixels in the desired spatial relation, then $S \subseteq R \times R$ and

$$P(i, j) = \#\{((r_1, c_1), (r_2, c_2,)) \in S | I(r_1, c_1) = i$$

and

$$I(r_2, c_2) = j\}$$

The gray level dependence matrix P can be normalized. One normalized form which produces a joint probability is given by

$$P_1(i, j) = \frac{P(i, j)}{\#S}$$

A second normalized form which produces a conditional probability is given by

$$P_2(i, j) = \frac{P(i, j)}{\sum_i P(i, j)}$$

163.   *Statistical texture measures* include the moments of the gray levels of the given region, typically the variance, the slewness and the kurtosis.

164.   The *gray level difference histogram* at a distance d of an image I is the histogram of values

$$\langle I(r, c) - I(r', c') | : (r - r')^2 + (c - c')^2 = d^2 \rangle$$

165.   A *structural texture description* is given by a set of primitives and placement rules which govern the stochastic spatial relation between them.

166.   *Fourier related texture descriptions* include the power spectrum and autocorrelation function.

13. SEGMENTATION

167. *Image segmentation* is a process which typically partitions the spatial domain of an image into mutually exclusive subsets, called regions, each one of which is uniform and homogeneous with respect to some property such as tone or texture and whose property value differs in some significant way from the property value of each neighboring region. Regions produced by an image segmentation process using image intensity as a property value produce regions which are called discrete tonal features.

168. *Region growing* refers to a sequential image segmentation process in which pixels are successively added to incomplete regions or initiate new regions when it is not appropriate to make them part of any of the existing incomplete regions. There are three basic kinds of region growing: region tracking, region aggregation, and region merging.

(a) 169. In *region tracking*, the image is scanned in raster scan order. The similarity of each pixel is compared with the regions to which the already processed 4- or 8-connected neighboring pixels belong. If one of these already processed neighboring pixels belongs to a region of sufficient similarity to the current pixel, then the pixel is added to the region. If the regions to which all the already processed pixels belong are dissimilar from the current pixel, then the current pixel initiates a new region.

(b) 170. In *region aggregation*, seed pixels are first found which serve as prototype pixels for the regions in the desired segmentation. Then in a sequential fashion, pixels having neighbors in any incomplete region join themselves into the region if they are similar enough. The aggregation process continues until all pixels are part of some region.

(c) 171. In *region merging*, the neighboring regions of an initial segmentation are successively merged together if they have similar enough properties. After each merging iteration, properties of the new regions are recomputed. The merging iterations continue until the properties of each pair of neighboring regions is sufficiently different from each other.

172. *Contour filling* is the process by which all pixels inside a blob defined by its bounding contour(s) are marked with the same unique label.

14. MATCHING

173. *Template matching* is an operation which can be used to find out how well a template subimage matches a window of a given image. The degree of matching is often determined by translating the template subimage all over the given image and for each position evaluating the cross-correlation or the sum of the squared or absolute image intensity differences of corresponding pixels. Tem-

plate matching can also be used to best match an observed measurement pattern with a prototype pattern.

174.   *Matched filtering* is a template matching operation done by using the magnitude of the cross-correlation function to measure the degree of matching.

175.   *Image matching* refers to the process of determining the pixel by pixel, arc by arc, or region by region correspondence between two images taken of the same scene but with different sensors, different lighting, or a different viewing angle. Image matching can be used in the spectral/temporal pattern classification of remote sensing, or in determining corresponding points for stereo, tracking, change analysis, and motion analysis. In one group of approaches, subimages of one image are translated over a second image. For each translation, the differences between appropriately transformed gray tone intensities and/or edges are measured. In this signal level approach, the unit being operated with is the pixel, since the measured difference is between values of pixels in two images.

176.   In *symbolic registration* or *symbolic matching*, higher level units are worked with. For example the scene can be segmented and a region matching then performed using segment features such as area, position, perimeter^2/area, orientation, length to width ratio, area/area of minimum bounding rectangle, area/area of bounding ellipse, gray tone intensity or color of segment, and number of corresponding neighbors. Because this matching uses a higher level unit than the pixel, it is called symbolic matching or symbolic registration.

177.   In *feature point matching*, selected points of each image are first determined, on the basis of the distinctive image values in a neighborhood or on the basis of an intersection between two feature lines. The location of each point can be to subpixel precision. After the locations of distinctive points are determined, a correspondence process associates as many as possible selected points of one image with selected points on the second image. The correspondence is based on similarity of the feature characteristics of the points.

178.   A *structural description* is a relational representation of a 2D or 3D entity. It consists of a set of primitives each having its own attribute description and a set of named relations which consist of tuples whose components are primitives which stand in the relation specified by the relation name.

A function $h : A \rightarrow B$ is a *relation homomorphism* from N-ary relation $R \subseteq A^N$ to N-ary relation $S \subseteq B^N$ if

$$R \bigcirc h \subseteq S$$

where $R \bigcirc h = \{(b_1, \ldots, b_N) \in B^N|$ for some $(a_1, \ldots, a_N) \in R, b_n = h(a_n), n = 1, \ldots, N\}$.

179.   The relation R is said to *match* relation S if there exists a relation homomorphism satisfying

$$R \bigcirc h = S \qquad \text{and} \qquad S \bigcirc h^{-1} = R$$

180. *Relational matching* refers to the process by which it is determined whether two relations match or do not match. *Structural matching* is a matching which establishes a correspondence or homomorphism from the primitives of one structural description to the primitives of a second structural description. In the ideal match a tuple of primitives which stand in a given relation in the first description will have its corresponding tuple of primitives stand in the same given relation in the second structural description.

181. The *local feature focus method* is a model based object recognition and location technique in which one feature, referred to as the focus feature, on the image is found and is used along with the object model to predict what other nearby features might be. After locating a set of features a relational matching is performed to infer a consistent correspondence between all the located image features and the object features. Once a consistent correspondence has been found, the object position orientation can be hypothesized. The hypothesized object position and orientation is then verified by template matching.

182. *Relaxation* refers to any computational mechanism which employs a set of locally interacting parallel processes, one associated with each image unit, which in an iterative fashion update each unit's current labeling in order to achieve a globally consistent interpretation of the image data. In *discrete relaxation*, the assessment of each unit's current state consists of that subset of labels not yet ruled out. In *probabilistic relaxation* the assessment of each unit's current state is a probability function associating with each possible label a probability of its being the correct state.

15. LOCALIZATION

183. An *interest operator* is a neighborhood operator which is designed to locate, with high spatial accuracy, pixel positions, or subpixel positions, whose central neighborhoods have distinctive gray tone patterns. Such neighborhoods are typically those whose autocorrelation functions falls off rapidly. Interest operators are usually used to mark pixels on a pair of images taken of the same scene but with some shift of either camera position or object from one image to another. The marked pixels are then the candidate pixels which are input to an algorithm which establishes correspondences between the selected pixels.

184. An *area of interest operator* is an operator which delineates regions of an image which have potentially interesting patterns. These delineated *regions of interest* are the ones which must be further processed.

185. *Screening* is an operation of selecting photographs or images containing areas of potential interest from those in a set of photographs, some or most of which contain no interesting areas.

186. In *area analysis*, the area of the image containing the objects or entities to be processed is located by some simple algorithm and a more complex

processing algorithm is only applied in the located area. This strategy of processing can often increase execution speed. The algorithm locating the area to be processed is called the *focus of attention mechanism*.

16. GENERAL IMAGE PROCESSING

187.   *Preprocessing* is an operation applied before pattern identification is performed. Preprocessing produces, for the categories of interest, pattern features which tend to be invariant under changes such as translation, rotation, scale, illumination level, and noise. In essence, preprocessing converts the measurement patterns to a form which allows a simplification in the decision rule. Preprocessing can bring into registration, bring into congruence, remove noise, enhance images, segment target patterns, detect, center, and normalize objects of interest.

188.   *Registering* or *registration* is the translation or translation and rotation alignment process by which two images of like geometries and of the same set of objects are positioned coincident with one another so that corresponding points of the imaged scene appear in the same position on the registered images. In this manner, corresponding image values can be made to represent the sensor output for the same object point over the full image frame.

189.   *Congruencing* is the geometric warping process by which two images of different geometries, but of the same set of objects, are spatially transformed so that the size, shape, position, and orientation of any object on one image is made to be the same as the size, shape, position, and orientation of that object on the other image.

190.   *Image compression* is an operation which preserves all or most of the information in the image and which reduces the amount of memory needed to store an image or the time needed to transmit an image.

191.   *Image restoration* is a process by which a degraded image is restored, as clearly or as best as possible, to its ideal condition. Perfect image restoration is possible only to the extent that the degradation transform is mathematically invertible. Common forms of restoration include *inverse filtering*, *Wiener filtering*, and *constrained least squares filtering*.

192.   *Image reconstruction* refers to the process of reconstructing an image from a set of its projections. The projections may be taken along a set of parallel rays, in which case they are called *parallel projections*, or they may be taken along a set of rays emanating from a point, in which case they are called *fan beam projections*. The most commonly employed reconstruction techniques are the *filtered back projection* and the *algebraic reconstruction* techniques. Image reconstruction techniques are important in computerized tomography, nuclear medicine, and ultrasonic imaging.

193.   *Image enhancement* is any one of a group of operations which improve the detectability of objects. These operations include, but are not limited to,

contrast stretching, edge enhancement, spatial filtering, noise suppression, image smoothing, and image sharpening.

194. *Image processing* or *picture processing* encompasses all the various operations which can be applied to image data. These include, but are not limited to, image compression, image restoration, image enhancement, preprocessing, quantization, spatial filtering, matching, and recognition techniques.

195. *Interaction image processing* is carried out by an operator or analyst at a console with a means of accessing, preprocessing, feature extracting, classifying, identifying and displaying the original imagery or the processed imagery for subjective evaluation and further interactions.

17. VISION

196. *Structured light* refers to a technique of projecting a carefully designed light pattern on a scene and viewing the scene from a different direction. Usually the pattern consists of successive planes of light at different positions and orientations. Those pixels which image a surface patch which is lit by a known light pattern have sufficient information to determine the 3D coordinates of the surface patch since the light pattern is designed so that the line of sight passing through the pixel and the lens will intersect the known light pattern in a unique point. For stereo matching purposes, the structured light pattern may be ''unstructured'' in the sense of being a texture pattern or consisting of random stipples.

197. *Light striping* refers to a simple form of structured lighting in which the light pattern consists of successive planes of light which are all parallel.

198. *Stereopsis* refers to the capability of determining the depth of a 3D point by observing the point on two perspective projection images taken from different positions.

199. A *stereo image pair* refers to two perspective projection images taken of the same scene from slightly different positions. The common area appearing in both images of the stereo pair is usually 40% to 80% of the total image area.

200. A point p on one image and a point q on a second image are said to form a *corresponding point pair* (p, q) if p and q are each a different sensor projection of the same 3D point. The *visual correspondence* problem consists of matching all pairs of corresponding points from two images of the same scene.

201. *Disparity* or *stereo disparity* refers to the difference in positions of the images of the same 3D point in two perspective projection images taken from different positions.

202. *Stereo matching* refers to the matching process by which corresponding points on a stereo image pair or identified.

203. *Triangulation* refers to the process of determining the (x, y, z) coordinates of a 3D point from the observed position of two perspective projections

of the point. The centers of perspectivity and the perspective projection planes are assumed known.

204. The *epipolar axis* of a stereo image pair is the line passing through the center of perspectivities of the image.

205. The two *epipoles* of a stereo image pair consist of one point on each of the perspective projection planes determined as the intersection of the image plane with the epipolar axis. For stereo image pairs having parallel perspective projection image planes, the epipoles are infinitely far to the left and right.

206. An *epipolar ray* is the line segment between an epipole and a point on the perspective projection image plane.

207. An *epipolar line* on one stereo image corresponding to a given point in another stereo image is the perspective projection on the first stereo image of the 3D ray which is the inverse perspective projection of the given point from the other stereo image.

208. An *epipolar plane* relative to a pair of stereo images is any plane determined by an observed 3D point, the position in 3D space of its perspective projection on the left stereo image and the position in 3D space of its perspective projection on the right stereo image. Every epipolar plane contains the epipolar axis and every plane which contains the epipolar axis is an epipolar plane.

209. An *occluding boundary* is a boundary appearing on an image due to a discontinuity in range or depth of an object in the observed scene.

210. For each fixed viewing position and point source of illumination, the *reflectance map* is a function defined on gradient space which specifies a surface's reflectivity. Thus, at each possible orientation of the surface normal (as encoded by the surface's first order partial derivatives) the surface's reflectance map specifies the reflection coefficient of the surface.

211. *Shape from shading* refers to the capability of determining the 3D shape characteristics of an object from the gray tone shading manifested by the object's surface on a perspective projection or orthographic projection image, that is, from its reflective map.

212. *Local shading analysis* refers to the capability of inferring the shape and predominant tilt of a section of an object's surface by the image intensities in a local neighborhood of a perspective or parallel projection view.

213. *Photometric stereo* refers to the capability of determining surface orientation by means of the shading variations present on two or more images taken of the same scene from the same position and orientation but with the light source in different positions.

214. The *motion field* or *image flow* is an image in which the value of each pixel is the projected translational velocity arising from a surface point of an object in motion relative to the camera. Each projected translational velocity vector is called an *optic flow vector*.

215. An *optic flow* or *optical flow* image is an image in which the value of each pixel is the estimated projected translational velocity arising from a surface point of an object in motion relative to the camera. Because the projected velocity may not be estimable at each pixel, there may be some pixels in an optic flow image having no optic flow information.

216. The *focus of expansion* of a motion field image arising from a moving camera and stationary scene is that point on the image at which the optic flow is zero and such that the optic flows of the neighboring points are directed away from it. In cases of relative motion toward the camera, there will be exactly one focus of expansion point in such a motion field image. In a motion field image arising from an object in relative motion to the camera, the focus of expansion is that point on the image having all the optic flow vectors arising from the moving object directed away from it.

217. The *focus of contraction* of a motion field image arising from a moving camera and stationary scene is that point on the image at which the optic flow is zero and such that the optic flows of the neighboring points are directed toward it. In cases of relative motion away from the camera, there will be exactly one focus of contraction point in such a motion field image. In a motion field image arising from an object in relative motion to the camera, the focus of contraction is that point on the image having all the optic flow vectors arising from the moving object directed toward it.

218. *Structure from motion* refers to the capability of determining a moving object's shape characteristics, and its position and velocity as well, from a sequence of two or more images taken of the moving object. Equivalently, if the camera is in motion, structure from motion refers to the capability of determining an object's shape characteristics, and its position and the camera's velocity, from a sequence of two or more images taken of the object by the moving camera. One fundamental kind of structure from motion problem is to determine the fixed position of M 3D points from a time sequence of N views and containing the 2D perspective projection of the M 3D points.

219. *Passive navigation* refers to the determination of the motion of a camera from a time varying image sequence.

220. *Dynamic scene analysis* refers to the analysis of time varying imagery. The purpose of the analysis may be to track moving objects, determine the motions of the objects, recognize the objects, determine the spatial positions of the objects at the time each image was obtained, or determine a shape description or characterization of one or more objects.

221. *Shape from texture* refers to the capability of determining the 3D shape characteristics of a homogeneously textured surface from the texture density variations manifested by the surface on a perspective projection or orthographic projection image.

222. *Surface reconstruction* refers to the process by which a 3D surface is analytically described by its boundary representation on the basis of processing a stereo image pair, a range map, or a time varying image sequence of the observed surface.

223. *Shape from contour* or *shape from shape* refers to the capability of inferring the 3D shape of an object from a 2D perspective projection view of a set of regularly marked contours on the object's surface.

224. The *2½D sketch* is a multiband image, each pixel providing information about the depth and surface orientation of the surface projected on it as well as providing an indication of the existence of a nearby depth discontinuity and an indication of the existence of a nearby surface discontinuity.

225. The *intrinsic scene characteristics* for each object surface point are: its depth from the image focal plane, its surface orientation, its reflectance, and its incident illumination.

226. An *intrinsic image* is a multiband image in which each pixel contains the predominant intrinsic scene characteristics of the surface patch projecting to its position. Hence each pixel of the intrinsic image specifies depth, surface orientation, reflectance, and incident illumination for the surface patch projecting to its position.

227. *Inverse optics* refers to the capability of inferring the 3D position and/ or the surface normal of each point on an object's surface and/or the surface shape from one or more perspective projection images of the object. Included in the techniques of inverse optics are stereo, photometric stereo, shape from shading, shape from texture, and structure from motion. Inverse optics techniques can be thought of as techniques which invert the perspective projection process and which, therefore, belong to the reconstructionist school of applied physics computer vision.

228. The *blocks world* refers to a world in which all objects have simple surfaces. The most common kind of objects in the blocks world are polyhedral objects.

229. A *view aspect* is a maximal connected region of viewpoint space having the property that when looking at a given object's center from any point of a view aspect, the resulting views are topologically identical.

230. An *aspect graph* is a graph in which the nodes are the view aspects and the arcs connect adjacent view aspects. The views represented by the nodes in the graph are all the stable views, characteristic views and principal views.

231. *3D vision* refers to the capability of a machine vision system to be able to infer some 3D characteristic of an object or object feature such as its position, dimensions, orientation or motion or some 3D characteristic of a point or an object surface such as its 3D position or surface normal orientation.

232. *Automatic visual inspection* or *automatic vision inspection* refers to an inspection process which uses a sensor producing image data and which uses

techniques from image processing, pattern recognition, or computer vision to measure and/or interpret the imaged objects in order to determine whether they have been manufactured within permitted tolerances. Automatic visual inspection systems usually integrate the technologies of material handling, illumination, image acquisition and special purpose computer hardware along with the appropriate image analysis algorithms into a system intended to be of practical use in the factory. Benefits from using automated visual inspection can include more accurate, reliable, repeatable, and complete quality assurance at a lower price and a higher speed of inspection than possible by manual labor.

233.   A *machine vision system* is a system capable of acquiring one or more images of an object, capable of processing, analysing and measuring various characteristics of the acquired images, and interpreting the results of the measurements in such a way that some useful decision can be made about the object. Functions of machine vision systems include locating, inspecting, gauging, identifying, recognizing, counting, and motion estimating.

234.   *Visual fixturing* or *visual pose determination* refers to the capability of inferring the position and orientation of a known object using a suitable object model and one or more cameras, range sensors, or triangulation based vision sensors.

235.   *Optical gauging* or *visual gauging* refers to the ability to measure specific positions or dimensions of a manufactured object by using non-contact light sensitive sensors to compare these measurements to preselected tolerance limits for quality inspection and sorting decisions. Gauging has wide application in manufacturing since it can determine the diameters of holes, openings, or cutouts, the widths of shafts, components, gaps, wires, or rods, and the relative locations of holes, folds, features, components, openings, or breaks.

236.   A gauging system can be either a *fixed inspection* or *flexible inspection* system. The fixed inspection system holds the part in a precision test fixture and has one or more sensors to take the required measurements. Flexible inspection utilizes sensors that are moved about the part being inspected, the motion being done along a programmed path trajectory. Flexible inspection gauging is sometimes called *robotic gauging*.

237.   A vision procedure is said to be *robust* if small changes in the assumed model on which the procedure or technique was developed produce only small changes in the result. Small fractions of the data which do not fit the assumed model and which in fact are very far from fitting the assumed model, constitute a small change in the assumed model. Data not fitting an assumed model may be due to rounding or quantizing errors, gross errors, or because the model itself is only an idealized approximation to reality.

238.   *Computer vision*, *image understanding*, or *scene analysis* is that combination of image processing, pattern recognition, and artificial intelligence technologies which focuses on the computer analysis of one or more images,

taken with a single/multiband sensor or taken in time sequence. The analysis recognizes, locates the position and orientation, and provides a sufficiently detailed symbolic description or recognition of those imaged objects deemed to be of interest in the three-dimensional environment. The computer vision process often uses geometric modeling and complex knowledge representations in an expectation or model based matching or searching methodology. The searching can include bottom up, top down, blackboard, hierarchical, and heterarchical control strategies.

239. The *bottom up* control strategy is an approach to problem solving that is *data driven*. It employs no object models in its early stages and only uses general knowledge about the world being sensed. In a computer vision system using a bottom up control strategy, the observed image data is interpreted and aggregated. The interpretations and aggregations are then successively manipulated and aggregated until a sufficiently high level description of the scene has been generated.

240. The *top down* control strategy is an approach to problem solving that is *goal-directed* or *expectation directed*. A form of solution is generated or hypothesized. Assuming the hypothesis is true and using the information in the knowledge data base, the inference mechanism then infers, if possible, some consistent set of values for the unknown variables or parameters. If a consistent set can be inferred, then the problem has been solved. If a consistent set cannot be inferred, then a new form of solution is generated or hypothesized. In a computer vision system using a top down control structure, the number or types of objects being sensed in the image is usually highly constrained and knowledge about the objects, relationships between objects, and object parts are all known. The system hypothesizes that the image shows a particular set of objects, infers values for parameters, and then tests to verify that the hypothesis is consistent with the observed data.

241. A *hierarchical* control strategy is an approach to problem solving in which the given problem is solved by dividing it up into a set of subproblems, each of which encapsulates an important or major aspect of the original problem. Then each subproblem is successively divided into more detailed subproblems. The refinement continues until the most refined subproblems can be solved directly.

242. A *blackboard* control strategy is an approach to problem solving in which the various components of the inference mechanism communicate with one another through a common working data storage area called the blackboard. When the blackboard has sufficient data to permit one component of the inference mechanism to make a deduction, the inference mechanism goes to work and writes its results on the blackboard where it becomes available for the other components of the inference mechanism. In this manner the inferred constraints are successively propagated and the required search is made more limited.

243. *Model based computer vision* is a computer vision process which employs an explicit model of the object to be recognized. Recognition proceeds in a top down manner by matching the object data structure inferred from the observed image to the model data structure.

244. *Knowledge based vision* refers to a computer vision process which has an image processing component, a reasoning or inference component, and a knowledge data base component. The knowledge data base stores information about the environment being imaged. The image processing component extracts primitive point, line, curve, and region information from an observed image. The reasoning or inference component is typically rule based and integrates the information produced by the image processing component with the information in the knowledge data base and reasons about what hypothesis should be next generated, what hypothesis should be next validated, what new information can be inferred from what has already been established, and what new primitives the image processing component should extract next.

18. PATTERN RECOGNITION

245. *Pattern recognition* techniques can be used to construct decision rules which enable units to be identified on the basis of their measurement patterns. Pattern recognition techniques can also be employed to cluster together units having similar enough measurement patterns. In *statistical pattern recognition*, the measurement patterns have the form of n-tuples or vectors. In *syntactic pattern recognition*, the measurement patterns have the form of sentences from the language of a phrase structure grammar. In *structural pattern recognition*, the measurements do not have the form of an n-tuple or vector. Rather, the unit being measured is encoded in terms of its parts and the relationships as well as properties of the parts.

246. *Pictorial pattern recognition* refers to techniques which treat the image as a pattern and either categorize the image or produce a description of the image.

247. The *unit* is the entity which is observed and whose measured properties constitute the measurement pattern. The simplest and most practical unit to observe and measure in the pattern recognition of image data is often the pixel (the gray tone intensity or the gray tone intensity n-tuple in a particular resolution cell). This is what makes pictorial pattern recognition so difficult, because the objects requiring analysis or identification are not single pixels but are often complex spatial formations of pixels.

248. A *measurement pattern* or *pattern* is the data structure of the measurements resulting from observing a unit.

249. A *measurement n-tuple* or *measurement vector* is the ordered n-tuple of measurements obtained from a unit under observation. Each component of the

n-tuple is a measurement of a particular quality, property, feature, or character-istic of the unit. In image pattern recognition, the units are usually picture el-ements or simple formations of picture elements and the measurement *n*-tuples are the corresponding gray tone intensities, gray tone intensity *n*-tuples, or prop-erties of formations of gray tone intensities.

250.   The *Cartesian product* of two sets *A* and *B*, denoted by $A \times B$, is the set of all ordered pairs where the first component of the pair is some element from the first set and the second component of the pair is some element from the second set. The Cartesian product of *N* sets can be defined inductively.

251.   *Measurement space* is a set large enough to include in it the set of all possible measurement patterns which could be obtained by observing some set of units.

252.   The *range set R_i* for the *i*th sensor, which produces the *i*th image in the multi-image set, is the set of all measurements which can be produced by the *i*th sensor. Simply, it is the set of all gray tone intensities which could possibly exist in the *i*th image. When the units are the pixels, measurement space *M* is the Cartesian product of the range sets of the sensors: $M = R_1 \times R_2 \times \ldots \times R_n$.

253.   Each unit is assumed to be one and only one given type. The set of types is called the set of *pattern classes* or *categories C*, each type being a par-ticular category.

254.   A *feature*, or *feature pattern*, or *feature n-tuple*, or *feature vector* or *pattern feature* is a *n*-tuple or vector whose components are functions of the ini-tial measurement pattern variables or some subset of the initial measurement pattern variables. Feature *n*-tuples or vectors are designed to contain a high amount of information relative to the discrimination between units of the types of categories in the given category set. Sometimes the features are predeter-mined and other times they are determined at the time the pattern discrimination problem is being solved. In image pattern recognition, features often contain information relative to gray tone intensity, texture, or region shape.

255.   *Feature space* is the set of all possible feature *n*-tuples.

256.   *Feature selection* is the process by which the features to be used in the pattern recognition problem are determined. Sometimes feature selection is called *property selection*.

257.   *Feature extraction* is the process by which an initial measurement pat-tern or some subset of measurement patterns is transformed to a new pattern feature. Sometimes feature extraction is called *property extraction.*

The word *pattern* can be used in three distinct senses: (1) as measurement pattern; (2) as feature pattern; and (3) as the dependency pattern or patterns of relationships among the components of any measurement *n*-tuple or feature *n*-tuple derived from units of a particular category and which are unique to those *n*-tuples, that is, they are dependencies which do not occur in any other category.

258. A *classifier* is a device or process that sorts patterns into categories or classes.

259. The *compactness hypothesis* states that the pattern measurements of a given class are nearer to other pattern measurements in the class than they are to pattern measurements from other classes.

260. The region of space occupied by pattern measurements coming from the same class or category is called a *class region*.

261. Two classes or categories are said to be *separable* if their class regions do not overlap. If for every class region there exists a hyperplane which separates the class region from all other class regions, the classes are said to be *linearly separable*.

262. A *prototype pattern* or *reference pattern* is the observable or characteristic measurement or feature pattern derived from units of a particular category. A category is said to have a prototype pattern only if the characteristic pattern is highly representative of the n-tuples obtained from units of that category.

263. A *data sequence* $S_d = \langle d_1, d_2, \ldots, d_j \rangle$ is a sequence of patterns derived from the measurement patterns or features of some sequence of observed units. d_1 is the pattern associated with first unit; d_2 is the pattern associated with the second unit; and d_j is the pattern associated with the jth unit.

264. A *decision rule* f usually assigns one and only one category to each observed unit on the basis of the sequence of measurement patterns in the data sequence S_d or on the basis of the corresponding sequence of feature patterns.

265. A *simple decision rule* is a decision rule which assigns a category to a unit solely on the basis of the measurements or features associated with the unit. Hence, the units are treated independently and the decision rule f may be thought of as a function which assigns one and only one category to each pattern in measurement space or to each feature in feature space.

266. A *hierarchical decision rule* is a decision rule in a tree form. In binary trees, each non-terminal node of the tree contains a simple decision rule which classifies patterns as belonging to its left child or to its right child. Each terminal node of the tree contains the assigned class or category of the observed unit.

267. A *compound decision rule* is a decision rule which assigns a unit to a category on the basis of some non-trivial subsequence of measurement patterns in the data sequence or in the corresponding sequence of feature patterns.

268. Provision can be made for a decision rule to *reserve judgment* or to *defer assignment* if the pattern is too close to the category boundary in measurement or feature space. With this provision, a deferred assignment is an assignment to the category of "reserved judgment".

269. A *category identification sequence* or *ground truth* $S_c = \langle c_1, c_2, \ldots, c_j \rangle$ is a sequence of category identifications obtained from some sequence of ob-

served units. c_1 is the category identification of the first unit; c_2 is the category identification of the second unit; and c_j is the category identification of the jth unit.

270. A *training sequence* is a set of two sequences: (1) the data sequence and (2) a corresponding category identification sequence. A training datum is the pair consisting of a pattern in the data sequence and the corresponding category identification in the category identification sequence. The training sequence is used to estimate the category conditional probability distributions from which the decision rule is constructed or it may be used to estimate the decision rule itself.

271. A *training procedure* is a procedure which uses the training sequence to construct a decision rule. It may operate by passing through the entire training sequence one time and construct the decision rule in a manner which is independent of the order in which the training data occurs in the training sequence. It may operate iteratively in which case it passes through the training sequence many times and after handling each training datum, it modifies or updates the decision rule. Such iterative training procedures may be affected by the order in which the training data occurs in the training sequence.

272. A *window training procedure* is an iterative training procedure in which each adjustment of the decision rule is made only when the training datum falls within a specified window, a subset of the pattern measurement space. Usually this subset or window contains the decision boundary.

273. An *error correcting training procedure* is an iterative sequential training procedure in which at each iteration the decision rule is adjusted in response to a misclassification of a training datum.

274. A classifier is said to *learn* if its iterative training procedure increases the classification performance accuracy of the classifier after each few iterations.

275. The *conditional probability* of a measurement or feature n-tuple d given category c is usually denoted by $P_c(d)$, or by $P(d|c)$, and is defined as the relative frequency or proportion of times the n-tuple d is derived from a unit whose true category identification is c.

276. A *distribution-free* or *non-parametric decision rule* is one which makes no assumptions about the functional form of the conditional probability distribution of the patterns given the categories.

277. A simple *maximum likelihood decision rule* is one which treats the units independently and assigns a unit u having pattern measurements or features d to that category c whose units are most probable to have given rise to pattern or feature vector d, that is, such that the conditional probability of d given c is highest.

278. A simple *Bayes decision rule* is one which treats the units independently and assigns a unit u having pattern measurements or features d to the category c whose conditional probability, given measurements d, is highest.

279. Let $\langle u_1, u_2, \ldots, u_j \rangle$ be a sequence of units with corresponding data sequence $\langle d_1, d_2, \ldots, d_j \rangle$ and known category identification sequence $\langle c_1, c_2, \ldots, c_j \rangle$. A simple *nearest neighbor decision rule* is one which treats the units independently and assigns a unit u of unknown identification and with pattern measurements or features d to category c_j where d_j is that pattern closest to d by some given metric or distance function.

280. A *discriminant function* $f_i(d)$ is a scalar function, whose domain is usually measurement space and whose range is usually the real numbers. When $f_i(d) \geq f_k(d)$, for $k = 1, 2 \ldots K$, then the decision rule assigns the ith category to the unit giving rise to pattern d.

281. A *linear discriminant function* f is a discriminant function of the form $f(d) = \sum_{j=1}^{n} a_i \delta_i + a_0$ where $d = (\delta_1, \delta_2, \ldots \delta_n)$ represents the measurement pattern.

282. A *quadratic discriminant function* f is a discriminant function of the form

$$f(d) = \sum_{i=1}^{n} \sum_{i=1}^{n} a_{11} \delta_1 \delta_1 + \sum_{i=1}^{n} a_i \delta_1 + a_0$$

283. A *decision boundary* between the ith and kth categories is a subset H of patterns in measurement space M defined by

$$H = \{d \in M | f_i(d) = f_k(d)\}$$

where f_i and f_k are the discriminant functions for the ith and kth categories.

284. A *hyperplane decision boundary* is the special name given to decision boundaries arising from the use of linear discriminant functions.

285. A *linear decision rule* is a simple statistical pattern recognition decision rule which usually treats the units independently and makes the category assignments using linear discriminant functions. The decision boundaries obtained from linear decision rules are hyperplanes.

286. The *pattern discrimination* problem is concerned with how to construct the decision rule which assigns a unit to a particular category on the basis of the measurement pattern(s) in the data sequence or on the basis of the feature pattern(s) in the data sequence.

287. *Pattern identification* is the process in which a decision rule is applied. If $S_u = \langle u_1, u_2, \ldots, u_i \rangle$ is the sequence of units to be observed and identified, and if $S_d = \langle d_1, d_2, \ldots, d_i \rangle$ is the corresponding data sequence of patterns, then the pattern identification process produces a category identification sequence $S_c = \langle c_1, c_2, \ldots, c_i \rangle$ where c_i is the category in C to which the decision rule assigns unit u_i on the basis of the j patterns in S_d. In general, each category in S_c can be assigned by the decision rule as a function of all the patterns in S_d. Sometimes pattern identification is called *pattern classification* or *classification*.

288. A *perceptron* or *neural network* is an interconnected network of nonlinear units or processing elements capable of learning and self organizing. The

response of a unit or a processing element is a non-linear monotonic function of a weighted sum of the inputs to the processing elements. The weights, called *synaptic weights* are modified by a learning or reinforcement algorithm. Typical nonlinear processing functions are sgn(x), $1/(1 + e^{-x})$, and tanh(x). When each processing element contributes one component to the output response vector and its inputs are selected from the components of the input pattern vector, the perceptron is called a *simple perceptron*. Processing units whose output only indirectly influences the components of the output response vector are called *hidden units*.

289.   An *error corrective reinforcement* for a perceptron is a learning or training algorithm in which the change in a synaptic weights is a function of the degree to which the output of the processing unit is not what it is desired to be. Error corrective reinforcement algorithms are also called *error back propagation* algorithms.

290.   A *forward coupled perceptron* is a perceptron in which the processing units are layered. The inputs to the processing units in layer n come from the outputs of processing units in layers prior to layer n. Single layered perceptrons can create linear decision surfaces. Two layered perceptrons can create convex decision regions. Three layered perceptrons can create almost arbitrarily shaped decision regions.

291.   A *series coupled perceptron* is a forward coupled perceptron in which the inputs to the processing center in layer n come from the outputs of processing units in layer $n - 1$.

292.   A *back coupled perceptron* is a perceptron which is not forward coupled. That is, there is some processor in layer n where output feeds back and is the input to a processor in some layer prior to layer n.

293.   A *cluster* is a homogeneous group of units which are very "like" one another. "Likeness" between units is usually determined by the association, similarity, or distance between the measurement patterns associated with the units.

294.   A *cluster assignment function* is a function which assigns each observed unit to a cluster on the basis of the measurement pattern(s) in the data sequence or on the basis of their corresponding features. Sometimes the units are treated independently. In this case the cluster assignment function can be considered as a transformation from measurement space to the set of clusters.

295.   The *pattern classification* problem is concerned with constructing the cluster assignment function which groups similar units. Pattern classification is synonymous with *numerical taxonomy* or *clustering*.

296.   The *cluster identification* process is the process in which the cluster assignment function is applied to the sequence of observed units thereby yielding a cluster identification sequence.

297.   A *misidentification*, or *misdetection*, or *type* I *error* occurs for category c_i if a unit whose true category identification is c_i is assigned by the de-

cision rule to category c_k, $k \neq i$. A misidentification error is often called an *error of omission*.

298. A *false identification*, or *false alarm*, or *type* II *error* occurs for category c_i if a unit whose true category identification is c_k, $k \neq i$, is assigned by the decision rule to category c_i. A false identification error is often called an *error of commission*.

299. A *prediction sequence*, or *test sequence*, or a *generalization sequence* is a set of two sequences: (1) data sequence (whose corresponding true category identification sequence may be considered to be unknown to the decision rule) and (2) a corresponding category identification sequence determined by the decision rule assignment. By comparing the category identification sequence determined by the decision rule assignment with the category identification sequence determined by the ground truth, the misidentification rate and the false identification rate for each category may be estimated.

300. A *confusion matrix* or *contingency table* is an array of probabilities whose rows and columns are both similarly designated by category label and which indicates the probability of correct identification for each category as well as the probability of type I and type II errors. The (*i*th, *k*th) element P_{ik} is the probability that a unit has true category identification c_i and is assigned by the decision rule to category c_k.

301. A unit is said to be *detected* if the decision rule is able to assign it as belonging only to some given subset A of categories from the set C of categories. To detect a unit does not imply that the decision rule is able to identify the unit as specifically belonging to one particular category.

302. A unit is said to be *recognized, identified, classified, categorized,* or *sorted* if the decision rule is able to assign it to some category from the set of given categories. In some applications, there may be a definite distinction between recognize and identify. In these applications, for a unit to be recognized, the decision rule must be able to assign it to a type of category, the type having included within it many subcategories. For a unit to be identified, the decision rule must be able to assign it not only to a type of category but also to a subcategory of the category type. For example, a small area ground patch which may be recognized as containing trees may be specifically identified as containing apple trees.

303. A unit is said to be *located* if specific coordinates can be given for the unit's physical location.

304. *Accuracy* refers to the degree of closeness an estimate has to the true value of what it is estimating.

305. *Precision* refers to the degree of closeness an estimate has to its expected value.

306. The *receiver operating characteristic* or the *receiver operating curve* of a pattern classifier is a function of its misdetection error rate against its false alarm rate.

307.   The *leave-K-out* method of evaluating a pattern classifier divides the training set into L mutually exclusive subsets each having K patterns. The classifier is successively trained using $L - 1$ of the subsets and tested on the Lth subset. The evaluation is then made on the accumulated performance tests of the experiments where in each experiment K patterns were omitted from the training set and then used in the testing set. Performance estimates obtained using the leave-K-out method are unbiased. However, for small K the estimates will have high variance.

308.   The *resubstitution method* of evaluating a pattern classifier uses the same set for training and testing. Performance estimates obtained using the resubstitution method are always biased high.

19.   INDEX OF TERMS

Absolute orientation   43
Accuracy   304
Acutance   6
Algebraic reconstruction   192
Ambient light   35
Analytic photogrammetry   39
Antiparallel   86
Area   128
Area analysis   186
Area of interest operator   184
Aspect graph   230
Automatic vision inspection   232
Automatic visual inspection   232
Back coupled perceptron   292
Backlighting   33
Band pass filter   79
Bayes decision rule   278
Bidirectional reflectance distribution function   30
Binary image   20
Binary image projection   144
Binary octree   150
Binary quadtree   148
Blackboard   242
Blob   113
Blob analysis   142
Block   106
Blocks world   228
Border   122

Bottom up 239
Boundary 122
Boundary delineation 84
Boundary detection 84
Boundary following 123
Boundary representation 154
Boundary surface description 154
Bounding contour 120
Bounding rectangle 136
Box filter 71
Brightness 3
Camera constant 47
Candela 24
Cartesian product 250
Categories C 253
Categorized 302
Category identification sequence 269
Center of perspectivity 46
Central moment 131
Centroid 129
Chain code 147
Change detection 65
Chromaticity 36
City distance 106
Classification 287
Classified 302
Classifier 258
Class region 260
Closing 97
Cluster 293
Cluster assignment function 294
Cluster identification 296
Clustering 295
Compactness 135
Compactness hypothesis 259
Compound decision rule 267
Computer vision 238
Concurve 124
Conditional probability 275
Confusion matrix 300
Congruencing 189
Connected 112

Connected component	113, 141
Connected component analysis	142
Constrained least squares filtering	191
Constructive solid geometry	153
Contingency table	300
Contour filling	172
Contour tracing	126
Contrast	7
Contrast difference	7
Contrast modulation	7
Contrast ratio	7
Contrast stretching	58
Convex	115
Convolving	73
Correlating	75
Corresponding point pair	200
Cross section function	152
Curve pyramid	149
Data driven	239
Data sequence	263
Decision boundary	283
Decision rule	264
Defer assignment	268
Density slicing	56
Depth map	15
Detected	301
Digital image	13
Digital photogrammetry	40
Digital picture function	13
Digital straight line segment	114
Digitized image	13
Dilating	94
Discrete Fourier transform	76
Discrete radon transform	100
Discrete relaxation	182
Discrete tonal feature	159
Discriminant function	280
Disparity	201
Distance transform	108
Distribution-free	276
Dynamic imagery	19
Dynamic scene analysis	220

Edge image	82
Edgel	10
Edge linking	83
Edge operator	81
Eight-connected	113
Elongatedness	139
Elongation	139
Epipolar axis	204
Epipolar line	207
Epipolar plane	208
Epipolar ray	206
Epipoles	205
Equal interval quantizing	60
Equal probability quantizing	61
Eroding	95
Error back propagation	289
Error correcting training procedure	273
Error corrective reinforcement	289
Error of commission	298
Error of omission	297
Euclidean distance	105
Euler number	134
Expectation directed	240
Extended Gaussian image	104
Exterior border pixel	117
Exterior orientation	42
Extremal pixel	137
Facet model	158
False alarm	298
False identification	298
Fan beam projections	192
Feature	254
Feature extraction	257
Feature *N*-tuple	254
Feature pattern	254
Feature point matching	177
Feature selection	256
Feature space	255
Feature vector	254
Figure *F*	109
Filtered back projection	192
Fixed inspection	236

Flexible inspection   236
Focus of attention mechanism   186
Focus of contraction   217
Focus of expansion   216
Foot candle   25
Forward coupled perceptron   290
Four-connected   111
Fourier descriptors   145
Fourier related texture descriptions   166
Frontlighting   34
Gaussian filter   72
Gaussian focal length   47
Gaussian image pyramid   80
Gaussian sphere   102
Generalization sequence   299
Generalized cone   152
Generalized cone axis   152
Generalized cylinder   152
Generalized cylinder spine   152
Goal-directed   240
Gradient operators   81
Gradient space   103
Gray level   3
Gray level co-occurrence matrix   162
Gray level dependence matrix   162
Gray level difference histogram   164
Gray level image   21
Gray scale image   21
Gray scale octree   150
Gray scale quadtree   148
Gray shade   3
Gray tone   3
Gray tone intensity   3
Ground truth   269
Hidden units   288
Hierarchical   241
Hierarchical decision rule   266
High pass filter   77
Histogram   23
Histogram equalization   61
Histogram hyperbolization quantizing   63
Hole   121

Homomorphic filtering 87
Hough transform 101
Hue saturation 37
Hyperplane decision boundary 284
Identified 302
Illuminance 26
Illumination 25
Image 1
Image compression 190
Image density 3
Image enhancement 193
Image flow 214
Image histogram 23
Image intensity 3
Image matching 175
Image operator 53
Image processing 194
Image pyramid 80
Image reconstruction 192
Image restoration 191
Image segmentation 167
Image smoothing 89
Image time sequence 19
Image transform 53
Image transform operator 53
Image understanding 238
Image value 3
Inner orientation 48
Interaction image processing 195
Interest operator 183
Interior border pixel 116
Interior pixel 118
Internal orientation 48
Intrinsic image 226
Intrinsic scene characteristics 225
Inverse discrete Fourier transform 76
Inverse filtering 191
Inverse optics 227
Irradiance 29
Iterative end point curve fitting 146
Kernel 70
Knowledge based vision 244

Lambertian surface   32
Laplacian image pyramid   80
Laplacian operators   81
Learn   274
Leave-K-out   307
Level slicing   56
Light striping   197
Limiting resolution   4
Linear decision rule   285
Linear discriminant function   281
Linearly separable   261
Linear quantizing   60
Linear spatial filter   69
Local feature focus method   181
Local shading analysis   212
Located   303
Low pass filter   78
Lumen   24
Luminous flux   24
Luminous intensity   24
Lux   25
Machine vision system   233
Masking   64
Mask matching   69
Match   179
Matched filtering   174
Mathematical morphology   92
Max distance   107
Maximum likelihood decision rule   277
Measurement N-tuple   249
Measurement pattern   248
Measurement space   251
Measurement vector   249
Medial axis   140
Median filter   88
Minimum perimeter polygon   125
Minimum variance quantizing   62
Misdetection   297
Misidentification   297
Model based computer vision   243
Moment   130
Morphological image pyramid   80
Morphologic edge operators   81

Motion field	214
Multi-band image	17
Multi-image set	17
Multi-level thresholding	57
Multi-spectral image	18
Multi-temporal image	19
Nearest neighbor decision rule	279
Neighborhood operator	67
Neural network	288
Non-parametric decision rule	276
Normalized central moment	132
Numerical taxonomy	295
Occluding boundary	209
Occluding edge	85
Octree	150
Opening	96
Optical flow	215
Optical gauging	235
Optic axis	44
Optic flow	215
Optic flow vector	214
Orientation histogram	104
Orientation map	16
Orthographic projection	50
Outer orientation	42
Parallax	51
Parallel	50
Parallel projections	192
Passive navigation	219
Pattern	248, 257
Pattern classes	253
Pattern classification	287, 295
Pattern discrimination	286
Pattern feature	254
Pattern identification	287
Pattern recognition	245
Pel	8
Perceptron	288
Perimeter	127
Perspective projection	49
Photometric stereo	213
Pictorial pattern recognition	246

Picture element   8
Picture processing   194
Pixel   8
Point operator   54
Precision   305
Prediction sequence   299
Preprocessing   187
Primal sketch   157
Principal axis   44
Principal distance   47
Principal point   45
Probabilistic relaxation   182
Projection index image   144
Property extraction   257
Property selection   256
Prototype pattern   262
Pyramid   80
Quadradic discriminant function   282
Quadtree   148
Quantizing   59
Radiance   27
Radiant intensity   28
Range image   12
Rangel   14
Range map   15
Range set   252
Raster scan order   11
Receiver operating characteristic   306
Receiver operating curve   306
Recognized   302
Reference pattern   262
Reflectance   30
Reflectance image   31
Reflectance map   31, 210
Reflection coefficient   30
Region aggregation   170
Region growing   168
Region merging   171
Region R   110
Regions of interest   184
Region tracking   169
Registering   188

Registration	188
Relational matching	180
Relation homomorphism	178
Relative orientation	41
Relaxation	182
Reserve judgment	268
Resolution	4
Resolution cell	5
Resolution limit	4
Resubstitution method	308
Robotic gauging	236
Robust	237
Rotation movement moments	133
Run length encoding	151
Scale space image	90
Scale space structure	91
Scene analysis	238
Screening	185
Second moment matrix	138
Separable	74, 261
Series coupled perceptron	291
Shape from contour	223
Shape from shading	211
Shape from shape	223
Shape from texture	221
Signature analysis	143
Simple boundary	119
Simple decision rule	265
Simple perceptron	288
Skeleton	98
Sorted	302
Spatial filter	68
Spatial frequencies	76
Spatial resolution	4
Spherical product	155
Square	107
Statistical pattern recognition	245
Statistical texture measures	163
Step edge operator	81
Stereo disparity	201
Stereo image pair	199
Stereo matching	202

Stereopsis   198
Strip tree   156
Structural description   178
Structural matching   180
Structural pattern recognition   245
Structural texture description   165
Structured light   196
Structure from motion   218
Structuring element   93
Subimage *F*   109
Superquadric   155
Surface reconstruction   222
Sweeping rule   152
Symbolic image   22
Symbolic matching   176
Symbolic registration   176
Symmetric axis   140
Synaptic weights   288
Syntactic pattern recognition   245
Template matching   173
Test sequence   299
Texel   161
Texture   160
Thickening operator   99
Thinning operator   98
Three-dee vision   231
Thresholding   55
Time varying image   19
Top down   240
Training procedure   271
Training sequence   270
Triangulation   203
Two and a half D sketch   224
Two-dimensional signal processing   66
Type I error   297
Type II error   298
Unit   247
Vanishing line   52
Vanishing point   52
Video image   2
View aspect   229
Visual correspondence   200
Visual fixturing   234

Visual gauging	235
Visual pose determination	234
Voxel	9
Wiener filtering	191
Window training procedure	272
X-parallax	51
YIQ coordinate	38
Y-parallax	51
Zero crossing operators	81

Index

Additive maximum operation, 146
Additive minimum operation, 147
Algorithms, for parallel processing,
   branching, 348
   convolution, 346
   rotation, 347
   transpose, 347
Aliasing, 375, 399
Antiextensive, 26, 36
ARIES, 380
Articulated object, 197

Benchmarks, 354–357
Base, 28
Basis, 29, 39
Blue noise, 367, 402

Census template, 153
Character recognition, 223
Chemical shift, 105
Closeness of fit, 158

Closing, 25
Clustered dot, 374, 388, 392
Clustering, 183
Coding
   arithmetic, 270
   hierarchical, 302
   Huffman, 266
   Lempel-Ziv-Welch, 272
   subband, 293
      quadrature mirror filters, 295
   transform, 285
Color halftone, 395
Compression
   lossless vs. lossy, 263
   of color images, 310
   of image sequences, 312
   redundancy and irrelevancy in,
      262
   standards, 313
     JBIG, 313
     JPEG, 314
     MPEG, 314

Coordinated graph (3D), 199
Critical pixel, 150

DCPM, 283
DCT, 288
Diffusion mask, 380, 403
Dilation, 22, 35
    geodesic, 77
Dispersed dot, 374, 388
Distance function, 73
Distance metric, 179
Dot growth pattern, 392
Dual filter, 24, 36

Early's parsing algorithm, 177
Echo, 110
    time, 110
Edge enhancement, 380, 407
Entropy, 268
Erosion, 21, 33
    geodesic, 79
    ultimate, 71
Error diffusion, 378, 402
Extensive, 26

Features, 232
    bar, 249
    cavity, 245
    direction value, 247
Field of view (FOV), 115
Flat filters, 39
Fractal-based compression, 300
Free induction decay (FID), 108

Generalized linear convolution,
        147, 153
Gradient descent, 199, 238
Grammar
    context-free, 174
    context-sensitive, 174
    phase-structure, 174
    tree, 184
    waveform, 189

Granulometry, 45, 57
    Euclidean, 58
    function, 65
    size distribution, 45
        local, 53
        normalized, 45
    statistical classification, 56
Gyromagnetic ratio, 105

Halftone, 363
    cell, 374, 388
    screen, 364, 378
        angle, 375, 395
        frequency, 397
Heuristic parsing algorithm, 201
High-pass filter, 407
Histogram, multidimensional, 124,
        128, 137
Hit-or-miss transform, 147
Homotopic, 150

Idempotent, 26, 36
Image algebra, 146
Image pyramids for compression,
        302
    Laplacian, 307
    mean, 305
    S-transform, 307
    subsampling, 304
Influence zone, 76
    skeleton by (SKIZ), 77
Invariant class, 28

Kernel, 28, 39

Larmor frequency, 106
Levenshtein distance, 179
Linearly separable, 15

Machine vision, 350–351
Magnetic field, 105, 107
    gradients, 111
        frequency-encoding, 112 ff

[Magnetic field]
    phase-encoding, 112 ff
    slice selection, 112 ff
Magnetic resonance imaging (MRI), 103
Matheron representation theorem, 28
Maximum-likelihood estimation, 3
Medial axis transform, 144, 148
Median, 2
  center-weighted, 16
  matched, 7
  two-dimensional, 8
  vector, 8
  weighted, 5, 14
    generating function, 18
    output distribution, 18
Minkowski addition, 22
Minkowski subtraction, 22
Moiré, 375, 399
Monotonically increasing, 24, 36
Morphological gradient, 89
Multiple-nuclei halftone, 375, 394

Neural network, 234
  back propagation algorithm, 240–243
  multilayer feedforward network, 234
  training, 235, 239
  weights, 237
90-FID pulse sequence, 109
Noise encoding, 366, 386
Non-terminals, 170
Nuclear magnetic resonance, 103

Opening, 24
Ordered dither, 372, 388
Orthographic gray scale, 366

Parallel matching algorithm, 206
Parallel processor architectures
  hypercube, 334, 337, 343

[Parallel processor architectures]
  linear array, 335, 338, 342
  mesh, 335, 337, 343
  MIMD, 330, 352
  pipeline, 331–333, 343
  pyramid, 336
  SIMD
    single bit, 334–336, 341–343
    word wide, 336–338
Partial dots, 366, 383
Partial volume effect, 110
Pattern representation, 199
Pattern spectrum, 45
  moments, 48
  representation, 50
Primitive patterns, 168
Processor design, 340–344, 353, 354
Production rules, 174
Pulse-density modulation, 380, 405

Quantization
  scalar, 275
    Lloyd-Max, 275
    entropy-constrained, 276
    bit allocation in, 277
  vector, 278
    codebook generation, 279
    tree-structured codebooks, 281
Quench function, 69

Reconstruction, 80
  gray-scale, 82, 83
Remainder image, 163
Repetition time, 109
RF pulse, 108

Saturation recovery pulse sequence, 109
Segmentation, 43, 226
  handwritten numeric field, 226
  multispectral, 117

[Segmentation]
  particle, 60
  texture, 54
Self-dual, 15
Sentence-to-sentence matching, 182
Serial processors, 329, 341,
      351, 352
Simulated annealing, 380
Skeletal lag, 157
Skeleton, 19, 143, 152, 161
  connectivity, 157
Spatial resolution, 382, 399
Spin, 104
  density, 111
    of brain tissues, 136
    calculation of, 123
Spin-echo pulse sequence, 109
Spin-lattice relaxation time, 106
  of brain tissues, 136
  calculation of, 121
Spin-spin relaxation time, 107
  of brain tissue, 136
  calculation of, 120
Stack filter, 12
  continuous, 16
    output distribution, 17
  representation via erosions, 39
String matching, 177
Structuring element, 20
Syntactic parsing, 177

Syntactic pattern recognition, 167

τ-opening, 27, 36
Template dot, 366
Terminals, 170
Thinning, 143, 150
  on nonrectangular domains, 161
  parallel algorithms for, 160
Threshold,
  array, 388
  decomposition, 10
Tone reproduction, 392
Translation invariant, 23, 35
Tree automata, 186

Visual perception, 381
VOXEL, 110
VQ-based compression, 298

Watershed segmentation
  binary segmentation procedure,
      83–85
  catchment basin, 86
  hierarchical, 92
  marker-driven   segmentation,
      90–91
  of gray-scale image, 87
  via conditional bisector, 96
Wavelets, 297

Lightning Source UK Ltd.
Milton Keynes UK
UKOW07n0853171214

243276UK00009B/163/P